Making Modern Medical Ethics

Basic Bioethics
Arthur Caplan, editor

Making Modern Medical Ethics

How African Americans, Anti-Nazis, Bureaucrats, Feminists, Veterans, and Whistleblowing Moralists Created Bioethics

Robert Baker

The MIT Press
Cambridge, Massachusetts
London, England

The MIT Press would like to thank the anonymous peer reviewers who provided comments on drafts of this book. The generous work of academic experts is essential for establishing the authority and quality of our publications. We acknowledge with gratitude the contributions of these otherwise uncredited readers.

This book was set in Sabon by Westchester Publishing Services. Printed and bound in the United States of America.

Library of Congress Cataloging-in-Publication Data

Names: Baker, Robert, 1937– author
Title: Making modern medical ethics : how African Americans, anti-Nazis, bureaucrats, feminists, veterans, and whistleblowing moralists created bioethics / Robert Baker.
Description: Cambridge, Massachusetts : The MIT Press, [2024] | Series: Basic bioethics | Includes bibliographical references and index.
Identifiers: LCCN 2023008794 (print) | LCCN 2023008795 (ebook) | ISBN 9780262547376 (paperback) | ISBN 9780262377409 (epub) | ISBN 9780262377416 (pdf)
Subjects: LCSH: Medical ethics—History. | Bioethics—History.
Classification: LCC R724 .B1846 2024 (print) | LCC R724 (ebook) | DDC 174.2—dc23/eng/20230315
LC record available at https://lccn.loc.gov/2023008794
LC ebook record available at https://lccn.loc.gov/2023008795

10 9 8 7 6 5 4 3 2 1

For Arlene

sine quo nihil

My wife, Arlene, suggested that instead of writing learned but dull commentaries about oaths and codes, I write a book about the scandals that led people to create them. Her suggestion was the nub around which this book was formed. As the manuscript evolved, however, she regretted the dark corners of history that my research led me to explore. She is also indirectly responsible for the prefatory material— the chronology of events, the list of names, and the explanations of abbreviations and foreign words. I included these after witnessing her frustration with a complex historical narrative that lacked these elements. As the prefatory Latin subtitle indicates, without her, this book would not exist.

Contents

Preface

The birth of bioethics is often presented as an origin tale in which morally disruptive medical technologies—dialysis machines, electroencephalographs, ventilators, and other innovations deployed in the latter part of the twentieth century—created a need to reconsider traditional conceptions of medical ethics. Such tales may mention whistleblowers, scandals, thinktanks, and government commissions, but they usually fail to focus on the insurrectionist aspects of the bioethical turn in modern medical ethics. As a function of this neglect, they understate, obscure, gloss over, elide, or ignore the contributions of African American civil rights activists and health care professionals who served in Allied armed forces in the world wars or in Vietnam, and they seldom analyze the impact of 1960s Great Society legislation that challenged structural classism, racism, and sexism endemic in mid-twentieth-century American health care. Standard histories also downplay the penalties played by moral change agents, such as the demotion of a stubborn bureaucrat who insisted on requiring and enforcing research subjects' informed consent, the systemic chastisement, suppression and ostracism of other change agents, and the postmortem rejection of various whistleblowing moralists.

Curiously, these histories offer only backhanded references to revelations at the Nuremberg Doctors Trials about medical complicity in the Holocaust and doctors' use of humans as guinea pigs. Such darker moments are discounted on the grounds that, with the onset of the Cold War (1945–1990), retrograde amnesia about such events became politic as the West sought to convert former Nazis into allies. In the words of British historian Duncan Wilson, "The Nuremberg Code . . . was routinely ignored by researchers in Britain . . . and elsewhere, who believed the guidelines were designed to prosecute 'barbarians' and did not apply to them."[1] American historian David Rothman concurs, reporting that "neither the horrors described at the Nuremberg Trial nor the ethical principles that emerged from it had a significant impact on the American research establishment. The trial itself did not receive extensive press coverage."[2] Similarly, influential American bioethicist Jonathan Moreno observes that, although "a number of newspaper articles . . . reported on the Nazi Doctors' Trial and its outcome," medical students "trained in that era saw little reaction to the [Nuremberg] Code among their US medical school professors."[3] Rothman concludes that "well into the 1960s, the American research community considered the Nuremberg findings, and the Nuremberg Code, irrelevant to its own work."[4] Wilson reports that this was true of the British research community as well.

I have no quarrel with these observations. As these historians and bioethicists correctly observe, during the 1960s and 1970s, Anglo-American medical and research establishments, including most medical school faculty, dismissed or ignored German physicians' complicity in the Holocaust, the Nuremberg Trials, and the ethics code issued by the Nuremberg Court. Yet moral revolutions, like their political and scientific counterparts, are not initiated by establishments but by the dissidents, critics of the established order. One of the hypotheses underlying this book is that modern medical ethics was initially created in response to Nazi medical ethics and medical practices and that it took what I call "a bioethical turn" after the 1970s when, in America, and to a greater or lesser extent elsewhere, values of transparency, shared decision-making, and egalitarian and antiracist views were integrated into official statements of medical ethics and became integral to modern perceptions of medical morality. Around this period, the East Coast American thinktanks that would play a vital role in establishing bioethics as an area of scholarship and research were founded. At the same time, various court cases and government commissions addressing issues raised by morally disruptive innovations inadvertently created occupational roles for trained bioethicists (i.e., "ethicists") to serve on hospital ethics committees (HECs), institutional review boards (IRBs), and institutional animal care and use committees (IACUCs). To date, most historians overemphasize the creation of the occupational aspects of the bioethical turn while ignoring, or glossing over, the pivotal role played by many of the whistleblowing moralists who publicized the ethics scandals of the 1960s and 1970s, thereby initiating the crisis of confidence in established Anglo-American medical institutions that triggered a reconception of both the researcher–subject and the clinician–patient relationship.

Perhaps the most overlooked or misrepresented players ingredient in the creation of modern medical ethics are the medical personnel who served in the Allied military during two worlds wars or in America's wars in Southeast Asia. Yet it was these veterans who founded an international medical organization that took as one of its prime objectives rescuing medicine and biomedical research from the scandal of Nazi medicine and its justificatory ethics. And it was also World War II and Vietnam military veterans who were the most prominent moralists alerting the media and, through it, the public to Anglo-American physicians' scandalously abusive exploitation of patients as unconsenting subjects of dangerous and sometimes deadly experiments.

In this book, I argue that the foundational documents of modern medical ethics were expressly designed to reject Nazi medical practices and the ethics that justified them. I also contend that if one distinguishes bioethics as a moral revolution from bioethics as a field of scholarship and a health care specialization, it becomes apparent that, as the chronology following this preface indicates, origin tales about the birth of bioethics as an artifact of American moralism and exceptionalism are myopic. Such tales focus on the founding of a few bioethics thinktanks—the Hastings Center and Kennedy Institutes—and on a series of US governmental commissions and US court cases. But by failing to acknowledge that modern medical ethics originated in Europe *before* the bioethics institutes were created and *before* American courts heard these cases, such histories erase the contributions of African American civil rights activists, female heroines, feminists, international organizations, Jews motivated by memories of the Holocaust, veteran military health care professionals, and others who led moral insurrections in America and Britain, thereby creating the bioethical turn in modern medical ethics.

The alternative origin narrative offered in this book unfolds in eleven chapters and an epilogue. The first chapter, "Moralists and True Scandals: Medical Ethics during the pre-and Interwar Periods," explores the relationship between moralists, scandals, and codes of ethics, focusing on an unduly neglected early twentieth-century German medical ethicist, Albert Moll. Chapter 2 addresses the development of *Rassenhygiene* (racial hygiene), the Nazi medical ethics that justified killing children with disabilities, mass sterilizations, and, ultimately, genocidal efforts to exterminate gays, Jews, and Roma. Chapter 3 briefly reviews the Nuremberg Trials and the origins of the Nuremberg Code to assess their influence on the creation of international medical oaths and codes. Chapter 4 analyzes the World Medical Association's (WMA's) Declaration of Geneva, which resurrected and modernized the Hippocratic Oath to rehabilitate German medicine and prevent any recurrence of Nazi medical ethics and medical practices.

Chapter 5 covers the creation of the WMA's Declaration of Helsinki, the first practical set of ethical standards designed for increasingly common forms of research on human subjects involving prospective multiarmed trials of innovative vaccines, drugs, or techniques. It also relates how Food and Drug Administration (FDA) bureaucrat Frances Kelsey slipped into US laws and implementing regulations an informed consent requirement originally formulated in a 1962 draft of the Declaration of Helsinki. Chapter 6, "Kelsey, Pappworth, and Beecher: Moral Awakenings," focuses on three influential moralists, examining the ways in which these moralists tried to make visible to their colleagues and to the public commonplace immorality in Anglo-American medical research that almost everyone else was either blind to, or knowingly turned a blind eye toward. Chapter 7 continues the discussion of moralists, some stymied, others effective, in protesting the US Public Health Service's infamous study of untreated syphilis in African American men. Chapter 8, "Scientistic Medical Paternalism and Its Discontents," explores various attempts to challenge scientistic medical paternalism. Chapter 9 focuses on reforms aimed at remedying some of the problems of medical paternalism. Chapter 10 analyzes the bioethical turn in modern medical ethics as an artifact of a successful moral revolution.

Chapter 11 offers a metahistorical review of prior histories, questioning why they were constructed in a way that ignores anti-Nazism, blacks, reformist bureaucrats, feminists, veterans, and most whistleblowing moralists. In a brief epilogue, I reflect on my role as a minor participant observer in the bioethics revolution and contend that it is important to the future of bioethics for bioethicists to understand that their field was founded as much by the heroic actions of insurgent dissidents as it was by the visionaries who founded thinktanks. Appended to the book's central narrative are two initial drafts and the final version of the document now known as the Nuremberg Code; also appended are two historically significant statements of patients' rights.

Notably absent from this book are analyses of the public policies and legal cases that helped to shape the field of bioethics, such as the Patient Self-Determination Act of 1990, the Health Insurance Portability and Accountability (HIPPA) Act of 1996, and the Oregon Death with Dignity Act of 1997. Also absent are analyses of such end-of-life cases as *Cruzan* (1983) and *Schiavo* (1990s–2000). As the dates of these policies and cases make clear, they occur after the bioethical turn in modern medical ethics in the 1960s and 1970s, which are the focal point of this narrative. I focus on legal cases and public policies only insofar as they influenced the bioethical turn in modern medical. For example, although I analyze the 1975–1976 *Quinlan* case, my focus is not on its precedent-setting establishment of a patient's or a surrogate's right to decline life-extending treatment but on a quirky turn of events involving a judge's misunderstanding that facilitated the spread of HECs, thereby inadvertently fashioning a role for bioethicists in the clinic contexts. Another factor is epistemic humility: I have no special expertise in health care law or government policy. And, to appropriate a line from the philosopher Ludwig Wittgenstein, whereof one cannot speak [expertly], one must remain silent.

Histories of bioethics   Two major histories of bioethics written in the 1990s serve as the standard histories of the field: David Rothman's *Strangers at the Bedside: A History of How Law and Bioethics Transformed Medical Decision Making* (Basic Books, 1991) and Albert Jonsen's *Birth of Bioethics* (Oxford University Press, 1998). Rothman's book is based on archival research and focuses on the regulatory history of biomedicine in the 1960s and 1970s and the research ethics scandals that changed biomedical research practices and clinical decision-making. He highlights Henry Beecher's (1966) whistleblowing article, "Ethics and Clinical Research." Jonsen's book is based on a 1992 conference he hosted at the University of Washington at which pioneering bioethicists contributed papers on the birth of bioethics. Jonsen supplements their analyses with some exploration of primary sources, augmenting them with his personal memories as a founding bioethicist present at major inflection points in the field's development.

Both histories were written from the perspective of the 1990s and present bioethics as a uniquely American phenomenon that was later internationalized. More recent scholarship, in contrast, indicates that the impetus for many of the moral innovations associated with bioethics originated as a fusion of European and American ideas, first in the Nuremberg Code and later in the World Medical Association's 1962–1964 Declaration of Helsinki. This international declaration served as a model for innovative American bureaucrats who, in the 1960s, implemented regulations focused on informed consent that were later unduly credited to bioethics movements of the 1970s. Both Rothman's and Jonsen's narratives focus on morally disruptive technologies, such as dialysis machines and ventilators, but they pay scant attention to the Great Society legislation creating the Medicaid and Medicare programs, even though these programs funded the purchase of those morally disruptive innovations and even as, over the decades, these programs served as a basis for challenges to the pervasive ableism, ageism, anti-Semitism, classism, exploitive condescension, racism, and sexism, endemic in post–World War II American health care. This book has a different emphasis. As its subtitle states, in this book I offer a more comprehensive view of the bioethics turn, expanding on earlier accounts to include "A History of How African Americans, Anti-Nazism, Bureaucrats, Commissions, Feminists, Thinktanks, Veterans, and Whistleblowing Moralists Turned Modern Medical Ethics into Bioethics."

Acknowledgments

This book could never have been written had Albert Jonsen and David Rothman not written their trailblazing histories of bioethics. Both authors were friends and I regret that neither is alive to review this manuscript and to respond to my alternative account of the bioethical turn in modern medical ethics. Research on this manuscript commenced in 2018 and continued through the COVID-19 pandemic. During this period, bioethics lost many pioneering voices, including Charles Bosk, Dan Callahan, Tris Engelhardt, Ken Kipnis, Bob Levine, Bob Orr, Ed Pellegrino, Kathy Powderly, and my very good friend and summertime colleague, Bob Veatch.

As I began research on this book, I recalled that Joseph D'Oronzio mentioned that A Patient's Bill of Rights had been negotiated by a civil rights leader. Searching Joe's name, I came across his 2001 article on A Patient's Bill of Rights. His work helped me to reconceive the focus of this book. I am also indebted to Ulf Schmidt and his colleagues, who invited me to a 2013 conference celebrating the fiftieth anniversary of the Declaration of Helsinki. Research for my presentation, which is published in Ulf's coedited book on the Declaration of Helsinki, *Ethical Research* (2020), forms the basis of chapter 4 of this book. A project for Lisa Rasmussen's edition of Mellanby's *Human Guinea Pigs* (2020) challenged my views on Pappworth, sending me to the Wellcome Library, where librarians graciously assisted my research. I also benefited greatly from supervising Melvin Wayne Cooper master's thesis, which was a translation of Rudolf Ramm's Nazi medical ethics textbook, *Ärztlizche Rechts- und Standeskunde Der Artz als Gesundheitserzieher*. A revised version has been published by Springer as *Medical Jurisprudence and Rules of the Medical Profession*. Closer to home: special thanks to Union College librarian Joseph Lueck for his assistance.

The COVID-19 pandemic precluded direct use of archives; consequently, many documents referenced in this book are quoted from the writings of other historians, and I gratefully acknowledge my use of their research.

This volume also draws on my prior research on the history of medical ethics in, among other works, *Before Bioethics* (2013), *The Structure of Moral Revolutions* (2019), and the *Cambridge World History of Medical Ethics* (2009, coedited with Laurence McCullough) and *The American Medical Ethics Revolution* (1999, coedited with Arthur Caplan, Linda Emanuel, and Steve Latham). I explored some of the points in this book in two essays published in 2022, "Erasing Blackness from Bioethics," which appeared in the *American Journal of Bioethics*, and "Principles and Duties: A Critique of Common Morality Theory," published in the *Cambridge Quarterly of Healthcare Ethics*. Commentators on drafts of these papers were helpful in sharpening my views.

Special thanks to Art Caplan for being open to the idea of publishing an alternative yet authoritative history of the bioethical turn in modern medical ethics in his Basic Bioethics series. My original manuscript has been improved by the many helpful comments offered by the external reviewers assembled by my editor at MIT Press, Philip Laughlin, and to the MIT Press's team of proof readers, led by Rashmi Malhotra, with thanks to Gillian Dickens, Emily Simon, Roger Wood, and those curiously identified as "CE," and "WPS eStyler." A special note of thanks to my friend and sometimes coauthor, Matt Wynia. Thanks to all the above for making this a better book. Finally, a shoutout and word of thanks to the baristas at the Barnes and Noble café in Colonie Center, New York, whose scones, green tea, and good will fueled this book. Cheers! bb

Chronology of Events Referenced

1605 Francis Bacon defines "euthanasia" as a physician-assisted painless passage from life

1735 Linnaeus classifies four races: *Africanus niger* (black), *Europaeus albus* (white), etc.

1803 Percival's *Medical Ethics* introduces this neologism to the Anglophone medical lexicon

1847 Newly formed American Medical Association (AMA) adopts code based on Percival's

1859 Darwin's *On the Origin of Species . . . or the Preservation of Favoured Races*

1852 Publication of Harriet Beecher Stowe's *Uncle Tom's Cabin; Life among the Lowly*

1861–1865 American Civil War fought over the issue of black slavery and the slave trade

1871–1918 Unification of German principalities with Prussia, later known as Germany

1871–1994 Sex between men criminalized in Section 175 of the new German Criminal Code

1883 Francis Galton coins "eugenics" as improving a race of people by selective breeding

1887 US National Institutes of Health (NIH) founded as the Hygienic Laboratory

1895 Jost, *Das Recht auf den Tod* (*A Right to Death*) by *euthanasie* (i.e., killing negative life)

1895 Ploetz introduces *Rassenhygiene* (racial hygiene) into the German medical lexicon

1898 Albert Neisser is fined by a Prussian court for unconsented experiments on patients

1900 Prussian regulations require *patients'* informed consent to experimental interventions

1900 *Subjects* in Walter Reed's yellow fever experiments in Cuba sign consent forms

1902 Albert Moll's *Ärztliche Ethik* (*Physicians Ethics*) documents unconsented human experiments

1914–1918 World War I: approximately 40 million casualties, 20 million deaths

1914–1923 Turkish genocide of Armenians: 1.5 million Armenians killed

1918 Armistice ends a world conflict between France, UK, US, and Austria, Germany, Turkey

1918–1933 German Weimar Republic founded on democratic values and human rights

1920 League of Nations founded to facilitate peaceful international cooperation (1920–1946)

1920 National Socialist German Workers Party, NSDAP, or "Nazi," party formed

1920 Hoche urges killing *lebensunwerten Lebens*: those living lives unworthy of being lived

1922–1939 German medical ethics society publishes *Ethik*, world's first medical ethics journal

1927 Fritz Jahr proposes a Kantian-Buddhist *bio-ethik* embracing all living entities

1929 Global stock market crash

1929–1932 Global economic depression

1930 Pope Pious XI's encyclical, *Casti Connubi*, condemns eugenics and birth control

1931 German Ministry of Health's "Guidelines for Human Experimentation on *Patients*"

1932–1972 US Public Health Service Study of "Untreated Syphilis in the Negro Male"

1933–1945 Nazi Germany nationalizes health care practitioners, de-professionalizing medicine

1938 US Congress passes the Federal Food, Drug, and Cosmetic Act (FD&C)

1939 Germany invades Poland; Britain and France declare war; World War II commences

1939–1945 Nazi *Kinder-euthanasie* (child euthanasia program)

1939–1945 Nazi *Sonderbehandlung* (special handling) mass killing of gays and disabled

1940–1945 Nazi Aktion T4 (*Gnadentod*/good death) for "incurably" mentally or physically ill

1941 November, Franz Büchner's anti-*euthanasie* public lecture on the Hippocratic Oath

1941 December 7, US enters World War II after Japanese bomb Pearl Harbor

1942 Nazis require Ramm's medical ethics textbook *Ärztlizche Rechts* (*Physicians' Rules*)

1944 June 6, Operation Overlord: Allied forces invade Nazi-occupied continental Europe

1945 May 7–9, German Reich surrenders to Allied forces, ending World War II in Europe

1946 United Nations founded to replace the defunct League of Nations

1946 November, UN Educational, Scientific, & Cultural Organization (UNESCO) founded

1946 August, Andrew Ivy writes first draft of the Nuremberg Code for the ISC

Fall, Ivy proposes a second draft of his code for adoption by the AMA

December, AMA adopts stripped-down version of Ivy's rules with no publicity

December 1946–April 1947, Leo Alexander revises Ivy's second draft of the code

1946 December–August 1947, Nuremberg Medical Trials (*US v. Karl Brandt, et al.*)

1946–1991 Cold War between US and allies and the Soviet Union (Russia) and allies

1947 August 20, Nuremberg verdict issued with "Code of Ethics for Permissible Research"

1947 World Medical Association (WMA) is founded

1947 AWA founded to represent physicians in the Western sector of occupied Germany

1948 WMA issues Declaration of Geneva, a modern version of the Hippocratic Oath

1948 World Health Organization (WHO) founded as a UN agency

1948 Bradford Hill establishes RCTs as the standard of evidence for innovative treatments

1948 UN issues Universal Declaration of Human Rights (UDHR) on December 10

1949 Council for International Organizations of Medical Sciences (CIOMS) founded

1949 WMA issues International Code of Medical Ethics (ICME)

1952–1954 Beecher-Lasagna CIA and US Army funded amphetamines and LSD experiments

1952 Pope Pius XII's Encyclical, *The Moral Limits of Medical Research and Treatment*

1953 US Secretary of Defense's memo: military research must conform to Nuremberg Code

1954 WMA issues *Principles for Those in Research and Experimentation*

1954–1975 Civil war between North Vietnam and South Vietnam

1955 Professor Count Gibson, MD, protests Tuskegee Study; Olansky represses him

1955–1970 Dr. Saul Krugman's hepatitis experiments at Willowbrook State School (NY)

1957 Pius XII's *Address to Anesthesiologists*, patient/family may cease "extraordinary means"

1958 Beecher's memo to Harvard Committee on ethical challenges in research on humans

1959 Beecher publishes "Experimentation in Man" in January issue of *JAMA*

1961 *British Medical Journal* reports that thalidomide causes congenital deformation

1961–1975 US enters Vietnam Civil War on the side of South Vietnam

1962 *Washington Post* crowns Frances Kelsey "Heroine" for preventing thalidomide disaster

1962 Senator Javits asks Kelsey, "Did women know thalidomide was an experimental drug?"

1962 Kefauver–Harris Amendment to FD&C law requires research *subjects'* informed consent

1962 Frances Kelsey receives Medal of Freedom from US President John F. Kennedy

1962 Maurice Pappworth publishes "Human Guinea Pigs—A Warning" in a popular magazine

1962 Clegg's unauthorized publication of a draft of the Declaration of Helsinki in the *BMJ*

1962–1965 Vatican II: Second Ecumenical Council of the Vatican reforms Catholic liturgy

1963–1965 Papal Birth Control Commission (Pontifical Commission on the Family)

1963 President John Kennedy assassinated, Lyndon Johnson,(1908–1973) is US President

1963–1969 President Johnson's: "Great Society" medicare/medicaid and civil rights laws

1963 Lawyer Hyman charges Southam's unconsented experiments violate "Nuremberg Code"

1964 US Civil Rights Act: forbids ethnic, racial, religious, and sexual/ gender discrimination

1964 WMA's Declaration of Helsinki, *Ethical Principles Medical Research on Human Subjects*

1964 FDA regulations formalize 1962 Helsinki requirements for subjects' informed consent

1964 Beecher objects to new FDA regulations on consent, challenges Kelsey's authority

1964 Dr. Louis Lasagna publishes a modernized version of the Hippocratic Oath

1964 Schatz writes letter to CDC protesting Tuskegee Study, Yobs resolutely ignores it

1965 July 30, Medicaid and Medicare enacted for President Johnson's Great Society program

1965–1966 Beecher–Pappworth correspondence on "outing" unethical experiments

1965 March, Beecher presents 18 examples of unethical experiments to media at Brook Lodge

1965 March, Beecher: "Guidelines for Clinical Investigation," published in *JAMA*

1966 June, Beecher's "Ethics and Clinical Research," in *New England Journal of Medicine*

1966 November: Buxtun reports to CDC comparing its Syphilis Study to Nazi experiments

1966 AMA endorses WMA's 1964 Declaration of Helsinki on ethics of research on humans

1966 Dr. Elisabeth Kübler-Ross's *On Death and Dying* inspires Death with Dignity movement

1967 Maurice Pappworth publishes *Human Guinea Pigs: Experimenta-tion on Man*

1967 Kelsey demoted to "bare desk" bureaucrat in response to industry and media pressure

1968 Media ignores Jenkins's CORE's *Drum* newsletter critique of the CDC's Syphilis Study

1968 April 4, American civil rights leader Reverend Martin Luther King Jr. assassinated

1968 African American riots in Baltimore, Chicago, Cincinnati, DC, Detroit, New York, etc.

1968 AMA Judicial Council authorized to expel member societies for racial discrimination

1968 June 6, presidential candidate, ex-US Attorney General Robert Kennedy assassinated

1968 Paul II's encyclical letter *Humanae Vitae* prohibits Catholics' use of contraception

1969 February 6, Blue Ribbon CDC panel discusses Tuskegee Study, votes to continue it

1969 Callahan and Gaylin found Hastings Center to focus on ethical issues instead of politics

1969 Medical school educators form the Society for Health and Human Values (1969–1997)

1969 George Wiley's negotiates A Patient's Rights agreement with the Joint Commission

1970 May 4, Kent State, National Guard kills black students protesting invasion of Cambodia

1970 Student protests against Vietnam War erupt at US colleges and universities nationwide

1970 *The Population Bomb: Population Control or Race to Oblivion* is a bestseller

1970 Beecher publishes *Research and the Individual: Human Studies*

1970 V. R. Potter uses the word "bioethics" in an article proposing an ethics for the biosphere

1971 Boston Women's Health Book's *Our Bodies, Our Selves: A Book by and for Women*

1971 André Hellegers founds the Kennedy Institute of Bioethics at Georgetown University

1972 Jay Katz and Alex Capron publish *Experimentation with Human Beings* (Russell Sage)

1972 July 25, AP reporter Jean Heller reports Buxtun's revelations on the Tuskegee Study

1972 August 28, DHEW forms the Ad Hoc Advisory Panel on the Tuskegee Syphilis Study

1972 October 25, Ad Hoc Panel recommends termination of Tuskegee Syphilis Study

1973 January, US Supreme Court decriminalizes abortion, *Roe v. Wade, 410 U.S. 113*

1973 April 24, DHEW Advisory Committee on Tuskegee publishes its final report

1973 Senator Ted Kennedy's "bioethics" hearings; Callahan, Gaylin, Veatch, etc. testify

1973 July 10, *Kaimowitz v. Michigan* incarcerated consent rejected, cites "Nuremberg Code"

1973 American Hospital Association "affirms" A Patient's Bill of Rights, but no enforcement

1973 Callahan publishes "Bioethics as a Discipline," in *Hastings Center Studies*, vol. 1, no. 1

1974 Library of Congress makes "bioethics" subject heading, citing Callahan's *Studies* article

1974 Gorovitz's Summer Institute on Medicine & Morals, trains philosophers as bioethicists

1974 July 12, National Research Act creates the first national "bioethics" commission

1974 August 9, Pres. Richard Nixon resigns over his burglary of Democratic headquarters

1974–1978 National Commission for Protection of Human Subjects of Biomedical Research

1975–1987 Kennedy Institute of Ethics publishes *Bioethics* bibliography, Walters is editor

1976 *In Re Quinlan*: disconnecting ventilators not homicide, requires ethics committee review

1976 *New England Journal of Medicine* publishes Harvard hospitals' non-resuscitation policies

1978 National Commission's *Belmont Report*: three ethical principles for research on humans

1978 International Committee of Medical Journal Editors (ICMJE) founded

1978 ICMJE: medical journals to reject research not compliant with Declaration of Helsinki

1978 Kennedy Institute of Ethics publishes *Encyclopedia of Bioethics*, Warren Reich editor.

1979–2019 Beauchamp and Childress publish *Principles of Biomedical Ethics*, eight editions

1980–1983 President's Commission Ethical Problems in Medicine and Biomedical Research

1982 *Clinical Ethics: A Practical Approach*, by A. Jonsen, M. Siegler, and W. Winslade

1983 President's Commission's report, *Deciding to Forego Life-Sustaining Treatment*

1986–1997 Society for Bioethics Consultation founded, incorporated into ASBH in 1997

1990 AMA's "Fundamental Elements of Patient-Physician Relationship" embraces bioethics

1991 *Strangers at the Bedside . . . How Bioethics . . . Transformed Decision Making*, David Rothman

1992 *The Nazi Doctors and the Nuremberg Code*, George Annas and Michael Grodin, editors

1992 *When Medicine Went Mad: Bioethics and the Holocaust*, Arthur Caplan, editor

1994–1997 American Association for Bioethics (AAB); incorporated into ASBH in 1997

1998 Pioneering bioethicist Albert Jonsen's (semi-autobiographical) *Birth of Bioethics*

1998 AAB, SBC, SHHV merge as American Society for Bioethics and Humanities (ASBH)

1999 AMA recognizes patient autonomy and patients' right to forego life-sustaining treatment

[Bioethics paradigm, discourse, and practices normalized as standard medical ethics]

Acronyms, Jargon, and Non-English Words

3R Rules: See Rules Regarding [the humane treatment of] Animals
ACP: American College of Physicians; Internal Medicine Society (founded 1915)
ADC: Aid to Dependent Children programs
AHA: American Hospital Association (founded 1898) or American Heart Association (1924)
Aktion T4: Nazi eugenics plan to eliminate *Lebensunwertes Lebens* by *euthanasie* (1939–1945)
AMA: American Medical Association, allopathic physicians' society (founded 1847)
AP: Associated Press: nonprofit cooperative news agency (founded 1846)
AWA: West German Physicians Professional Organization (founded 1947)
Bioethical turn: creation of a biomedical ethics recognizing patients and subjects' rights
BMA: British Medical Association (founded 1832)
BMJ: *British Medical Journal* (founded 1840)
BWMA: *Bulletin of the World Medical Association* (1949–1953)
Calling a code: Emergency notice to attend to a patient experiencing cardiopulmonary arrest
CBE: Commander Order of the British Empire: highest British award excluding knighthood
CDC: USPHS branch, known after 1980 as Centers for Disease Control and Prevention
CIOMS: Council for International Organizations of Medical Sciences (founded 1949)
CMA/AMC: Canadian Medical Association/Association Médicale Canadienne (1867)

Coding: Call for emergency care for a patient in cardiopulmonary arrest (US hospital argot)

CORE: Congress on Racial Equality (1942), nonviolent direct action civil rights group

CPR: Medical procedure of nonsurgical (closed chest) cardiopulmonary resuscitation

DHEW or HEW: Department of Health Education & Welfare, Federal Agency (1953–1979)

DNR: Do Not Resuscitate; order not to initiate CPR or other resuscitative interventions

Durchschnittsmoral: Everyday commonsense morality

Euthanasia: English: mercy killing to alleviate a person's pain or suffering

Euthanasie: Pre-1945 German: mercy killing those burdening themselves, a family, or society

FDA: US Food and Drug Administration (founded 1906)

FD&C: US Food, Drug, and Cosmetic Act of 1938; enables FDA to police drugs for safety

FRCP: Fellow of the Royal College of Physicians

Führerprinzip: Nazi principle of authoritarian charismatic leadership (e.g., Adolf Hitler)

HEC: Hospital ethics committee (sometimes, "health care ethics consultant")

HEW: See DHEW

HQ: Headquarters

IACUC: Institutional Animal Care and Use Committee; satellite regulators for lab animals

ICD: WHO's International Classification of Diseases, Diagnoses, and causes of death

ICME: WMA's 1949 International Code of Medical Ethics

ICMJE: International Committee of Medical Journal Editors (founded 1978)

Investigator: US bureaucratic terminology for medical researchers

IOM: Institute of Medicine (1970–2015), now called the National Academy of Medicine

IRB: Institutional review board; satellite regulator policing research ethics standards

ISC: International Scientific Commission on War Crimes of a Medical Nature (World War II)

JAMA: *Journal of the American Medical Association* (founded 1883)

Joint Commission: Accrediting commission for health care organizations (founded 1951)

Lebensunwertes Lebens: Lives unworthy of life, or lives unworthy of being alive

LGBTQ+: Lesbian, gay, bisexual, transgender, queer (or questioning)

MAID: "Medical Aid in Dying," prescribing lethal drugs to capacitated terminally ill patients

National Academies (US): National Academies of Science, Engineering, and Medicine

NIH: US National Institutes of Health, founded in 1887 as the Hygienic Laboratory

NSDAP, NAZI: National Socialist German Workers Party (1920–1945)

NWRO: National Welfare Rights Organization, George Wiley, founder (1966–1975)

Ombudsman: Scandinavian for a (traditionally male) independent conflict mediator

Operation Overlord: Code name for June 6, 1944, Allied invasion of Nazi-occupied Europe

PRIM&R: Public Responsibility in Medicine & Research, society for IRB members (1974–)

PAS: Physician-assisted suicide; legal in many US states; euthanasia is not legal in US

RAMC: Royal Army Medical Corps of the United Kingdom

Rassenhygiene: Racial hygiene, 1895 neologism for eugenic social medicine for *Volkskörper*

RCT: Randomized controlled trials: a method of limiting experimenters' or statistical biases

RCP: Royal College of Physicians (founded 1518)

R3 Rules Regarding Animals: Reduce number, Refine to mitigate, Replace with alternatives

Salus aegroti suprema lex: Latin for "the health of the patient is the supreme law"

Satellite regulators: Local institutional regulators enforcing federal policies (IACUCs, IRBs)

Scientistic paternalism: "Doctors as lords almighty" by virtue of their scientific expertise

Shell shock: Now called "posttraumatic stress disorder" (PTSD)

Slow/show coding: Feigning CPR to appear to adhere to regulatory or legal requirements

SNCC: Student Nonviolent Coordinating Committee, civil rights action group (founded 1960)

Social medicine: Public health policies to improve social conditions affecting people's health

SS: *Schutzstaffel*, paramilitary Nazi "protection squad" used for violent intimidation

STI: Sexually transmitted infection (formerly called "venereal diseases")

Total institutions: Organizations with regulated populations cut off from society (e.g., armies)

UN: United Nations, a global organization representing almost all the world's nations

UNESCO: United Nations' Educational, Scientific and Cultural Organization

USPHS, or PHS: US Public Health Service

USAF or USAAF: US Air Force, US Army Air Force

Vivisection: Experiments on humans or other animals to advance biomedical knowledge

***Volk*:** People connected culturally, genetically, geographically, and linguistically

***Volkskörper*:** Literally, "body politic": a conception of a people, as a single organic entity

VRS: Vivisection Reform Society, patient protection movement founded in Chicago (1903)

***Weltanschauung*:** A group or person's comprehensive view of the world

WHO: World Health Organization, United Nations' international health care organization

WMA: World Medical Association, international association of national medical societies

Woke: Morally awakened to unethical conduct or social injustice

Who's Who: People Referenced

Abram, Morris (1918–2000), prosecutor Nuremberg Trials, President's Ethics Commission Chair

Alexander, Leo (1905–1985), Austrian American physician, coauthor of the Nuremberg Code

Annas, George J. (1945–), US lawyer, patients' rights advocate, Boston University professor

Appiah, Kwame Anthony (1954–), American Ghanaian philosopher of moral revolutions

Aristotle (384–322 BCE), Greek philosopher, often cited as "The First Scientist"

Asperger, Hans (1906–1980), Austrian physician, discovered "Asperger's syndrome"

Bacon, Francis (1561–1626), English philosopher urged euthanasia, end-of-life palliative care

Baker, Robert (1937–), Bronx-raised US philosopher, bioethicist, historian of medical ethics

Baxter, Richard (1615–1691), English preacher, preached about "true scandals"

Beauchamp, Tom (1939–), US philosopher, coauthor of *Principles of Biomedical Ethics*

Beecher, Henry K., née Unangst (1904–1976), US anesthesiologist, veteran, whistleblower

Beecher, Henry Ward (1813–1887), Brooklyn, New York; pastor, anti-slavery activist

Bentham, Jeremy (1748–1832), English philosopher, founder of modern utilitarianism

Bigelow, Henry (1818–1890), US physician, Harvard professor, anti–animal vivisectionist

Booth, Sir Christopher (1924–2012), English naval frogman, physician, historian, author

Bosk, Charles (1948–2020), US medical sociologist, analyzed clinical morality

Brady, Joseph (1922–2011), US World War II veteran, psychologist, and neuroscientist

Brandt, Alan (1953–), US historian of medicine and science, Harvard University

Brandt, Karl (1904–1948), German physician, director Aktion T4 *euthanasie* program

Bronowski, Jacob (1908–1974), British-Polish mathematician and historian

Büchner, Franz (1895–1991), German pathologist who publicly criticized Nazi *euthanasie*

Buxtun, Peter (1937–), Czech-born US lawyer, Vietnam veteran, USPHS whistleblower

Cannon, Walter (1871–1945), US physician, Harvard physiologist, founder of AMA's CDMR

Caplan, Arthur (1950–), US philosopher (NYU), bioethicist, *When Medicine Went Mad*, editor

Capron, Alexander (1944–), US lawyer, bioethics pioneer, founding Hastings Center Fellow

Carpenter, Daniel, (1967–), US social scientist, Harvard University professor

Carroll, James (1854–1907), US physician, researcher in Reed's yellow fever experiment

Chalmers, Thomas (1917–1995), US physician at Harvard, Mt. Sinai, headed NIH and VA

Chamberlain, Neville (1869–1940), British prime minister, Conservative Party, 1937–1940

Childress, James (1940–), US moral theologian, coauthor of *Principles of Biomedical Ethics*

Cibrie, Paul (1881–1965), French, secretary general of a French medical association.

Clegg, Hugh (1900–1983), UK physician, editor of *British Medical Journal* (1947–1965)

Cooper, Cynthia (1962–), US accountant, whistleblower, *Time* "Person of the Year," 2002

Cutler, John (1915–2003), director of USPHS's STI Division, led Study of Untreated Syphilis

Darwin, Charles (1809–1882), UK biologist, theorist of evolution by natural selection

Downer, Carol, (1933–), US lawyer, feminist, founder of women's self-help health movement

Edelson, Paul (1943–), US physician, author, educator, historian of medicine

Edelstein, Ludwig (1902–1965), German-Jewish classicist, translated Hippocratic Oath

Ehrlichman, John (1925–1999), US lawyer, Air Force veteran, counsel to President Nixon

Eisenhower, Dwight D. (1890–1969, US president 1953–1961), commander, Operation Overlord

Engelhardt, H. Tristram, Jr., MD, PhD (1941–2018), US founding bioethicist, libertarian theorist

Evans, John H. (1965–), US sociologist, studied bioethics discourse, U. California, San Diego

Fitzgerald, A. Ernst (1926–2019), US engineer, whistleblower about USAF's $640 toilet seats

Fleming, Arthur (1905–1996), US Commissioner on Aging, funded patient representatives

Fletcher, Joseph F. (1905–1991), US Christian theologian, educator, situational ethics founder

Fox, Renée (1928–2020), US sociologist, tracked development of bioethics from its origins

Freedman, Benjamin (1951–1997), Canadian bioethicist, coins "theoretical equipoise"

Gallinger, Jacob (1837–1918), US senator, homeopath, Vivisection Reform Society founder

Galton, Francis (1822–1911), UK, coined "eugenics," improving humans through breeding

Garland, Joseph (1893–1973), US, editor, *New England Journal of Medicine* (1947–67)

Gibson, Count (1921–2002), US physician, first to protest USPHS's untreated Syphilis Study

Gisborne, Thomas (1758–1846), English Anglican priest, poet, antislavery crusader, ethicist

Gleeson, Geraldine A. (1911–), US CDC statistician, discouraged Jenkins's Tuskegee protests

Gore, Martin, CBE (1951–2019), English physician, pioneering oncologist

Gorovitz, Samuel (1938–), US philosopher, founds bioethics training camp for philosophers

Green, Dwight (1897–1958), US governor, state of Illinois, convened ethics commission

Gregory, John (1724–1773), Scottish physician, medical educator, and moral philosopher

Gustafson, James (1925–2021), influential US professor of Christian ethics at Yale

Halberstam, Michael (1932–1980), US cardiologist, author, critic

Heidegger, Martin (1889–1976), German philosopher, educator, championed Nazism

Hellegers, André (1926–1979), Dutch, "Pope's Biologist," founds, KennedyEthics Institute

Hill, Sir Austin Bradford (1897–1991), UK epidemiologist, pioneers RCT methodology

Himmler, Heinrich (1900–suicided 1945), German SS *Führer*, architect of the Holocaust

Hitler, Adolf (1889–1945), Austro-German Chancellor, Nazi party *Führer*/leader (1921–1945)

Hobbes, Thomas (1588–1679), British empiricist philosopher and contractarian theorist

Hoche, Alfred (1865–1943), German psychiatrist, urges killing of lives unworthy living

Hufeland, Christoph (1816–1898), German physician, advocates palliative care for the dying

Hughes, Justice Richard (1909–1992), New Jersey judge, ruled in Karen Quinlan case

Hyman, William (1893–1966), New York lawyer and patients' rights advocate (Southam case)

Ivy, Andrew C. (1893–1978), US physician, coauthor of the Nuremberg Code

Javits, Jacob (1904–1986, NY senator 1957–1981), introduces FDA consent requirement

Jahr, Fritz (1895–1953), German theologian, urges a Kantian-Buddhist "*bio-ethik*" in 1927

Jenkins, William (1945–2019), African American CDC statistician, protests Syphilis Study

John XXIII, (1881–1963), Pope 1958–1963, initiates Vatican II to reform the Catholic Church

Jones, James (1943–), US Arkansas historian, wrote first monograph on PHS's Syphilis Study

Jonsen, Albert (1931–2020), US theologian, founding bioethicist, and historian of bioethics

Jost, Adolf (1874–1908), Austrian psychologist, author of *The Right to Death* (1895)

Kamowitz, Gabe (1935–), US independent lawyer, lifelong advocate for the disadvantaged

Kant, Immanuel (1724–1804), German philosopher, redefined Greek concept of autonomy

Katz, Jay (1922–2008), German American USAF veteran, psychiatrist, founding bioethicist

Kefauver, Estes (1903–1963), US senator (1949–1963) amended the US Food & Drug Act

Kelsey, Frances (1914–2015), Canadian American FDA official, champions informed consent

Kennedy, Edward, "Ted" (1932–2009), US senator, holds first hearings on "bioethical" issues

Kennedy, John (1917–assassinated 1963), US president 1961–1963, gave Kelsey medal

Kennedy, Robert, (1925–assassinated 1968), US ex-attorney general, civil rights reformer

King, Martin Luther, Jr. (1929–assassinated 1968), US minister, nonviolent rights activist

King, Patricia (1942–), African American, Belmont commissioner, Georgetown Law School

Koestler, Arthur CBE (1905–1983), Hungarian British anticommunist author

Kroprowski, Hilary (1916–2013), Polish American virologist, developed a polio vaccine

Kübler-Ross, Elisabeth (1926–2004), Swiss-US psychiatrist, *On Death and Dying* (1969)

Kuhn, Thomas (1922–1996), US philosopher, author of *Structure of Scientific Revolutions*

Lasagna, Louis (1923–2003), Beecher's assistant for Army and CIA drug experiments

Lazear, Jessie (1866–1900), US Army physician, died as a yellow fever research subject

Lebacqz, Reverend Karen (1945–), US Christian moral theologian, founding bioethicist

Lederer, Susan E. (1955–), US historian of medicine and medical ethics, U. of Wisconsin

Leibbrand, Werner (1896–1974), anti-Nazi German psychiatrist and medical historian

Lenz, Fritz (1887–1976), German, author of textbook on *Rassenhygiene*, Widukind's father

Lenz, Widukind (1919–1995), German pediatrician, showed thalidomide causes birth defects

Leopold, Nathan (1904–1971), incarcerated US "Nietzschean" murderer, Jolliet prison

Lifton, Robert Jay (1962–), US psychiatrist, analyzed psychology of Holocaust perpetrators

Linnaeus, Carl (1707–1778), Swedish botanist, invented "scientific" racial classifications

Luther, Martin (1483–1546), German priest, theologian, catalyzed the Protestant Reformation

Lysaught, M. Therese (1963–), US, moral theologian, Neiswanger Institute, Loyola University

Macklin, Ruth (1938–), US philosopher, feminist, bioethicist, Einstein Medical College

Maehle, Andreas-Holger (1957–), German historian of medicine, Durham University, UK

Marcuse, Herbert (1898–1979), German-Jewish refugee, Frankfurt school philosopher

Mellanby, Kenneth CBE (1908–1993), UK entomologist, experiments on conscientious objectors

Mill, John Stuart (1806–1873), English utilitarian moral philosopher, Bentham's godson

Miller, Franklin (1948–), US philosopher, bioethicist, National Institutes of Health (NIH)

Moll, Albert (1862–1939), German physician, wrote *Ärztliche Ethik* (*Physicians Ethics*, 1902)

Moore, George E. (1873–1958), UK philosopher, Cambridge, wrote *Principia Ethica* (1903)

Moreno, Jonathan (1952–), US influential philosophically trained founding bioethicist

Munk, William (1816–1898), British physician, advocated palliative care for the dying

Myerson, Abraham (1881–1948), Lithuanian American neurologist, anti-eugenicist

Neisser, Albert (1855–1916), famous German physician punished for unconsented research

Nixon, Richard M. (1913–1994), US president 1969–1974, resigned facing impeachment

Nowell-Smith, Patrick (1914–2006), English philosopher of epistemological moral relativism

Nuland, Sherwin (1930–2014), US surgeon, writer, medical educator, bioethicist, Yale

Olansky, Sidney (1914–2007), US director of CDC's STI Research and its Syphilis Study

Pappworth, Maurice (1910–1994), English physician and whistleblowing moralist

Paul VI, Pope (1897–1978, Pope 1963–1978), 1968 *Humanae Vitae* forbids birth control

Percival, Thomas (1740–1804), UK physician, wrote *Medical Ethics* (1803), coined the term

Perkins, Maxwell (1884–1947), US, famous editor for Scribner Publishers (New York City)

Plato (circa 429–347 BCE), Athenian philosopher, student of Socrates, teacher of Aristotle

Pius XII, Pope (1876–1958, pope 1939–1958), OKs disconnecting "extraordinary" life support

Ploetz, Alfred (1860–1940), German eugenicist, wrote *Foundations of Racial Hygiene* (1895)

Potter, Van Rensselaer (1911–2001), US physician, dubs "bioethics" ethics of the biosphere

Pridham, John (1891–1965), English physician, World War I veteran, cofounder of WMA

Quinlan, Karen (1954–1985), persistently comatose woman, subject of 1976 legal case

Ramm, Rudolf (1887–1945), German physician, SS, author of Nazi medical ethics textbook

Ramsey, Paul (1913–1988), influential Methodist moral theologian at Princeton University.

Rascher, Sigmund (1909–1945), German physician convicted of unethical experiments

Ravich, Ruth (1921–2001), founder-director, Mount Sinai Patients' Advocacy program

Reed, Walter (1851–1902), US Army physician, yellow fever researcher, used consent forms

Reich, Warren (circa 1932–), KIE scholar, edits first edition of the *Encyclopedia of Bioethics*

Reverby, Susan (1946–), US historian, expert on Tuskegee Syphilis Study, Wellesley College

Rockwell, Donald (1931–2013), Lt. Colonel USPHS, researcher on USPHS's Syphilis Study

Rothman, David (1937–2020), US historian of medicine, wrote first history of bioethics

Rutstein, David (1909–1986), US physician, Harvard University, hosted popular TV series

Sabine, Albert (1906–1993), Polish American developer of a live virus oral polio vaccine

Salk, Jonas (1914–1995), US virologist, developer of a killed polio virus vaccine

Schatz, Irwin (1931–2015), Canadian American cardiologist, protests PHS's Syphilis Study

Schilling, Klaus Karl (1871–1946), German physician, experimented on Dachau prisoners

Schmidt, Ulf, (1967–), German historian, director of ethics center, Hamburg University

Schweitzer, Albert (1875–1965), Alsatian physician, philosopher, won Nobel Peace Prize

Scribonius Largus (circa 1–50 CE), Greek physician to Romans, invoked Hippocratic Oath

Seldon, Joanna (1954–2016), UK educator, Maurice Pappworth's daughter and biographer

Serpico, Frank (1936–), New York City police officer and whistleblower

Silkwood, Karen (1946–1974), US chemical technician, union organizer, whistleblower

Southam, Chester (1919–2002), US oncologist, implanted cancer cells in uninformed patients

Spencer, Herbert (1820–1903), English bio-philosopher, coined "survival of the fittest"

Spinelli, A. (fl. 1960s), Italian, chair WMA Committee drafting 1964 Helsinki Declaration

Sternberg, George (1838–1915), US physician, researcher, surgeon general 1893–1902

Stowe, Harriet Beecher (1811–1896), US author of antislavery novel, *Uncle Tom's Cabin*

Swain, Margaret (1906–1973), US wife of Harvard whistleblowing physician, Henry Beecher

Swazey, Judith (1928–2020), US sociologist, tracked development of bioethics from origins

Taber, Sydney (1862–1930), US lawyer, founder the Vivisection Reform Society (VRS, 1883)

Tacitus, Publius Cornelius (56–120 CE), Roman historian and politician

Talbott, John (1902–1990), US WW II colonel, *JAMA* editor, rejected Beecher's 1966 article

Taylor, Telford (1908–1998), US general and lead prosecutor at Nuremberg Trials

Toulmin, Stephen (1922–2009), UK moral philosopher and founding bioethicist

Trapasso, Reverend Thomas (1924–2012), spiritual advisor to Karen Quinlan's parents

Turnberg, Barron Leslie (1934–), UK president of Royal College Physicians, 1992–1997

Veatch, Robert (1939–2020), founding bioethicist, member of Hastings and Kennedy institutes

Virchow, Rudolf (1821–1902), German physician-pathologist, founds social medicine

Voncken, Jules (1887–1975), Belgian surgeon general, founds Military Medicine Conference

Walters, Leroy (1940–), moral theologian, first director, Kennedy Institute's Bioethics Center

Watkins, Sharron (1959–), US Enron whistleblower, *Time* Person of the Year 1962

Whewell, William (1794–1866), English philosopher, coined expression "scientific revolution"

Wiley, George (1931–1973), African American NWRO leader, negotiats Patients' Bill of Rights

Wilhelm II, Kaiser (1859–1941), emperor of Germany and king of Prussia (1888–1918)

Williams, Bernard (1929–2003), English moral philosopher, Cambridge University

Williams, Samuel (fl. 1870s), English teacher who redefined euthanasia as mercy killing

Wittgenstein, Ludwig (1889–1951), influential Austrian-English mystic-logician-philosopher

Yesley, Michael, US lawyer, director of National Commission Protection of Human Subjects

Yobs, Anne (1929–2010), CDC physician, censors Schatz's protest letter on Syphilis Study

1

Moralists and True Scandals: Medical Ethics during the Pre- and Interwar Periods

Mistake not (with the vulgar) the nature of scandal, as if it lay in offending men, which is nothing but grieving or displeasing them; or in making yourself to be of evil report; . . . So that all men must avoid whatever a censorious person will call scandal, when he means nothing else himself by scandal than a thing that is of evil report with such as he . . . *true scandals* are those "which tempteth [people] into sin . . . and maketh [them] stumble, or occasioneth them to think evil of a holy [or good] life."
—Richard Baxter, 1673[1]

On Nazi Medical Ethics: A *True* Scandal

Modern medical ethics arose as a reaction to a *true* scandal, the sort of scandal that seventeenth-century Protestant minister Richard Baxter (1615–1691) preached about when he warned his congregants, "Mistake not (with the vulgar) the nature of scandal," *true scandals* are "a murdering of souls . . . a stumbling block or temptation by which a man is in danger of falling into sin . . . or occasioneth them to think evil of a holy [or good] life."[2] Physicians' complicity in the Holocaust is a *true* scandal, but perhaps the most scandalous aspects of Nazi doctors' complicity in genocide was not just their lack of contrition; it was replacing a medical ethics of caring, healing, and curing, with an ethics in which, to use the language of Isiah 5:20, they "call[ed] evil good, and good evil; put darkness for light, and light for darkness; put bitter for sweet, and sweet for bitter!" In this book, I relate how, on discovering this true scandal, veteran military physicians tried to ensure that "never again" would medical professionals lose sight of what constitutes good medical practice, inventing, in the process, the foundations for modern medical ethics.

Among the change-agents who sought to prevent a recurrence of the truly scandalous conduct of the German physicians who engaged in unethical experiments and facilitated genocide was a one-eyed horseback-riding

World War I British colonel who revived an ancient oath to save the soul of modern medicine. Another change-agent was a former eugenicist who rejected eugenics as a cult rather than a science. He collaborated with a religious Christian who, in defiance of St. Paul's warning not to do evil that good may come of it, perjured himself, or came perilously close to doing so, for a greater good. Some decades later, bioethical reform was championed by a guilt-ridden bible-reading Methodist who, in seeking absolution from his own abusive conduct, risked rejection by his colleagues to champion moral reform and, in the process, transformed himself from villain to hero— although some believe his villainy outweighed his heroism. Another heroic moral reform change-agent was an afro-wearing African American scholar who led single women with dependent children to protest academic medical centers' condescending exploitation of African Americans, older Americans, single mothers, and other welfare recipients. The book introduces readers to their stories and relates stories of numerous Jews who, time and again, played the role of coal mine canaries, sounding the alarm about abuses reminiscent of those suffered by their co-religionists during the Holocaust.

Another theme in this book is the scandal-to-ethics code cycle. It might seem odd to think of codes of ethics as responses to scandals, but if, like Baxter, one looks to the Hebrew bible for guidance, one will find that the most influential ethics code in Western civilization, the Ten Command- ments, was written in response to a true scandal. As told in the Book of Exodus, when Moses first descended from Mount Sinai, he cradled in his arms two tablets inscribed with Jehovah's commandments to the Israelites. However, on entering the Israelites' camp, Moses discovered them worship- ping a false god, the golden statue of a calf. As reported in Exodus 32–33, Moses' "anger waxed hot, and he cast the tablets out of his hands and brake them beneath the mount." After a period of conflict, Moses again ascended Mount Sinai. He later descended with a second set of tablets on which are engraved the Ten Commandments as we now know them. Exodus is silent about what was inscribed on the first set of tablets, but the second set is a post-scandal code of ethics designed not as a code for a good God-fearing people but as a code for disobedient calf-worshiping ingrates so morally obtuse that they had to be told that it was wrong to kill fellow humans.

The foundational oaths, codes, principles, and rights statements of mod- ern medical ethics, from the Nuremberg Code to A Patient's Bill of Rights to the *Belmont Report*, follow a similar scandal-to-code pattern: they too were formulated in response to conduct so truly scandalous that two physicians inscribed new commandments setting forth ten requirements for morally permissible experiments on fellow humans. Later dubbed "The Nuremberg

Code," this formulation of the ethics of experimenting on humans was inspired by German physicians' lack of contrition for treating humans as if they were mere experimental biological material. Their lack of contrition for this and for their participation in genocidal attempts to eliminate Jews, gays (LGBTQ+), people with disabilities, and Roma in the Holocaust so profoundly appalled the founders of the World Medical Association that they rewrote an ancient oath to ensure that future physicians would never engage in similar moral scandals. Their efforts laid the foundation for what could be called its bioethical turn in modern medical ethics—that is, a post-1970s version of medical ethics that fuses ideals of transparency with egalitarian values and new norms recognizing patients' rights.

On Moralists

Moses was a religious moralist. The defining characteristic of moralists, whether secular or religious, is that they see as immoral or sacrilegious acts or practices that their community views as morally or religiously acceptable. According to John Stuart Mill (1806–1873), his godfather, Jeremy Bentham (1748–1832), founder of the utilitarian reform movement and of modern utilitarian philosophy, became a moralist—and later a moral philosopher—because he was shocked by the legal community's blind acceptance of what he saw as immoral conduct. As Mill tells the tale, while starting his career as a lawyer, the young Bentham encountered a "particular abuse that first gave shock to his mind ... the custom of making the client pay for three attendances in the office of a Master in Chancery, when only one was given."[3] What impressed Mill about Bentham's reaction to his fellow lawyers' routine practice of overcharging their clients for three visits when only one visit was performed was that it

> was known to every lawyer that practiced, to every judge that sat on the bench, and neither before nor for a long time after did they cause any apparent uneasiness to the consciences of these learned persons. ... During so many generations, in each of which thousands of educated young men were successively placed in Bentham's position and with Bentham's opportunities, he alone was found with sufficient moral sensibility and self-reliance to say to himself that these things, however profitable they might be, were frauds and that between them and him there should be a gulf fixed. To this rare union of self-reliance and moral sensibility we are indebted for all that Bentham has done.[4]

Mill puts his finger precisely on what distinguishes moralists, like Bentham, from those who merely have moral qualms: Bentham recognized, condemned, and publicly refused to participate in conduct that everyone in his community either engaged in or found morally tolerable. For generations, hundreds of young lawyers had witnessed and profited personally by systematically overcharging their clients, yet none objected to this bold-faced fraud. Bentham's ability to recognize, renounce, and decline to participate in this practice served, according to his protégé Mill, as a prologue to his later practice of publicly denouncing as immoral such then commonplace and near universally accepted practices as animal abuse and slavery. Bentham publicly condemned and lobbied for abolition of the slave trade and of slavery, two highly profitable practices accepted throughout the British Empire. He also condemned animal abuse, anti-Semitism, women's unequal standing in the law and he championed voting rights for ordinary people, female and male alike.[5]

Bentham also condemned the execution of "sodomists" (i.e., people we would today describe as having an LGBTQ+ sexual orientation). But, fearful of the consequences, he did so only in conversations with his disciples and in unpublished writings.[6] Since his critique was *sub rosa*, on this issue, Bentham was merely a moral philosopher. He was not a "moralist." As I use the word in this book, "moralists" are people who *publicly* decry as immoral practices that their community does not recognize as immoral, even though this community will later condemn these practices as immoral.

One more linguistic note: although Bentham went public with most of his critiques, he was not a "whistleblower," at least not in the sense that this word is currently used. The word "whistleblower" gained currency in the nineteenth century because it was the custom of London police officers to, quite literally, blow whistles to alert fellow officers to criminal acts. The meaning of the term changed in the twentieth century when it came to be viewed as analogous to referees in football and other sports who blew whistles to stop play when they spotted a rule violation. The word gained traction in the media after 1968 when A. Ernst Fitzgerald (1926–2019) "blew the whistle" on a $2.3 billion cost overrun in the Lockheed C-5 aircraft program that had been covered up by Air Force officials.[7] After that, media and the public came to call people who exposed information about immoral or illegal activities by organizations or institutions they were affiliated with "whistleblowers." Whistleblowing is not a matter of moral grandstanding or virtue signaling; it involves publicly exposing information about actual institutional or organization wrongdoing.

These exposés often have significant consequences for both organizations and whistleblowers. Fitzgerald's whistleblowing, for example, led to years of congressional investigations and changes in governmental practices. But, like many other whistleblowers, Fitzgerald paid a price: President Richard M. Nixon (1913–1994, president 1969–1974) retaliated by ordering his assistant, John Ehrlichman (1925–1999), to "get rid of that son of a bitch."[8] Fitzgerald was fired in fall 1969. After years of litigation, Fitzgerald got his old job back and was soon informing media that the US "Airforce was paying $ 400 for hammers and $ 600 for toilet seats. Fitzgerald retired from the Defense Department in 2006."[9]

Other famous whistleblowers from this era were New York City police officer Frank Serpico, who exposed corruption in the city's police department in the 1960s and 1970s. Fellow officers turned against him, but the New York Police Department gave him the department's highest award, the Medal of Honor—and his story became the subject of films, novels, and television shows. Another well-known whistleblower from that period, Karen Silkwood (1946–1974), exposed health and safety violations at a nuclear fuel plant that employed her. She died shortly thereafter under suspicious circumstances. Twenty-first-century accountants Cynthia Cooper and Sharron Watkins "blew the whistle" on malfeasance and corporate fraud in their respective companies, the Enron Corporation and WorldCom, which earned them public acclaim and the accolade "Persons of the Year 2002" from *Time* magazine.

As these vignettes indicate, the word "whistleblowing" gives a positive spin to a person whom others in the community might deem "disloyal," an "informer," or a "snitch." Political scientist Allison Stanger characterizes whistleblowers, as "risk takers for the truth and what they risk, above all, is their own well-being."[10] Not surprisingly, therefore, whistleblowing does not always end happily for those who blew the whistle. Even as a whistleblower like Ernst Fitzgerald was lauded by Congress, the media, and the public, he was denounced, fired, and ostracized by his employers and colleagues as "not a team player." "An Office of Research Integrity report on whistleblowing in science . . . showed that two-thirds of whistleblowers experience negative consequences from speaking out. One in four loses her or his job. Despite this, in a 1995 study, close to 80 percent [of the whistleblowers interviewed] said they would do it again. [In another study] 86 percent of those who had been victims of retaliation claimed they would be willing to report again, versus 95 percent who did not suffer retaliation."[11] Worse yet, studies show that "only a small percent of federal whistleblowers succeed in having their assertions believed—5 to 20 percent."[12]

The situation is even worse for those I characterize as "moralists." Whistleblowers can be believed as much as 20 percent of the time because they are exposing conduct that, although tolerated or encouraged by their organizations, is nonetheless deemed immoral or otherwise improper by the wider community, in the law, or by others in their profession. Public exposure of $2.3 billion cost overruns or of $400 hammers and $600 toilets, for example, invites criticism and ridicule by other officials, media, and the public. Safety violations at nuclear plants are dangerous, conjuring up nightmare visions of Chernobyl or Three Mile Island. However, when moralists like Moses publicly denounce calf worshiping, they condemn a practice that many in the wider community do not view as wrong or at least find tolerable. Consequently, moralists cannot rely on the wider community for support because, as Baxter observed, the "nature of scandal . . . is nothing but . . . displeasing" the community.[13] Thus, communities are likely to reject not only a moralist's condemnation of customary conduct but also the moralists themselves. As seen from the perspective of the prophet-moralist Jeremiah, it is as if the community consisted of "foolish people . . . without understanding; which have eyes, and see not; which have ears, and hear not" (Jeremiah 5:21). Or, in the language of an Evangelical apostle, "they [who] seeing, see not; and hearing they hear not, neither do they understand" (Matthew 13:13).[14]

Albert Moll and the Prussian Ethics Code of 1900

Insofar as the wider community neither sees, hears, nor understands moralists, it may simply ignore them, as the English legal community ignored Bentham's refusal to overcharge his clients, or it may disparagingly dismiss the moralist as a crank or noisome nuisance; alternatively, the community, or its authorities, influencers, and powerbrokers, may seek to silence the moralist. An illustrative example is that of German physician, medical ethicist, and moralist Albert Moll (1862–1939), whose case also offers a glimpse of German physicians' understanding of research ethics during the Weimar Republic (1918–1933)—that is, just prior to the Nazi era (1933–1945).

Born to a Jewish family in Chancellor Bismarck's newly emerging German state, Moll converted to Protestantism in 1895, during the period when Germany was ruled by Kaiser Wilhelm II (1859–1941), whose Wilhelmine reign lasted from 1888 until his abdication in 1918. In 1902, Moll published a massive 650-page book, *Ärztliche Ethik* (*Physicians Ethics*, or *Medical Ethics*), specifying physicians' ethical duties in various aspects of their practice.[15] His conception of medical ethics was founded on a traditional maxim, *Salus aegroti suprema lex*, the health of the patient is the supreme law.[16] Taking this maxim in a new direction, Moll envisioned a tacit social contract specifying physicians' and patients' respective duties to each other viewed from the perspective of *Durchschnittsmoral* (i.e., everyday commonsense morality). This, in turn, led him to believe that, since patients consulted physicians seeking remedies for their ailments, physicians had a reciprocal obligation to promote their patients' health. He then drew the logical conclusion that nontherapeutic medical experiments that might endanger patients' health or life would be ethical only if patients were informed of the possible effects that the proposed experiments might have on them and freely consented to participate. Thus, experimental interventions are only moral if a patient or a patient's surrogate is informed and voluntarily consents to them.

Although this may seem self-evident in our bioethics era, it was a relatively new concept in Germany of the Wilhelmine period. Hospital physicians of that period saw themselves as "absolute rulers over their 'clinical material' [i.e., the charity patients in their wards and thus used them] . . . as needed for their research interests. The question of consent was viewed to be for the most part irrelevant."[17] Reflecting this attitude, a researcher could unabashedly mention in a medical publication that he tested a new vaccine "on children in a foundling hospital . . . because the planned [subjects], calves, [had] been too expensive in price and maintenance; [another researcher] reported on 'feeding' round worm cultures to children, observing in a passing remark that it is much easier to induce the roundworm [infestation] than to eliminate it completely from the body"[18]—thus implying that his experiments may have left his child subjects with lifelong worm infestations.

Not content to make purely philosophical pronouncements, anticipating the methods of empirical bioethics, Moll decided to measure the extent of unethical research by examining articles published in leading German and international medical journals. He eventually amassed "a collection of approximately six hundred examples [of unethical experiments] from the national [i.e., German] and international professional literature."[19] Curious about why unethical research was so commonplace, Moll examined "thank you notes" in the prefaces and endnotes of published papers. Reading them revealed that "clinic directors made the 'material' [i.e., patients] 'kindly available' to young researchers [who] stated 'their humble thanks' in their publications."[20] He also found that "dying patients were viewed as 'corpses vile'; and thus 'appropriate material' for experiments . . . [and] that patients were apparently viewed as objects, animals, or slaves which the physician believed he could use with impunity at his discretion."[21]

Moll reluctantly concluded "with increasing surprise that some medics, obsessed by a kind of research mania, have ignored the areas of law and morality in a most problematic manner. For them, the freedom of research goes so far that it destroys any consideration for others. The borderline between human beings and animals is blurred for them. The unfortunate sick person who has entrusted herself to their treatment is shamefully betrayed by them, her trust is betrayed, and the human being is degraded to a guinea pig. . . . There seem to be no national or political borders for this aberration."[22] One point worth noting is that at that time, neither the editors of medical journals nor their readership and, presumably, none of the researchers conducting these experiments seemed to recognize these experiments as immoral.

Moll's objective in documenting the abuse of human patients as guinea pigs was not to name and shame malefactors but to change the regulations governing medical research, and—mindful of the consequences of alienating his medical colleagues—he never published the names of the miscreant researchers whose experiments he condemned, the names of the institutions at which they conducted their research, or the journals from which he had culled his data. His reticence to ascribe blame to fellow medical professionals was challenged in the aftermath of a headline-grabbing scandal involving Albert Neisser (1855–1960), an internationally renowned German physician and researcher who isolated the bacterium that causes gonorrhea (gonococcus) and who co-discovered the causative agent of leprosy. Neisser's research became controversial after he published a paper in 1898 in which he described inoculating prostitutes and young girls with an experimental vaccine designed to immunize them against syphilis.[23] After word of the experiment attracted the attention of the media, the case found its way onto the floor of the Prussian parliament and into the law courts. It eventually became entangled with anti-vivisectionist and anti-Semitic agitation (since Neisser was Jewish).

Testimony by Neisser and other physicians revealed their blatant dismissal of the necessity of obtaining a research subject's consent. Consent was, Neisser testified, "a mere matter of form of completely secondary importance."[24] "If it had been a matter of formally covering up for something," Neisser explained, "I would certainly have gotten their consent, since there is nothing easier than getting ignorant persons to agree to what you want by means of friendly persuasion when it concerns such a harmless, everyday matter as an injection."[25] The females he used as subjects in his experiment were simply told that the injection "was going to cure their illness." This "corresponded to contemporary practice, as becomes clear in many documents, as well as to the views of most of his contemporaries."[26]

Nonetheless, the court found Neisser guilty of a breach of professional responsibility for his failure to acquire the informed consent for his nontherapeutic interventions on the young girls. They fined him an amount equal to about two-thirds of his annual salary.[27]

Scandal playing its characteristic reform-inducing role, in 1900, the Prussian ministry responsible for health care issued an edict to all public hospitals prohibiting nontherapeutic interventions (*Eingriffe*) on minors and noncompetent people and requiring researchers to obtain "unambiguous consent" from "human subjects" who have received "a proper explanation of the possible negative consequences of the intervention."[28] Researchers would also have to obtain a medical director's approval before initiating any nontherapeutic intervention on patients and would have to document their compliance with these requirements. Therapeutic or preventive interventions were exempt:[29] this exemption mirrors the court's findings in the Neisser case where the unconsented injection of experimental serum into prostitutes was deemed legally acceptable because the intent was preventative (i.e., Neisser was inoculating them against a disease that they were exposed to as an occupational hazard). However, since the injection was presumed not to be preventive for well-behaved young girls, it required informed consent from the girls' parents.

Published two years after the 1900 Prussian health ministry regulations, Moll's *Ärztliche Ethik* attracted the attention of the authorities, who asked him to share his data on the unethical experiments. Moll complied. After reviewing Moll's data, the ministry concluded that Moll's cases were "totally harmless." Sharing the views of Neisser and the wider German medical community of that period, the health ministry official could see nothing immoral in Moll's examples of unconsented experiments. What the ministry clearly discerned, however, was that Moll was a threat to the reputation of German experimental medicine. So, the minister ordered a police investigation of Moll that, Moll believed, quashed his appointment to a university professorship.[30] Refusing to be intimidated, Moll continued to play the role of moralist, speaking out against such abuses as the "patient trade"—a fee-splitting scheme in which specialists financially compensated those who referred patients to them (i.e., specialists gave them kickbacks). Historian Andreas Holger-Maehle observes that "Moll ... established himself as an 'expert' in medical ethics ... at a reputational cost. [Fellow physicians] ridiculed Moll as 'God's ethicist' [*Ethiker von Gottes Gnaden*] and as a 'medical Sherlock Holmes' [whose] campaign against the ... [medical] professors had damaged the international reputation of German medicine and was 'anti-German' [*deutschfeindlich*]. The latter remark may well have been understood by informed readers as an anti-Semitic gesture towards Moll's Jewish descent."[31]

Moll's experience illustrates the moralist's dilemma: insofar as the wider community and its authorities do not share the moralist's perception of immorality, when moralists speak out, they risk ostracism and censure by their immediate community and the wrath of its authorities. In the German medical community of the early 1900s, researchers subscribed to the view that consent was only required for potentially physically harmful "injury [that] . . . probably would result in long-term damage or death."[32] Consequently, as Moll discovered, since "it was impossible [for dying patients] to incur long-term injury . . . nothing stood in the way of [infecting them] with serious disease from the [accepted] moral point of view."[33] Not surprisingly, therefore, given the German medical community's understanding of what constituted ethical research, during the dozen years that the German health ministry conducted official surveys of the German medical literature (1901 to 1913), they found only six ethically suspect cases: fewer than one per year. On further investigation, however, viewed through the eyes of medical officialdom, not one patient had been abused during the dozen years surveyed.[34] In the words of Jeremiah and Matthew, they were morally blind; they had eyes but saw not. This was also true of most of the scandals that led to the formation of modern medical ethics. They arose in contexts in which both the medical community and the officialdom that had oversight responsibilities saw nothing unethical or scandalous in practices that would later be deemed immoral. Moreover, since neither the community nor its powerbrokers shared the moralist's vision of morality, they were more likely to ignore, ridicule, silence, or persecute the moralist than to change customary conduct.

2

Nazi Medical Ethics

During the powerful upheaval of the intellectual and moral structure of the German *Volk* ... the National Socialist revolution ... there was also a fundamental
rearrangement of the ideal conception of medicine. The overpowering individualism of the liberal age had also influenced the thinking of the physician and produced a purely individualistic professional conception of the physician and the
entirety of medical science ... with the inexorable ... racial decline and stepped
with continuously growing clarity towards the day of the death of the *Volk*. There
was however no way for them to stop this catastrophe. ...

   A complete change first occurred when Adolf Hitler succeeded in snatching
the German *Volk* back from the brink of decay ... to show the way from the doctrine of the individual to that of becoming the physician to the nation. ... Fulfilling
these new duties presupposes that each individual physician must change his attitude, and that the entire medical community must undertake a moral-intellectual
renewal.
—Rudolf Ramm, 1943[1]

The above prefatory quotation confirms John Stuart Mill's observation
that "all political revolutions not effected by foreign conquest originate
in moral revolutions."[2] Like other early proponents of National Socialism, Rudolf Ramm (1887–1945) viewed himself as leading a moral and
political revolution to reverse the decay of the German *Volk* caused by "the
overpowering individualism of the liberal age."[3] What they were rejecting as "the liberal age" was Germany's post–World War I Weimar period
(1918–1933), during which a democratic republic survived a period of
hyperinflation to prosper in a brief "golden era" from 1924 to 1929. The
stock market crash of 1929 and the ensuing global depression and socioeconomic collapse (1929–1932) led to a period of paramilitary clashes
undermining Weimar's legitimacy, eroding faith in democracy in Germany
(and elsewhere) and paving the way for the 1933 electoral victory of the
Nationalsozialistische Deutsche Arbeiterpartei (National Socialist German
Workers Party, or NSDAP), a right-wing nationalist and socialist party
commonly referred to in English as "the Nazi party."

It was not unreasonable for the German electorate and its intellectuals, influencers, and powerbrokers to reject "liberal individualism." National Socialism appealed to them because the dominant liberal democratic model had failed to provide political order or to offer the prospect of a socioeconomic recovery. When a community comes to believe that its current forms of moral or political governance have failed to serve such basic functions as mediating conflict and facilitating cooperation, protests and insurgencies naturally erupt. Insurgencies alone, however, would be insufficient to culminate in a successful revolution without the support of an active dissident intelligentsia (actors, artists, journalists, philosophers, playwrights, religious figures, scientists, writers, etc.) supplying alternative justificatory concepts and principles that promise to resolve problems seeming unresolvable on the currently accepted morality or form of governance. Thus, leading German intellectuals, like Martin Heidegger (1889–1976), raised questions about what should be done if "the West breaks down and cracks at the seams, if that worn out make-believe culture collapses, expends all its greatness in confusion, and smothers in its lunacy?" To Heidegger and many other German intelligentsia, the National Socialist party seemed to provide an answer. Heidegger joined the Nazi party in 1933 and, in his role as rector of Freiberg University, proclaimed to his students that "the Führer himself [i.e., Adolf Hitler], and he alone, is the German reality of today."[4]

Nazi Medical Ethics

As a populist socialist party of the right, the National Socialists acted as if their electoral victory legitimized a revolutionary moral, social, and political transformation of German society, in much the same way that the French Revolution sanctioned the transformation of French society. In their view, this revolution involved a "breakthrough of a new attitude of mind" in which the formerly dominant liberalistic-materialistic attitude would be replaced by the National Socialist philosophical *weltanschauung* (worldview) that embraced top-down governance based on the *Führerprinzip* (the "leadership principle": one people, one government, *one leader*) dedicated to promoting the interests of the German *Volk* (i.e., the German race), defined as a people sharing common cultural, genetic, geographical, and linguistic heritage, sometimes referred to as common "blood and soil." These ideals fused with two older conceptions of medicine, Rudolf Virchow's (1821–1902) concept of medicine as a social endeavor (social medicine) and a race-based eugenicist conception of preventive medicine extended to include a race's future gene pool, *Rassenhygiene* (racial hygiene).[5] In effect, this new *weltanschauung* replaced Moll's *Salus aegroti suprema lex* ("the health of the patient is the supreme law") with something closer to Cicero's original, *Salus populi suprema lex esto*, "the welfare of the people (*Volk*) shall be the supreme law."[6] Or, as Ramm's textbook on National Socialist medical ethics puts this point, "Even though the ultimate responsibility goes to the healing of patients and the perpetuation of life," the new *weltanschauung* requires "an essential expansion" in physicians' duties "through coming to grips with biological thinking in the National Socialist state," in which the physician's role is to promote the health of the *Volkskörper*, literally the "body politic," conceived as a single organic entity.[7]

Before discussing Ramm's textbook in greater detail, it is important to clarify what is meant by the words "moral" and "ethics" in this book. As I use it, the word "moral" designates standards that members of a community internalize and use to appraise the actions and character of fellow community members and themselves as praiseworthy or blameworthy. The word "ethics" characterizes justifications or critiques of these moral standards—that is, explanations of why some character traits or actions are deemed morally virtuous or praiseworthy, while others are deemed moral vices or morally blameworthy. Communities create moralities for the functional purpose of facilitating cooperation and mediating or preventing conflict. A community's intelligentsia (philosophers, priests, theologians), in turn, develop ethics standards and codes of ethics to justify and explain or critique a community's morality or to defend it against criticisms.[8]

These terms are descriptive. Thus, when I describe and analyze the phenomenon of "Nazi medical ethics," this expression should not be dismissed as not an oxymoron. The Nazis clearly recognized medical ethics and offered ethical justifications of why it is praiseworthy and virtuous for medical practitioners to prevent, mitigate, and cure disease and disability of the *Volkskörper*. What is distinctive about Ramm and other Nazi ethicists is that unlike most commentators on medical ethics, they applied their medical ideals to the German people conceived as one organic entity, the *Volkskörper*, prioritizing its health, and that of its future gene pool, over the health of the present generation and individuals in it. It is important to recognize, describe, and analyze Nazi medical ethics for the same reason that pathologists should study deceased patients, their organs, and tissues: to understand, prevent, and remedy pathological processes. Similarly, historians of immorality need to study moral pathologies to correct misdiagnoses and use the information to recognize and, if possible, prevent its recurrence. As a historian of Nazi medical ethics, Robert Procter observed it is a misdiagnosis, an error, to believe that

> the Nazis abandoned ethics . . . that Nazi doctors' overzealous scientific curiosity led them to abandon all moral sense in the pursuit of medical knowledge. The image is of the unfettered quest for knowledge, a kind of scientific zealotry reminiscent of the Faustian bargain, of science practiced without limits or of an overly aggressive search for the truth. The problem with this view is that there were in fact ethical standards at this time. Medical students took courses on medical ethics, and medical textbooks from the time treated medical ethics. There are discussions in German journals of the obligations of physicians to society, to the state, and sometimes even to the individual. Nazi medical philosophers were critical of the ideal of "neutral" or value-free science, which was often equated with apathetic ivory-tower liberal-Jewish "science for its own sake." Science was supposed to serve the German *Volk*, the healthy and productive white races of Europe. We have to distinguish between no ethics and a lot of bad ethics, between chaos and evil. Surprisingly, there never has been a systematic study of medical ethics under the Nazis.[9]

This chapter does not offer the systematic study of Nazi medical ethics that Proctor envisioned. That would involve, among other things, a detailed study of the Nazis' takeover of the world's first medical ethics journal, *Ethik* (1922–1938).[10] Instead, it focuses on a neologism that Alfred Ploetz (1860–1940) added to the German medical lexicon in 1895, *Rassenhygiene*:[11] a form of social medicine that blends a social Darwinist conception of racial conflict with Mendelian and eugenicist ideas. In the first decades of the twentieth century, *Rassenhygiene* gained widespread acceptance in German medical, scientific, and intellectual circles. Facilitating its spread was the rediscovery of Mendelian genetics in 1900, which supported the conception of race as "hard genetics" (i.e., as a genetic trait not susceptible to environmental or cultural influences). As Proctor points out, "Most of the 20-odd university institutes for racial hygiene [*Rassenhygiene*] were established at German universities before the Nazi rise to power, and by 1932 racial hygiene had become an orthodox fixture in the German medical community . . . [moreover] most of the 15-odd journals of racial hygiene . . . were established long before the rise of National Socialism. . . . Racial hygienists were convinced that many human behaviors were at the root genetic—crime, alcoholism, wanderlust, even divorce."[12] Initially, Proctor remarks, the *Rassenhygiene* movement "was primarily nationalistic and autocratic rather than anti-Semitic or Nordic Supremacist. Eugenicists worried more about the indiscriminate use of birth control among 'the fit,' and the provision of inexpensive medical care (to the unfit). . . . In fact, for Alfred Ploetz, Jews were to be classed along with Nordics as one of the superior 'cultured' races of the world."[13] By the 1920s, however, the National Socialists had appropriated the concept of *Rassenhygiene* and were beginning to call their movement "applied biology": a moral-political movement based on racial interpretation of social Darwinist survival-of-the-fittest biology dedicated to restoring the health and vigor of a racially purified German *Volkskörper*, a social organism composed of multigenerational people serving as the cells that kept the vital organs of this body politic functional.

Another concept that the National Socialists appropriated was "*euthanasie*." At the onset of the seventeenth century, English philosopher Francis Bacon (1561–1626) resurrected the word "euthanasia" from ancient texts and characterized "Outward Euthanasia" as a form of medically assisted dying in which physicians use their "skill . . . whereby the dying may pass more easily and quietly out of life . . . the easy dying of the body [in contrast to] Inward Euthanasia, which regards the preparation of the soul."[14] By the eighteenth century, Bacon's palliative conception of medical care for terminal patients had become a standard medical usage. The 1708 edition of Blanchard's *The Physical Dictionary* defined "euthanasia" as "a soft and easy Passage out of the World without Convulsions or Pain." This ideal of palliative end-of-life care was also disseminated in the works of influential British physician-ethicists like John Gregory (1724–1773) and Thomas Percival (1740–1804) and, later in the nineteenth century, by William Munk (1816–1898). The idea of using opiates to induce palliative care of the dying crossed over into Germany in the writings of the eminent physician Christof Wilhelm Hufeland (1816–1898), who argued that physicians' duties of prolonging life should never override their duty to relieve the suffering of terminal patients by administering opiates. Mercy killing, however, was a step too far. Hufeland believed that it "annihilates the vocation of the physician [who] is bound in duty to do nothing but what tends to save life, whether existence be fortunate or unfortunate, whether life be valuable or not, is not for the physician to decide. If he once permits such considerations to influence his actions, the consequences cannot be estimated, and he becomes the most dangerous person in the community. For if he once trespasses his line of duty and thinks himself entitled to decide on the necessity of an individual's life, he may by gradual progressions apply the measure to other cases."[15]

The Hufeland–Munk conception of euthanasia as palliative care for the dying that stopped short of mercy killing dominated the English-language and German-language medical literature until the 1870s.[16] In 1870, however, definitional discord disrupted the purely palliative conception of "euthanasia." The disrupter, an English schoolteacher, Samuel D. Williams Jr., gave a lecture on "euthanasia" in which he recommended "that in all cases of painful and hopelessness it should be the recognized duty of the medical attendant, whenever so desired by the patient, to administer chloroform . . . so as to destroy consciousness at once and put the sufferer to a quick and painless death; all needful precautions being adopted to prevent any possible abuse of such duty."[17] Williams's essay was soon reprinted as a small book, and within three years, medical journals were filled with denunciations of "euthanasia" as mercy killing. Scandal eclipsing tradition, in a linguistic counterpart to Gresham's law, the deviant mercy-killing usage became so common that it displaced the original palliative-care-for-the-dying usage in the Anglo-American medical lexicon.

Consequently, in 1895, when Austrian philosopher-psychologist Adolf Jost (1874–1908) imported "euthanasia" into German medical lexicon as "*euthanasie*," in his book, *Das Recht auf den Tod* (*The Right to Death*), the word meant "mercy killing." Jost, whose father suicided when he found life offered him no happiness, and who suffered from bouts of mental illness himself, argued that since "there really are cases, in which, mathematically considered, the value of a human life is negative" to offer relief in such cases, the state ought to have a right to kill these people because of the burden they place on others as well as on themselves. The state has this right, Jost argued, because it is analogous to the state's right to conscript citizens into the military where they also face death to protect others.[18]

In the aftermath of World War I, law professor Karl Binding (1841–1920) and liberal psychiatrist Alfred Hoche (1865–1943) coauthored an influential 1920 treatise, *Die Freigabe der Vernichtung lebensunwerten Lebens* (*The Release and Destruction of Lives Unworthy of Being Lived*).[19] Employing Jost's concept of lives of negative value or, as they characterize them, *lebensunwerten Lebens* (lives unworthy of being lived, or lives unworthy of life), they reiterated Jost's claim that the state has a right to kill those whose lives were of negative value. Jost may have been influenced by German philosopher Friedrich Nietzsche's (1844–1900) characterization of the "the sick man as a parasite on society. In certain cases, it is indecent to go on living. To continue to vegetate in a state of cowardly dependence upon doctors and special treatments, once the meaning of life, the right to life, has been lost, ought to be regarded with the greatest contempt by society. . . . A new responsibility should be created, that of the doctor—the responsibility of ruthlessly suppressing and eliminating degenerate life."[20] The National Socialists fused these concepts and lines of argument into a more or less coherent philosophy of applied biology in which, as a matter of *Rassenhygiene*, it was morally permissible to employ sterilization or *euthanasie* to purge the germ pool of the German *Volk* of *lebensunwerten Lebens*.

To implement these ideals, the Nazis had to reject traditional professional and entrepreneurial conceptions of medical occupations. These, in their view, had bankrupted German physicians financially, intellectually, and morally,[21] leading to a decline in the physical, genetic, and demographic viability of the German *Volkskörper*, "to the point of a serious biological crisis. [Had not the National Socialists taken charge] at the end of this process there would ultimately have been a drop off in the number of *Volk*, collapse of the race and finally the death of the *Volk*."[22] The new National Socialist *weltanschauung* promised "a complete psychological change of mind in the German *Volk*, leading to a reawakening, hardening, and a strengthening of their will to live."[23] This new scientifically advanced application of biology, *Rassenhygiene*, would support the German *Volk* by improving social conditions conducive to public health, rather than by prioritizing individual treatment of the sick or the dysgenic dispensation of medical care to the physically or psychologically disabled. Moreover, since the unfit and racially or sexually different threatened to pollute the purity of the *Volk*'s gene pool, they should be treated as "parasites on society"[24] and, like other parasites, quarantined, isolated, or eliminated to facilitate the future health of the gene pool of the German *Volkskörper*.[25]

The lines quoted or summarized in the previous paragraph are from Rudolf Ramm's official National Socialist textbook on governmental health care regulations and Nazi medical ethics. Ramm (1887–1945) was a physician who served in the German medical corps during World War I and who later joined the NSDAP in 1930, a date that is noteworthy because he joined before the National Socialists became the governing party. He then served as a physician to the Nazi's paramilitary "protection squad," the notorious Schutzstaffel (SS). After 1933, Ramm would have sworn the SS oath: "I swear to you, Adolf Hitler—as the Führer and Chancellor of the Reich—loyalty and bravery. I pledge to you and to my superiors, appointed by you, obedience unto death, so help me God."[26] Ramm's textbook, *Ärztlizche Rechts - und Standeskunde Der Artz als Gesundheitserzieher* (*Physicians' Duties and the Rules of the Medical Profession*, hereafter referred to as *Rules of the Profession*), was first published in 1942; a second edition followed immediately thereafter in 1943. The book was at once a National Socialist revolutionary manifesto, a comprehensive presentation of the National Socialist's biomedical *weltanschauung*, and a summary of the governmental regulations affecting health care practitioners in the newly nationalized health care system that converted medical practitioners from self-regulating professionals into (de-professionalized) government employees.

Ramm's book immediately became a required textbook for all German medical students and for postgraduates seeking advanced credentials or degrees. Thus, the book's section on medical ethics became an authoritative primary source for anyone seeking to analyze the National Socialist's revolutionary conception of biomedical ethics; some scholars deem it "the most important known historical source pertaining to the instruction of Nazi medical ethics."[27] Textbooks are often derided as mere pedagogical instruments. Ignoring such academic snobbery, the philosopher of science and historian of scientific revolutions Thomas Kuhn (1922–1996) observed that since a textbook's function is to introduce students to whatever the current generation deems important, they also reflect a profession's understanding of its past. Consequently, textbooks ignore those parts of the past its authors' regard as irrelevant, relating only those elements believed to validate the field's current beliefs and practices. To accomplish this, Kuhn observes,

> Textbooks ... begin by truncating [the reader's] sense of [a] discipline's history and then proceed to supply a substitute for what they have eliminated. Characteristically textbooks ... contain just a bit of history, either in an introductory chapter, or more often, in scattered references to the great heroes of an earlier age. From such references students and professionals come to feel like participants in a long-standing historical tradition. Yet the textbook-derived tradition from which [readers] come to sense their participation is one that, in fact, never existed. ... Partly by selection and partly by distortion ... earlier ages are implicitly represented as having worked on the same set of fixed problems and in accordance with the same set of fixed canons that the most recent revolution ... has made seem [relevant].[28]

Ramm's *Rules of the Medical Profession* offers precisely this sort of truncated selective history of medicine and medical ethics: that is, a history designed to valorize National Socialist medical ethics. Thus, Ramm deems both Hippocratic medicine and the Hippocratic Oath forms of Aryan medicine[29] precursors to the medical ethics of *Rassenhygiene*.[30] Ramm's textbook also transubstantiates famous German physician Rudolf Virchow (1821–1902),[31] who urged physicians to be advocates for social reforms benefiting the health of the poor, into a National Socialist hero because Virchow also proclaimed, "Medicine is a social science and politics is nothing else but medicine on a large scale."[32] Appropriating and reinterpreting the past, Ramm cites the reformation of medical practice in the context of the French Revolution as precedent for the NSDAP's transformation of the German medical profession from relatively autonomous professional organizations into state-run public services directed by the NSDAP.[33] This transformation, he claims, will provide physicians with a more stable source of income, even as it offers the NSDAP control over medical practitioners and their organizations.[34]

Ramm lays out in detail the implications of the National Socialists' view that the medical professions should serve as guardians of the health of the *Volkskörper*. Thus, although physicians have the "responsibility for the healing of patients and the perpetuation of life," given "an essential expansion . . . with [the] biological thinking in the National Socialist state,"[35] they will also be responsible for promoting healthy eating (of whole grain breads, for example), for discouraging the use of tobacco (as carcinogenic),[36] and for obeying "law[s] for the reestablishment of German blood . . . for Prevention of Genetically Ill Offspring . . . for the Defense of the Genetic Health of the German *Volk* . . . which prohibits for all time a further mixing of pure-blooded German people with the Jewish and lower races. *The Sterilization Law* preclude[s] . . . genetically ill and morally inferior people from transmitting their genes."[37] Consequently, health care practitioners were required to report children and adults with disabilities to hereditary courts. (These courts would refer children and, later, adults with disabilities to specialized institutions, such as Hadamar Psychiatric Hospital, where they would be covertly killed.[38]) During this period, even famous physicians, like Johann Asperger (1906–1980), discoverer of the eponymous "Asperger's syndrome," reported children with disabilities to the authorities.[39]

Rules of the Medical Profession was not one of those textbooks whose dicta are dutifully memorized but forgotten after exams have been passed or a certificate or diploma conferred. It is a comprehensive, coherent, but succinct explanation and justification of medical ethics in National Socialist Germany. Thus, when reporters pressed Dr. Karl Brandt (1904–1948) to justify his actions as director of the Aktion T4 program for killing children with disabilities, he responded with a statement that could have been taken directly from the pages of *Rules of the Medical Profession*. "We German physicians look upon the state as an individual to whom we owe prime obedience. We therefore do not hesitate to destroy an aggregate of, for instance, a trillion cells in the form of a number of individual human beings if we believe they are harmful to the total organism—the state."[40] Brandt's explanation was not idiosyncratic. When psychiatrist Robert J. Lifton interviewed physicians who had staffed concentration camps (like Auschwitz and Dachau), he too found that they expressed no signs of remorse because they accepted "the principle of 'racial hygiene' [*Rassenhygiene*, and were] working toward a noble vision of the organic renewal of a vast 'German biotic community' [*Volkskörper*] . . . with a positive mission involving the principle of 'the necessity to sweep clean the world' . . . in the words of their leader, Adolf Hitler, '*to see to it that the blood is preserved pure and by preserving the best of humanity, to create the possibility of a nobler development.*'"[41]

Moral innovation often hinges on pivotal concepts—in this case, *Rassenhygiene* and *Volkskörper*—that support a reinterpretation of established moral norms. Thus, many of the duties Ramm describes are like the well-established norms found in conventional statements of 1940s Western medical ethics. Physicians, for example, are held to have a duty to respond to the medical needs of the poor and rich equally, to preserve medical confidentiality, and to perform abortions "only if there is a danger to the life of the pregnant woman."[42] What transforms these otherwise standard views of 1940s Western medical ethics into Nazi medical ethics is the commitment to *Rassenhygiene* and *Volkskörper*, concepts that expand the scope of physicians' responsibilities to include as yet unconceived future generations of the German *Volk*. Thus, for Ramm, the wrongfulness of aborting a German fetus is not an assertion of the value of unborn life; it is a condemnation of an assault on the *Volk*'s gene pool. Ramm emphasizes this point using italics, "*Whoever weakens the Volk community through abortion of a fetus is to be placed on the same plane as a traitor to the country and Volk.*"[43] To reiterate for clarity and emphasis, whereas (except in Soviet Russia, which legalized abortion in 1920)[44] conventional Western medical ethical pronouncements on the wrongfulness of abortion in the 1940s emphasize a commitment to preserve fetal life from the moment of conception, Ramm emphasizes that abortion is an assault on the *Volk*'s gene pool. A similar line of reasoning leads to an absolute prohibition against practicing euthanasia on any member of the *Volk*.[45] Infamously, everything is different with respect to non-*Volk* who are perceived as contaminating the *Volk*'s gene pool.

Like Brandt, Ramm underscores the idea that *Rassenhygiene* justifies eugenic prohibitions for the prevention of genetically ill offspring and requires defending the genetic health and purity of the *Volk*'s germline.[46] Neither saw themselves as rogue physicians or unethical sadists; instead, they envisioned themselves as agents of a new and superior morality. Brandt's moral idealism was evident from his youthful efforts to work with Albert Schweitzer's medical mission in Africa—which were ultimately prevented by bureaucratic obstacles.[47] Brandt had also studied with liberal psychiatrist Alfred Hoche, coauthor of *The Release and Destruction of Lives Unworthy of Being Lived*. Brandt's biographer, Ulf Schmidt, writes that "Hoche—and later Brandt—applied British sociologist Herbert Spencer's (1820–1903) concept of the social organism to the mentally ill." They "saw the state as an organic entity in which the mentally ill were parts of the 'body politic' (*Volkskörper*) that had been damaged, useless, or harmful and needed to be removed . . . [and this] . . . provided the intellectual and moral basis from which Brandt would later argue his case after Hitler asked him to implement [Aktion T4] and also during the Nazi Doctor's Trial."[48]

Brandt was not the only German physician to unapologetically proclaim the superiority of Nazi medical ethics. Lack of remorse was so widespread among German health care professionals in the immediate postwar era that when the World Medical Association (WMA) made legitimization of the German medical profession contingent on a statement of apology and contrition, the organization representing (West) German physicians, the *Arbeitsgemeinschaft Westdeutscher Artzekammern* (AWA), refused to apologize or act contrite. The WMA was "astonish[ed] . . . that no sign whatever had come from Germany [i.e., from the AWA] that the doctors were ashamed of their share of the crimes, or even that they were fully aware of the enormity of their conduct."[49] Eventually, the AWA apologized, but it did so under duress.[50,51]

Many German physicians complicit in the Holocaust or who used concentration camp inmates as "human material" for their experiments subscribed to National Socialist medical ethics not despite morality but because they were committed to living a moral life: to them, the Aktion T4, mass sterilization and genocide, and other aspects of the Holocaust were morally permissible. As Schmidt puts this point, "In the worldview of Brandt and other Nazi physicians . . . they genuinely believed that their actions could be justified on the basis of what they perceived as their noble motivation. . . . As [Brandt] later defended his actions at Nuremberg [he] never felt that it was not ethical or was not moral."[52] In contrast, when a Turkish physician was asked whether he had violated his medical calling by participating in the genocide of Armenians (1914–1923, 1.5 million killed), he replied, "My Turkishness prevailed over my medical calling."[53] The physician then analogized his genocidal acts to preventing the spread of pathogens. "Armenian traitors had found a niche for themselves in the bosom of the fatherland: they were dangerous microbes. Isn't it the duty of a doctor to destroy these microbes? . . . I shut my eyes and surged forth without reservation."[54]

The Turkish physician's need to "shut his eyes" indicated that his remarks about pathogens were metaphorical; they were not part of his *weltanschauung*. The Nazi's conception of the German race as an organism, the

Volkskörper, transubstantiated the Turkish doctor's metaphor into a literal "truth" in their way of thinking. This "truth" reconciled physicians' medical ethics with the use of deviant human material for experiments and permitted Aktion T4 and other medical eliminations of deformed, homosexual, Jewish, and Roma contaminants to the *Volkskörper* organism. Brandt and Ramm and a generation of German physicians conceived the German people as a *Volkskörper* and so they acted on the ethics of *Rassenhygiene* with a clear conscience. In doing so, they created what Baxter would have denounced as a *true scandal*, one "which tempteth [people] into sin . . . and maketh [them] stumble, or occasioneth them to think of evil [as good]." As explained in the next chapter, reacting to that true scandal in which evil masqueraded as a good, Anglo-American and European military veterans attempted to prevent any recrudescence of this perversion of evil as a good by creating the foundational documents of modern medical ethics.

3

The Nuremberg Doctors Trial and the Nuremberg Code

Over an acre of ground lay dead and dying people. You could not see which was which. . . . The living lay with their heads against the corpses and around them moved the awful, ghostly procession of emaciated, aimless people, with nothing to do and with no hope of life, unable to move out of your way, unable to look at the terrible sights around them. . . . This day at Belsen was the most horrible of my life.
—Richard Dimbleby, BBC News, April 15, 1945[1]

I accepted the invitation to serve at the Nuremberg Trials only because I had in mind the objective of placing in an international judicial decision the conditions under which human beings may serve as subjects in a medical experiment, so that these conditions would become the international common law on the subject. Otherwise, I would have had nothing to do with the nasty and obnoxious business. I believe in prevention, not a "punitive cure". . . . The Judges and I were determined that something of a preventative nature had to come out of the "Nuremberg Trial of the Medical Atrocities.
—Andrew C. Ivy, Letter to Maurice Pappworth, April 6, 1966[2]

Corpses Living and Dead: The Genesis of the Nuremberg Trial

The National Socialist moral-political revolution was centered on an applied eugenic biological theory, *Rassenhygiene*, that transformed medicine and medical ethical norms in ways that justified use of inmates as experimental material, Aktion T4, and the Holocaust. Yet while this rhetoric was public, the experiments, Aktion T4 and other genocidal *euthanasie* programs, and extermination camps like those at Auschwitz-Birkenau were thinly veiled from a German public that preferred not to see and opaquely veiled from the rest of the world, which preferred not to know. So, Allied troops were shocked and horrified by the near-skeletal figures and stacks of corpses they saw on entering German concentration camps at Auschwitz, Bergen-Belsen, Dachau, and elsewhere. A member of the military film crew at Bergen-Belsen recalls that "the bodies were a ghastly sight.

Some were green. They looked like skeletons covered with skin—the flesh had all gone. There were bodies of small children among the grown-ups. In other parts of the camp there were hundreds of bodies lying around, in many cases piled five or six high."[3] A BBC reporter described

over an acre of ground lay dead and dying people. You could not see which was which. . . . The living lay with their heads against the corpses and around them moved the awful, ghostly procession of emaciated, aimless people, with nothing to do and with no hope of life, unable to move out of your way, unable to look at the terrible sights around them. . . . A mother, driven mad, screamed at a British sentry to give her milk for her child, and thrust the tiny mite into his arms, then ran off, crying terribly. He opened the bundle and found the baby had been dead for days. This day at Belsen was the most horrible of my life.[4]

On April 12, 1945, General Dwight D. Eisenhower (1890–1969), commander of the Allied forces, visited a concentration camp with US generals Bradley and Patton. He reported that "the things I saw beggar description. . . . The visual evidence and the verbal testimony of starvation, cruelty and bestiality were so overpowering as to leave me a bit sick." Anticipating what would later be called "Holocaust denial," Eisenhower remarked that he "made the visit deliberately, in order to be in a position to give first-hand evidence of these things if ever, in the future, there develops a tendency to charge these allegations merely to 'propaganda.'"[5]

In August 1946, an International Scientific Commission on War Crimes of a Medical Nature (ISC) met in Paris to set up war crimes tribunals. By then, victory over the Nazis and tensions between allies had eroded the desire for collaborative trials, and war crimes hearings began to be conducted separately by each of the formerly Allied nations—Britain, France, the USSR (Russia), and the US—in territories under their control. The 1946–1947 Nuremberg Doctors Trial, *US v. Karl Brandt, et al.*, was part of a series of American trials designed to hold Germans, in this case twenty-three German physicians and administrators, accountable under international law for crimes against humanity, war crimes, and membership in a criminal organization (the SS, or Schutzstaffel). Brandt and fifteen other physicians were found guilty, and seven received the death penalty, including Brandt.

Opposition to Holding the Nuremberg Doctors Trial

Many in diplomatic, medical, and military circles objected to the idea of putting German doctors in the dock at a public trial because it might undermine efforts to preserve useful data from valuable Nazi experiments, like the Dachau experiments on treating people in shock from extreme cold (hypothermia).[6] Others feared that holding a public trial might revitalize the antivivisectionist movements that threatened to ban, or radically restrict, medical experimentation on humans.[7] One eminent figure voicing these concerns was Major Kenneth Mellanby, CBE (1908–1993), an English entomologist who had directed experiments on British conscientious objectors and who served as the *British Medical Journal*'s (*BMJ*'s) foreign correspondent at the Nuremberg Doctors Trial.[8] In a December 7, 1946, letter to *The Lancet*, Mellanby condemned the idea of discarding data from the Nazi experiments as "pernicious sentimentality,"[9] even as he confessed "a great deal of sympathy for some of those who were responsible for carrying out the experiments ... especially serious research workers."[10] Mellanby explained further that "given the chance of using prisoners for experiments, which one believed to be of great importance and value to mankind, what would one do if ... the victims were dangerous criminals who were anyhow condemned to death and likely to die in some particularly abominable manner?" Indeed, Mellanby remarked, he was "not sure what I myself should have done."[11] Mellanby concluded by reemphasizing the need to preserve the Nazi data.[12] He seemed unconcerned by the prospect that recognizing these experiments as valuable might validate the morally deplorable methods employed. Seeking precisely such *ex post facto* moral rehabilitation, indicted German scientists, like Dr. Klaus Karl Schilling (1871–1946), pleaded for publication of his experiments because "it would be an enormous help ... for my colleagues, and a good part to rehabilitate myself."[13]

Tu Quoque: "What about What You Did?"

As it turned out, there were few objections to publicly condemning the doctors and administrators directly involved in *euthanasie*/Aktion T4 programs, or those involved in so-called thanatological experiments (i.e., experiments designed to discover efficient means of mass sterilization or mass killing).[14] Nor were there any challenges to condemning those complicit in killing 112 Jews at Auschwitz for a skeleton collection that was part of an anatomical project at the State University of Strasbourg designed to display Jewish racial inferiority. Experiments on human subjects directed at biomedical issues related to the war itself, however, were not as easily condemned. Armies confronting each other on the same battlefields encountered many of the same biomedical issues. Quite naturally, therefore, scientists from combatant countries on both sides conducted seemingly similar experiments seeking to improve treatments for wounded soldiers, for pilots downed in freezing waters, and for finding better treatments for tropical diseases, like malaria and typhus.

Thus, American prosecutors at the Nuremberg Trials faced the challenge of condemning as "unethical" experiments conducted by German researchers without, in the process, indicting seemingly similar experiments conducted by American scientists. Mellanby exploited this conundrum in his articles for the *British Medical Journal* and in a 1973 edition of his book, *Human Guinea Pigs*. There Mellanby defended his own morally questionable experiments on British conscientious objectors and echoed the German defense attorneys' charge that "America criminals in gaol [jail] did take part in tests involving malaria infection not unlike those done at Dachau by Schilling." Mellanby continues with the observation that "it is somewhat ironical that Schilling was . . . found guilty of mass murder, and summarily hanged."[15]

Mellanby's *Tu Quoque* (What about what you did?) charge was an *ad hominin* designed to discredit the American prosecutors by citing experiments with antimalarial drugs conducted by the University of Chicago's Malaria Research Unit at Jolliet-Stateville prison from 1944 to 1946. During that period, researchers infected 500 healthy white male prison inmates with malaria to test the efficacy of various antimalaria treatments.[16] Yet, although superficially like Schilling's experiments, as was pointed out at the trial, the Jolliet-Stateville experiments were conducted differently. As a result, no deaths were associated with the Jolliet-Stateville experiments.[17] In contrast, according to such witnesses as Schilling's assistant, Czech prisoner Dr. Franz Blaha, hundreds of deaths were associated with Schilling's Dachau experiments.

> During my time at Dachau, [Dr. Blaha testified], I was familiar with many kinds of medical experiments there with human victims. These persons were never volunteers but were forced to submit to such acts. Malaria experiments on about 1200 people were conducted by Dr. Klaus Schilling between 1941 and 1945. . . . The victims were either bitten by mosquitoes or given injections of malaria sporozoites taken from mosquitoes. Different kinds of treatment were applied, including quinine, pyrifer, neosalvarsan, antipyrine, pyramidon and a drug called 2516 Behring. I performed autopsies on bodies of people who died from these malaria experiments. Thirty or forty died from the malaria itself. Three hundred to four hundred died later from diseases which proved fatal because of the physical condition resulting from the malaria attacks. In addition, there were deaths resulting from poisoning due to overdoses of neosalvarsan and pyramidon. Dr. Schilling was present at the time of my autopsies on the bodies of his patients.[18]

Dr. Blaha pointedly testified that none of Schilling subjects were volunteers.[19] Other witnesses confirmed this claim. In contrast, all the subjects in the Jolliet-Stateville experiment signed the following "waiver and release" form attesting to their volunteer status. The function of these forms, which were standard for US government-supported research involving human subjects at that time and during much of the Cold War, was to have subjects waive their legal rights to sue and to release the researchers from any personal responsibility.

> I . . . , N[umber]. . . . aged . . . , hereby declare that I have read and clearly understood the above notice [describing the experiments and associated risks], as testified by my signature below, and I hereby appeal to the University of Chicago, which is at present engaged in malarial research at the orders of the Government, for participation in the investigation of the life cycle of the malarial parasite. I hereby connect all risks connected with the experiment and on behalf of my heirs and my personal and legal representatives, I assume all the risks of this experiment. I hereby absolve from such liability the University of Chicago and all the technicians and researchers who take part in the above-mentioned investigations. I similarly absolve the Government of the State of Illinois, The Director of Department of Public Security of the State of Illinois, the warden of the State Penitentiary at Joliet-Stateville, and all employees of the above institutions and Departments, from all responsibility, as well as from all claims and proceedings or Equity pleas, from any injury or malady, fatal or otherwise, which may ensue from these experiments.
>
> I hereby certify that this offer is made voluntarily and without compulsion. I have been instructed that if my offer is accepted, I shall be entitled to remuneration amounting to [100] dollars, payable as provided in the above notice.[20]

Although, like most consent forms of that period, this one focuses on minimizing the researchers' legal liability for harms suffered by experimental subjects, it nonetheless states that the prisoner who signed it was informed about the experiments and gave written consent to the researchers. To reiterate: prior to the bioethics moral revolution, such "waiver and release" forms were standard and used in US government-sponsored or government-funded research during World War II and in the early years of the subsequent Cold War period. During this period,

> When experiments were hazardous, federal sponsors had [their] agency lawyers review the wording of waiver statements and required researchers to have subjects sign approved documents. The prevalence of waiver provisions was consistent with a general pattern, scientists and their government patrons distanced themselves from all but immediate research-related maladies. . . . Investigators provided medical care for enrollees' illnesses while the studies were ongoing, but participants were on their own if their symptoms did not resolve or if they developed lasting disabilities.[21]

In 1948, Illinois Governor Dwight Green (1897–1958) commissioned a review of the Jolliet-Stateville malaria study.[22] Andrew Conway Ivy, MD

(1893–1978), coauthor of the Nuremberg Code, headed the commission.[23] Assessing the Jolliet-Stateville experiments in terms of the criteria for ethical experiments that Ivy himself had recommended to the Nuremberg Tribunal,[24] the committee cited these experiments as "an example of human experiments which were ideal because of their conformity with ethical rules."[25] A later 2013 retrospective review by National Institutes of Health (NIH) bioethicist Franklin Miller (1948–) reached a similar conclusion.[26] Pivotal to Miller's analysis were autobiographical comments by an infamous Jolliet-Stateville prisoner, Nietzschean murderer Nathan Leopold (1904–1971),[27] one of the subjects in the malaria experiment. Leopold had been incensed by the German defense lawyers' claim that the Jolliet-Stateville prisoners' participation was "involuntary." "That was absolutely false,"[28] Leopold wrote in his 1958 autobiography, *Life Plus 99 Years*: "The young docs and Dr. Alving leaned over backward in handling the matter of volunteering in a scrupulous and ethical manner."[29] Leopold also remarks that he and other inmates were informed of the experiment by "an announcement [that] was read over the institutional radio, asking for volunteers. The first day 487 men volunteered."[30] Based on these and similar statements, Miller concluded that "the voluntariness of consent . . . [in] principle was satisfied in the recruitment and consent process for the Stateville malaria experiments. Although we have no record of exactly what the prospective subjects were told about the nature of the experiments and what to expect, if Leopold's description is accurate, it appears that they were sufficiently informed to give valid consent."[31] What Miller finds most persuasive was that "according to Leopold, a major motivating factor for at least some of the prisoners was the opportunity to contribute to research that might aid the war effort. The incentives to participate in the Stateville malaria experiments seem far from irresistible, especially as many prisoners did not volunteer."[32]

The Nuremberg Code

"Success," it is said, "has many fathers, failure is an orphan." The scholarly consensus is that parentage of the ten principles for morally permissible research on humans now known as the "Nuremberg Code" belongs to two fathers: Leo Alexander (1905–1985) and Andrew C. Ivy. Their memos, reports, questions, and testimony were midwifed by the Nuremberg justices to live on as a statement of ten principles for morally permissible research on human subjects, later christened "The Nuremberg Code."

Prior to and after the Nuremberg Trials, Alexander and Ivy lived in different worlds. Ivy was raised as an all-American baseball- and piano-playing

Christian boy who grew up to become a medical media star. Alexander, in contrast, was a Jewish refugee from the Holocaust trying to make his way as a stranger in a strange land. He was born in 1905 to an affluent, well-connected Viennese Jewish family and came of age in a mansion where such eminent Jewish intelligentsia as Sigmund Freud and Gustav Mahler joined the family for dinner. Following in his father's footsteps, Alexander studied medicine in Vienna and then at the prestigious Kaiser Wilhelm Institute for Brain Research in Berlin, a major center for eugenics research. In 1933, Alexander took leave from his position at a psychiatric hospital in Frankfurt to serve as a visiting Rockefeller lecturer in neurology and pathology in Beijing, China. While Alexander was working in China, the National Socialists took power in Germany, and it became too dangerous for him to return to Germany or Austria. Grasping his only option, Alexander accepted a position at Worcester State Hospital in Massachusetts, receiving only room and board as compensation.[33]

When Alexander arrived in America in January 1934, he likely held his German professors' views on eugenics and medical ethics. Recall that Moll described this moral milieu as one in which "clinic directors made the 'material' [i.e., patients] kindly available to young researchers,"[34] and where "patients were . . . viewed as objects, animals, or slaves which the physician believed he could use with impunity at his discretion."[35] Alexander's biographer, Ulf Schmidt, indicates that the situation remained unchanged in the 1920s and 1930s, when German medical school "professors [still held] utilitarian and paternalistic belief systems [that] led them to regard patients as little more than objects of scientific curiosity and research. It is almost certain that Alexander failed to obtain consent from the members of the . . . family who formed the basis of his schizophrenia research."[36]

However, Schmidt continues, "The process of emigrating, first to China in 1933 and then to the United States, had a defining impact on Alexander's life and character . . . [and] also on his view of medical ethics and eugenics."[37] Evidence of a change in Alexander's perspective emerged two years later, in 1935, when he volunteered to become a research assistant for a fellow neurologist, Abraham Myerson (1881–1948), son of a Lithuanian Jewish immigrant. Myerson led a group of eminent experts writing a report to the American Neurological Association that was critical of American eugenics. In the United States, as in Germany, the eugenics movement was fueled by alarmist literature about the decline of a dominant ethnic group, white Anglo-Saxon Protestants, and its fear of replacement by "inferior" races. Rebutting these alarmist claims, Myerson and colleagues reported that "there is nothing to indicate that mental disease and defect are increasing . . . there is no evidence of biological deterioration of the race."[38] They also noted that "most of the [eugenics] legislation which has been enacted so far is based more on a desire to elevate the human race than upon proven facts." However, they continue, "Our knowledge of human genetics has not the precision nor the amplitude which would warrant the sterilization of people *who themselves are normal* in order to prevent [mental disease] in their descendants. . . . [Moreover] There is at present no sound scientific basis for sterilization on account of immorality. . . . Any law concerning sterilization passed in the United States should be voluntary and regulatory rather that compulsory. . . . Your committee . . . can only recommend sterilization in selected cases of certain diseases and [then] only with the consent of the patient."[39] Alexander's willingness to sign this report is notable because he now associated himself with critics of eugenics who contended the consent was a prerequisite to ethical medical interventions.

When America entered World War II, Alexander joined the military where he treated soldiers with "shell shock" (now called "posttraumatic stress disorder," or PTSD). After the war, one of Alexander's former students was assigned the task of choosing a war crimes investigator. Spotting Alexander's name near the top of an alphabetical list of qualified candidates, as grateful students often do, he nominated his former professor for the position.[40] On his return to Europe, Alexander spent months in fruitless investigation until he came upon evidence of SS Dr. Sigmund Rascher's (1909–1945) hypothermia experiments at Dachau. In these experiments, "human material" (i.e., inmates involuntarily serving as human subjects) was submerged in vats of ice water to determine how long they survived and how best to revive them. The data proved valuable, but the accounts of human suffering made Alexander even more determined to bring these physicians to justice.[41] Years later, musing about those times in a conversation with fellow psychiatrist Robert Lifton, Alexander remarked, "What anguish I did not feel [in studying the abuse of human subjects] was probably prevented by the satisfaction of having been victorious over so much evil. Therefore, I was able to view this evil no longer as a threat . . . but also with the hope of even preventing it in the future."[42] Although Alexander was originally motivated to bring abusive researchers to justice, his daughter recalls that he later "felt . . . when he looked back on it, that the whole point, was to create a new standard [for ethical research on humans] not just to convict those people."[43]

Alexander's colleague, Andrew Conway Ivy, was a religious pacifist born in Farmington, Missouri, to a devout Christian family in 1893. He attended college at Southeast Missouri State Normal School in Cape Girardeau, a teachers' training school where his father taught science. In 1918, he left

Missouri to study medicine at the University of Chicago. He later chaired the Division of Physiology and Pharmacology at Northwestern University in Chicago, where he headed a large well-equipped laboratory that was often featured in *Life Magazine* (a leading media influencer in the 1940s and 1950s).[44] Having established a reputation as a scholar, researcher, writer, and media celebrity, Ivy was elected president of the American Physiology Society (1939–1941). In addition, although a religious pacifist, Ivy became director of the Naval Medical Research in 1942–1943—but he declined any military rank.

In 1946, the American Medical Association (AMA) chose Ivy as its representative at the Nuremberg Doctors Trial. As Ivy explained in a letter to Maurice Pappworth (1910–1994), "I accepted the invitation to serve at the Nuremberg Trials only because I had in mind the objective of placing in an international judicial decision the conditions under which human beings may serve as subjects in a medical experiment, so that these conditions would become the international common law on the subject. Otherwise, I would have had nothing to do with the nasty and obnoxious business. I believe in prevention, not a 'punitive cure'. . . . The Judges and I were determined that something of a preventative nature had to come out of the 'Nuremberg Trial of the Medical Atrocities.'"[45]

Ivy was also motivated by his concern that "the publicity of the medical trial [does] not stir public opinion against the ethical use of human subjects for experiments."[46] To forestall such an outcome, Ivy "felt that some broad principles should be formulated by this meeting enumerating the criteria for the use of humans as subjects in experimental work."[47] He presented his idea at a July 1946 conference in Paris that he attended as special consultant to the Secretary of War. Ivy envisioned three principles governing morally permissible experiments on humans, and these became, in effect, the first draft of the Nuremberg Code. Historian Ulf Schmidt comments that "here . . . we find the rationale for the creation of principles for permissible experimentation on humans in embryonic form . . . which he subsequently produced, [and that were subsequently] adopted and publicized for the purpose of contrasting ethical and unethical experimentation on human subjects."[48] Ivy's idea was manifest in the opening line of the Nuremberg Code, which states, "The great weight of the evidence before us is to the effect that certain types of medical experiments on human beings, when kept within reasonably well-defined bounds, conform to the ethics of the medical profession generally."

As should be evident from Moll's 1902 report on unethical experiments on patients, this opening statement perpetuates a fictive narrative that "the involved Nazi physicians and scientists ignored . . . ethical principles and rules, which have been well established by custom, social usage, and the ethics of medical conduct, and which are necessary to ensure the human rights of the individual and to avoid the debasement of a method for doing good, and the loss of faith in the medical profession."[49] In point of fact, as Moll's survey indicates, no such ethical principles and rules were "well established by custom, social usage," prior to the Nuremberg Trials. Moreover, the concept of human rights (i.e., rights humans should recognize in fellow humans)—as opposed to rights conferred by a deity or creator (as specified in the American Declaration of Independence)—first became prominent in the aftermath of World War II.[50] Ivy hoped that the war crimes trial would establish ethical principles and rules for morally permissible experiments on humans based on the concept of human rights, but to attest this as a historical fact was pure fabrication. Ivy repeated this falsehood in his testimony at the trial. Consequently, as bioethicist Jonathan Moreno observes, "When the defense put contemporary US practices under a microscope to press . . . Dr. Andrew Ivy, [was put] into a position in which he virtually perjured himself, asserting that there was US ethics standards in place when in fact he himself had rushed an [AMA] ethics code into print just in time for his testimony for the prosecution."[51]

Ivy's near perjury was motivated by his fear of admitting that neither the AMA nor any other American health care organization had endorsed ethics standards for conducting experiments on humans prior to, or during, World War II. Ironically, Ivy perpetrated the falsehood that "all agree" to common ethical standards to inspire such ethics standards. It bears repeating that no professional codes of ethics operant during World War II, or previously, addressed the ethics of experimenting on human subjects. Ironically, there had been a 1931 German Health Council (*Reichgesundheitsrat*) regulation stating the ethical conditions for using experimental treatments on *patients*[52] and (in response to the Neisser episode) a 1900 Prussian precursor regulating experimental interventions on *patients*.[53] However, there is no evidence that the 1931 regulations were ever enforced before the Nazis assumed power in 1933, and they were almost instantly invalidated because they were not part of National Socialists' new rules of medical conduct.[54] Furthermore, since the scope of the earlier German regulations had been limited to innovative therapeutic interventions on *patients*, had they been operative after 1933, they would not have protected healthy prisoners in concentration camps.[55]

Thus in 1946–1947 there were no formal professional ethical standards, regulations, or laws regulating experiments on human subjects in America, Europe, or elsewhere. Ivy either lied, knowingly perpetuated a falsehood, or, perhaps, he exaggerated the status of a hoped-for truth—but he did so under oath. Consequently, he either perjured himself or came close to doing so; however one characterizes what he did, it is important to recognize that he did it with the noble intention of protecting both the rights of scientific researchers to experiment on humans and the human rights of people subjected to their experiments.

Debates over the moral propriety of such "noble lies" extend back to the earliest history of Western ethics and moral theology (since lying to God about eating the fruit of the tree of knowledge was associated with Eve and Adam's original sin). Saint Augustine (354–430 CE) famously declared lying morally wrong, even when lying appears necessary to prevent a greater evil (*De Mendacio*, On Lying, circa 395). Augustine's arguments buttressed the sentiments of Saint Paul (née Saul of Tarsus 5–64/65 CE), who had previously stated that doing evil is never morally permissible, even if good may come of it (Romans 3:8). Plato (circa 429–347 BCE), on the other hand, held that physicians and politicians may tell well-intended or noble lies, the former to minimize patients' fears, the latter to protect communities from enemies or to otherwise benefit a community.[56]

The idea that deceit, noble lies, and placebos (i.e., fake treatments) were morally permissible if they induced positive psychophysiological reactions dates to the Hippocratic corpus, and discussions of using deceit to prevent or relieve pain or anxiety were commonplace in the Renaissance and early modern medical literature.[57] Thomas Percival, author of *Medical Ethics*, the work that introduced the expression "medical ethics" into the English medical lexicon, defended a physician's right to tell white lies in ministering "hope and comfort to the sick." Anglican clergyman Thomas Gisborne (1758–1846) criticized Percival on this point, arguing that physicians' attempts to administer hope and comfort are constrained by a higher duty to limit one's comforting words to "whatever truth and sincerity will admit."[58] Percival replied that insofar as a white lie or minor deception was intended to promote a moral good, if that good outweighed the harm of lying, the lie was morally excusable.

Just shy of a century and a half later, another physician, Andrew Ivy, came to the same conclusion. He was probably ignorant of Percival (interest in Percival's ethics was first rekindled in the context of the bioethics revolution of the 1970s), but as a religious Christian, Ivy, like Percival, believed that it was morally excusable to tell a white lie to establish "ethical principles and rules necessary to ensure the human rights of the individual and to avoid the debasement of a method of doing good and the loss of faith of the public in the medical profession."[59] So Ivy knowingly fabricated a myth in the expectation of promoting a greater good: to establish the first international code of ethics for experiments on humans. And his fabrications did just that.

However, just as the scandal-induced set of tablets that Moses cradled in his arms as he descended Mount Sinai for a second time was not a code for good God-fearing Israelites, so too, the Nuremberg Code that Ivy and Alexander created in the aftermath of the Nazi medical experiments and the Holocaust was not a code of ethics for moral researchers. It was a set of precepts designed to prevent immoral researchers from engaging in bad conduct. Thus, Principle 1 prohibits researchers use of "force, fraud, deceit, duress, over-reaching, or other ulterior form of constraint or coercion" to recruit human subjects. Similarly, Principle 2 restricts "experimentation to [those experiments intended to produce] results for the good of society [rather to puff up resumés] unprocurable by other methods or means of study, and not random and unnecessary in nature" (Ivy's drafts and the Nuremberg Code as officially issued are available in the Appendix).

These admonitions might seem self-evident today, but if one recalls Barbara Elkeles's report of pre-Nazi researchers who unabashedly mention that they tested a new vaccine "on children in a foundling hospital . . . because the planned [subjects], calves, [had] been too expensive in price and maintenance," their relevance becomes evident: in this example, there were other means of conducting the experiment; it was just cheaper to use orphaned human children.[60] The code was specifically fashioned to prevent Nazi-type abuses. Principle 5, for example, would have prohibited Rascher's hypothermia experiments at Dachau because it forbids any experiment where there is an "a priori reason to believe that death or disabling injury will occur." Principle 7 reinforces this prohibition, emphasizing that "proper preparations should be made, and adequate facilities provided to protect the experimental subject against even remote possibilities of injury, disability, or death." Note that the use of the expression "experimental subject[s]" expands the code's scope to include people who were not patients.

To reiterate, this was not a code designed to refine the inclinations of moral experimenters; it was a code for experimenters tempted to engage in the scandalously immoral, and it was designed to prevent future scandals. Curiously, Principle 5 permits potentially lethal experiments on other people but only if "the experimental physicians also serve as subjects." Ivy introduced this exception in his second draft of the code (the AMA version). He based it on Walter Reed's experiments on "yellow fever where the experimenters serve as subjects along with non-scientific personnel."[61] Yet

Ivy appears have been unaware of the details of that incident. Reed's team agreed to serve as experimental subjects because, in the words of one member, Dr. James Carroll (1854–1907), it was anticipated that "several lives [will be] sacrificed in the course of the [experiment], [so] the best atonement we could offer for endangering the lives of others was to take the same chances ourselves."[62] After Carroll and his colleague, Jessie William Lazear (1866–1900), were stricken with yellow fever, however, Carroll admitted to Surgeon General George M. Sternberg (1838–1915) that his work as a researcher unsuited him to serve as a subject because he was not in the same controlled environment as other subjects. Moreover, Dr. Lazear's death depleted the staff of medically qualified researchers. Thereafter, Sternberg forbade researchers to serve as subjects in their own experiments.

It is also important to recognize the flawed logic of the claim that self-sacrifice legitimizes or permits similar sacrifices from others. Just as self-flagellation does not permit one to whip others, researchers' willingness to submit to the same dangers that subjects undergo does not entitle them to experiment on others without their consent. Different people value things differently, and (except in instances of surrogate or guardian decision-making) one person's consent reaches no further than that person herself or himself. Self-sacrifice may alleviate a person's feelings of guilt, but as Carroll admitted, it is typically incompatible with the controlled conditions required for experimenting on human subjects. It also undermines researchers' claims to objectivity, and as Dr. Lazear's death demonstrates, it may diminish researchers' ability to complete an experiment. Researchers' feelings of atonement, no matter how deeply or sincerely felt, confer no right to place other people at risk of indignity, harm, or death. Just as suicide bombers' self-murders in the name of some presumed social good does not justify murdering others in the name of that good, researchers' willingness to risk their health or lives to achieve some scientific or humanitarian good does not justify placing other people's lives at risk to achieve this same good—unless, of course, those people, or appropriate surrogates, having been fully informed, having understood and appreciated the potential inconveniences and harms, consent to serve as subjects.

Nuremberg Principles 9 and 10 establish subjects' liberty rights and researchers' responsibilities: that is, subjects' rights to discontinue an experiment and researchers' correlative duty to terminate experiments if they believe "that continuation of the experiment is likely to result in injury, disability, or death to the experimental subject." This means that research subjects' initial consent is not irrevocable; it is not a Ulysses contract from which subjects may not withdraw. Subjects may withdraw if they fear injury, disability, or death—or even inconvenience. Thus, when Dr. Rascher's experimental subjects saw the ice vats awaiting them, they had the right to withdraw from the experiment, even had they agreed to serve as subjects. Moreover, researchers have a duty to terminate experiments likely to have disabling or lethal results. Thus, researchers conducting Rascher-like experiments that risk subjects' injury, disability, or death have a duty to discontinue the experiment if at any time they become aware that the experiment is likely to cause subjects injury, disability, or death. There were no such duties in any code operant in 1945, but Alexander and Ivy campaigned to ensure future researchers would accept these ethical constraints—and, correlatively, that future subjects would have rights protecting them against unethical researchers.

4

The Declaration of Geneva: Old Things Become New

Old things are passed away; behold, all things are become new.
—King James translation of *The Holy Bible* 2, Corinthians 5:17

While awaiting the signal to commence Operation Overlord, the invasion of Nazi-occupied continental Europe, an international assemblage of physicians used the headquarters of the British Medical Association (BMA, founded 1832) as their home base.[1] After Germany surrendered on May 7–9, 1945, one of them, a British physician, John Pridham (1891–1965), a World War I veteran who had lost an eye during that conflict, initiated an effort to transform the informal multinational fellowship of physicians who served in Operation Overlord into an international medical society to be called the "World Medical Association" (WMA, founded 1947). As he and his colleagues initially conceived it, the WMA would coordinate efforts to provide medical assistance to war-ravaged continental Europe. However, after vague mentions of atrocities were replaced by news reports of medical war crimes, mass murder, and unethical experiments, Pridham and his colleagues took up a secondary goal: restoring "the honor of medicine."

To quote Pridham, medicine had been stained by "the failure of German doctors to attempt to combat the Nazi ideology which led to both professional and national debasement, and their own profession becoming disintegrated."[2] Consequently,

> At the WMA's first official meeting the question of the censure of medical men who had been [implicated] in war crimes was raised. This was the first occasion when doctors from the enemy occupied countries had met representatives of the free countries. Much evidence was offered of crimes against humanity committed by medical men which shocked the whole profession. The fact that such horrors could be perpetrated by doctors underlined the need for the formation of a world medical authority and it was proposed that a full statement on the subject be drawn up. . . . The statement was accepted at the First General Assembly of the WMA in the following year . . . and in the course of time a modern version

of the Hippocratic oath named, at the suggestion of Dr. Pridham, "The Declaration of Geneva."[3]

This succinct account neglects to mention the Nuremberg Medical Trials (*US v. Karl Brandt, et al.*) by name, although a summary of the tribunal's findings had been published in the *World Medical Association Bulletin* (1949–1953). That summary concludes that

> it is clear that certain doctors carried out their inhumane experiments both for the furtherance of the war and for research in disease. In the course of the experiments and in their application of the findings, they deliberately killed persons politically undesirable to the régime in power. They misused their medical knowledge and prostituted scientific research. They ignored the sanctity and importance of human life, exploiting human beings both as individuals and in mass. They betrayed the trust society had placed in them as a profession. The doctors who took part in these deeds did not become criminals in a moment. Their amoral methods were the result of training and conditioning to regard science as an instrument in the hands of the State to be applied in any way desired by its rulers. It is to be assumed that initially they did not realize that the ideas of those who hold political power would lead to the denial of fundamental values on which Medicine is based. Thus, the care of the individual patient ceased to be the doctor's primary aim and the humanitarian purpose of medical science was subordinated to the needs of war.[4]

In disseminating the findings of the Nuremberg Doctors Trial, this summary propagates a misleading narrative that implies that since only "certain doctors" allowed themselves to become "an instrument in the hands of the [Nazi] state," only those doctors "betrayed the trust that society had placed in them as a profession." An unstated implication is that since only "certain doctors" allowed themselves to be seduced by a rouge state, the rest of the German medical profession was redeemable. So, by indicting a dozen or so "Nazi" doctors, and by focusing on medical experiments, rather than the more widespread medical complicity in the Holocaust, the Nuremberg narrative, as faithfully conveyed in the WMA's *Bulletin*, facilitated the rehabilitation of the German medical profession, since West German physicians not complicit with the Nazis could be readmitted into the international fellowship of physicians and serve the medical needs of the public in now-defeated postwar Germany—provided that they made a formal apology and reasserted their loyalty to traditional ideal of *Salus aegroti suprema lex*.

To this end, the WMA made recognition of the West German medical profession (represented at that time by the *Arbeitsgemeinschaft Westdeutscher Artzekammern* [AWA], the Medical Chamber of the West German Working Group) contingent on the AWA's acknowledgment of "the participation of *certain* German medical doctors, in numerous acts of cruelty and oppression, and in the organization and perpetration of numerous acts of brutal experimentation on human beings without their consent." A further requirement was West German physicians admission that "by not denouncing those physicians whose acts and experiments . . . have resulted in the deaths of millions of human beings, the German medical profession has violated the ethical tradition of Medicine, has debased the honor of the medical profession, and has prostituted medical science in the name of war and political hatred."[5] If only the West German physicians admitted this and "promis[ed] . . . future good behavior,"[6] they could be accepted as members of the WMA.

Yet, to the WMA's leadership's "astonish[ment], . . . no sign whatever had come from Germany that the doctors were ashamed of their share of the crimes, or even that they were fully aware of the enormity of their conduct."[7] Independently, Andrew Ivy reported to the AMA that "at present, there isn't any information showing that the organized medical profession in Germany protested or that it plans to condemn medical crimes. The Church appears to be the only medical group which protested, and their protests pertained specifically to the killings at Hadamar Asylum." Ivy also noted that organized medicine's silence could not have been for lack of knowledge because Heinrich Himmler (1900–1945), director (*Reichsführer*) of the SS (*Schutzstaffel*), complained that "in . . . 'Christian Medical Circles,' . . . a young German aviator should be allowed to risk his life, but the life of a criminal . . . is too sacred and one should not stain oneself with this guilt . . . they were strongly opposed to Dr. Rascher and his experiments." Ivy's point was that since this was common knowledge in German Christian medical circles, the rest of the German medical community must also have been aware of Rascher's experiments; thus, its silence was tantamount to complicity.[8]

After the Nuremberg Doctors Trial ended, the AWA grudgingly accepted the role of penitent scripted for them by the WMA. On October 18, 1948, the AWA passed a resolution accepting the findings of the Nuremberg court and distributed copies of its findings to West German physicians. The resolution also accepted the narrative that "compared to the number of doctors working in Germany only a very small number of members of the medical profession shared in these crimes."[9] The German medical profession now claimed to be "aghast to learn of the actions which were the basis of the trial," and they "deeply regret that men of their own rank committed such horrifying crimes." They "mourn for the victims sacrificed by a despotic régime which availed itself of science as one of its instruments and was assisted in so doing by doctors."[10] They also claimed that "the German medical profession ... [is] aware of the wider dangers which are engendered in *the errors of a few*" (italics added) and thus the need to "formulate basic principles for the present and the future."[11]

The AWA also proposed a new foundation for German research ethics. It argued that since the research process had been corrupted by "the private interests of medical scholars ... where new investigative methods are being tried ... experiments [should be] submitted to a body of experts for their consent." If the experiment is approved, this committee of experts "should also explain the significance of these experiments to the public before they were tried on human beings."[12] Thus, in the view of the AWA, peer review by fellow medical experts, accompanied by a public statement of the rationale for planned experiments, would suffice to prevent unethical experimentation. More insightfully, considering the National Socialists' drive to de-professionalize health care delivery, the AWA also insisted that professional independence from governmental regulation was "fundamental to preserve the individuality and independent responsibility of doctors in medical activities. Society must guarantee the doctor sovereignty in his own field. In giving medical care, a doctor should not be subjected to any orders or directives from government but should live up to his knowledge of science and professional standards of good conduct."[13]

In effect, the lesson that the AWA claimed to have learned was that past abuse of human research subjects would be curtailed through professional self-regulation if medical peers were given the power to review and approve proposed research on humans and then explain the value of the research to the public. Although transparency might inhibit some unethical conduct, the more fundamental issue, well illustrated by Moll's reports on unethical medical experiments, is that although physicians may profess Hippocratic loyalty to patients' health and welfare, as well as mouth slogans like *Salus aegroti suprema lex*, their loyalties to institutions, to grants conferring organizations, to scientific advance, or to a national or racial interest (as in *Rassenhygiene*) may combine with self-serving interests in career advancement and professional status to blind researchers and fellow professionals to morally problematic experiments they proudly and willingly publicize.

One of the primary functions of any code of ethics, and thus for any code addressing morally permissible research on human subjects, is to recognize conflicting interests and loyalties and to set priories. The first line in Percival's *Medical Ethics*, for example, makes it clear to hospital physicians and surgeons that in the hospital, their conduct should "reflect that the ease, health, and lives of those committed to their charge depend on their skill, attention, and fidelity."[14] That is, when in the hospital, their priority is to attend to the patients in their charge faithfully, attentively, and skillfully. The AMA restates this obligation in its 1847 Code of Ethics. Perhaps because professional codes of ethics had not played a significant role in Germany, nothing comparable is in the AWA proposals.[15] More significantly, the AWA made no mention of subjects' informed consent, or to the 1931 guidelines or its 1900 Prussian precursor, or to the 1947 Nuremberg principles for morally permissible research on humans. Yet, without guidance from formal codes of ethics or regulatory standards, peer review readily degrades into a form of club regulation in which morally blind members validate the morally compromised research of their peers as "ethical."

Diplomatically, the WMA deemed the proposal a "preliminary statement" and entered serious negotiations with the AWA. Explaining that "research in Medicine as well as in its practice must never be separated from eternal moral values . . . [the WMA insisted that German physicians must reject] policies that degrade or deny fundamental human rights." Furthermore, they stressed that "medical knowledge and progress, unless governed by humanitarian motives, may become the instruments of wanton destruction."[16] Nonetheless, mindful of the fact that in 1948, Europe lay in ruins as the Soviet Army inched westward, the WMA found it prudent to accept the membership of West German physicians, because "present political and social conditions may imply the return of dangers similar to those of the Nazi régime."[17] After the West German physicians were admitted into the WMA, by parity of reasoning, in May 1951, the WMA also admitted a Japanese physicians' organization.[18]

Yet the WMA still insisted that the AWA publicly commit itself to some statement of the "fundamental values of medicine." As is evident from Ramm's textbook, German medical practitioners held the Hippocratic tradition in such high regard that Ramm tried to fashion a Hippocratic heritage for *Rassenhygiene*. Thus, one statement of the fundamental values of medicine that all parties could readily agree on was some form of the Hippocratic Oath. As it happened, Moll had invoked the Oath in condemning *euthanasie*,[19] and his invocation inspired Freiberg pathologist Franz Büchner (1895–1991) to do something similar in a public lecture on the Oath. In this November 1941 lecture, Büchner (echoing Hufeland) declared, "Life, is the only master the physician must serve. From the medical viewpoint death is the great opponent of life, and of the physician. If the physician would be expected to carry out the killing of the incurably ill, however, this would mean forcing him to make a pact with death. But if he makes a pact with death, he stops being a physician."[20] Büchner's lecture was one of the rare occasions on which a German physician publicly challenged Nazi medical ethics during the Nazi era.[21]

After the war, Büchner's singularly courageous act suggested that the Oath could serve as a talisman of medical morality. Apparently mindful of this, the lead prosecutor at the Nuremberg Medical Trial, Brigadier General Telford Taylor (1908–1998), invoked the Oath in his opening statement. This is "no mere murder trial," he proclaimed, "because the defendants were physicians who had sworn to 'do no harm' and to 'abide by the Hippocratic Oath.'"[22] In point of fact, none of the German physicians on trial had sworn the Hippocratic Oath.[23] The last vestiges of Hippocratic Oath swearing had faded into obsolescence throughout Europe in the interwar period, during which communist and fascist regimes filled the vacated ceremonial space with pledges of allegiance to a leader, a party, a social class, a people, or a race.

Undaunted by such inconvenient facts, Nazi physicians' violations of their Oath became a "recurring theme" at the Nuremberg Doctors Trial. Prosecution witnesses repeatedly discussed "the relevance of Hippocratic ethics to human experimentation and whether Hippocratic moral ideals could be an exclusive guide to the ethics of research [on] human subjects."[24] Three prosecution witnesses cited the Oath as a universally recognized statement of the ethics of research on human subjects: Leo Alexander, Andrew Ivy, and a German psychiatrist and medical historian, Werner Leibbrand (1896–1974). An outspoken critic of Nazi medical ethics, Leibbrand condemned the Nazis de-professionalization initiatives, which, in his view, debased medical professionals into mere "biological state officers" who treated their patients as if they were simply biological "objects." These biomedical state employees, Leibbrand charged, neglected Hippocratic ethics, and embraced in its place an ideology that "lack[ed] morality and reverence for human life."[25] He also claimed that if physicians adhered to the ethics of the Hippocratic Oath, no patient would be subject to potentially harmful experimental treatment. Leibbrand also "referred to Moll who had insisted that the morality of a physician is to hold back his natural research urge in order to maintain his basic medical attitude that is laid down in the Oath of Hippocrates and [prohibits experiments] which may result in doing harm to his patient."[26] As historian Ulf Schmidt observes, "For Leibbrand human experimentation was only permissible if the person had given voluntary informed consent, was mentally capable of giving consent, and was not forced or in any way coerced. He categorically ruled out experiments on prisoners, infants, or the handicapped."[27] Ivy concurred, claiming that "every physician should be acquainted with the Hippocratic Oath, [which] represents the Golden Rule of the medical profession . . . throughout the world."[28] Linking the Oath to the then nascent concept of human rights, Ivy contended that the Oath requires physicians to "have respect for life and the human rights of his experimental patient."[29]

Ceremonial invocations of the Hippocratic Oath in German medicine date to sixteenth-century Wittenberg,[30] but by the twentieth century, oath-swearing had gone out of fashion in Germany. Nonetheless, to a German medical profession seeking to rehabilitate itself in the eyes of the medical world, the reiterated invocation of the Oath to criticize the Nazi medical ethics overdetermined their decision to accept the Oath as a symbol of German medicine's renewed commitment to traditional medical ethics. Necessity requiring the resurrection of a long-dead tradition, the AWA required that "every doctor taking his license" in West Germany swear a revised version of the Hippocratic Oath.[31] In the words of scripture, "Old things are passed away; Behold, all things are become new"—and indeed, a long-abandoned oath became new again.

The 1948 Declaration of Geneva

The oath also became relevant to the WMA, but for a different reason.[32] In their eyes, a revised and modernized version of the Oath could teach future generations of medical students "to honor the traditions of Medicine and to absorb its humanitarian purposes—the succor of the bodily and mental needs of the individual irrespective of class, race or creed; the cure of disease; the relief of suffering; the prolongation of human life; and the prevention of disease."[33] Openly acknowledging that the tradition of oath swearing had "fallen into disuse in many countries," the WMA promulgated its modernized version "to impress on newly-qualified doctors the fundamental ethics of medicine and to raise the general standard of medical conduct," not only in West Germany but also in "every age and every country."[34]

The WMA understood the fundamentals of Hippocratic Oath to be: "The brotherhood of medical men; The motive of service for the good of patients; The duty of curing, the greatest crime being cooperation in the destruction of life by murder, suicide, or abortion; Purity of living and honorable dealing; Professional secrecy for the protection of patients; Dissemination of medical knowledge and discovery for the benefit of mankind."[35] "In view of the recent war crimes and continued troubled state of the world," the WMA promulgated an updated version of the Hippocratic Oath, now renamed "The Declaration of Geneva," hoping that, if sworn "by every newly qualified doctor [it] should have a beneficial effect.... afford[ing] a world-wide bond uniting [physicians] in a common service to humanity."[36]

In the 1940s, the ancient Hippocratic Oath was read in the German- and English-speaking cultural spheres as a strong prolife statement that was likely to have been written by Hippocrates himself.[37] This reading of the Oath, as Moll, Büchner, and Pridham would have understood it, is captured by a 1943 English translation by Ludwig Edelstein (1902–1965), a renowned Berlin- and Heidelberg-educated German classicist who found refuge at Johns Hopkins University in Baltimore after the National Socialists took power in Germany.[38]

Edelstein's Translation of the Hippocratic Oath

1943

I swear by Apollo Physician and Asclepius and Hygeia and Panacea and all the gods and goddesses, making them my witnesses, that I will fulfill according to my ability and judgment this oath and this covenant:

To hold him who has taught me this art as equal to my parents and to live my life in partnership with him, and if he is in need of money to give him a share of mine, and to regard his offspring as equal to my brothers in male lineage and to teach them this art—if they desire to learn it—without fee and covenant; to give a share of precepts and oral instruction and all the other learning to my sons and to the sons of him who has instructed me and to pupils who have signed the covenant and have taken an oath according to the medical law, but no one else.

I will apply dietetic measures for the benefit of the sick according to my ability and judgment; I will keep them from harm and injustice.

I will neither give a deadly drug to anybody who asked for it, nor will I make a suggestion to this effect. Similarly, I will not give to a woman an abortive remedy. In purity and holiness, I will guard my life and my art.

I will not use the knife, not even on sufferers from stone, but will withdraw in favor of such men as are engaged in this work.

Whatever houses I may visit, I will come for the benefit of the sick, remaining free of all intentional injustice, of all mischief and in particular of sexual relations with both female and male persons, be they free or slaves.

What I may see or hear in the course of the treatment or even outside of the treatment in regard to the life of men, which on no account one must spread abroad, I will keep to myself, holding such things shameful to be spoken about.

If I fulfill this oath and do not violate it, may it be granted to me to enjoy life and art, being honored with fame among all men for all time to come; if I transgress it and swear falsely, may the opposite of all this be my lot.[39]

To reiterate, Edelstein's translation of the Oath, or some similar English translation,[40] represents the Oath as would have been understood in

1946–1948, when the WMA used it as the basis for a post-Holocaust oath that it published as an appendix to its International Code of Medical Ethics (ICME).[41]

The International Code of Medical Ethics (1949)

Jules Voncken (1887–1975), surgeon general of the Belgian Medical Corps, was one of the physicians searching for a practical method for implementing the Nuremberg Code. Shocked by the revelations at the Nuremberg Trial, he suggested drafting "*un Enseignement du Droit International Medical*," that is, an international code of medical ethics that would, among other things, offer practical resolutions to issues of permissible research on human subjects. Acting on Voncken's suggestion, a committee was formed to draft such a code, and in 1949, the WMA adopted an ICME: a list of doctors' general duties posted side by side with their specific duties to the sick. According to this list, doctors were obligated to comply with "the highest standards of professional conduct," to eschew the "profit motive" in professional conduct, to forego unauthorized forms of advertising, and were not to surrender their professional independence by joining a union or a group practice. Doctors should not engage in fee-splitting or in any act that would weaken a patient's physical or mental resistance (except in that person's own interest). They were also to use "great caution" in "divulging discoveries or new techniques or treatments."

Listed side by side with these general admonitions was a correlative list of a doctor's duties to *his* patients. Western medicine in the 1940s was a man's world, and the choice of masculine pronouns reflects this fact. As stated in the ICME, a physician's primary duty was to "preserve human life." Thus, *he* must refer *his* patients to specialists "when an examination or treatment is beyond *his* capacity." *He* must also offer "emergency care as a humanitarian duty."[42] The ICME also states that doctors should follow the Golden Rule by behaving toward "*his* colleagues as *he* would have them behave towards *him*" and so were "not [to] entice patients from *his* colleagues." Yet, nothing in the ICME addresses Voncken's original concerns about implementing a practical version of the Nuremberg Code. The ICME did, however, require physicians to "observe the principles of the Declaration of Geneva," which was published in English, French, and Spanish in 1949. The ICME code has since that time had changes in format and revisions in content in 1968, 1983, 2006, and most recently in 2022. It remains the least frequently invoked or cited of the WMA's three core ethical statements: the other two are its Declarations of Geneva and Helsinki.

The Declaration of Geneva

Declaration of Geneva

Serment de Geneve
Declaracion en Genebra
1947–1949[43]

(i)   AT THE TIME OF BEING ADMITTED AS A MEMBER OF THE MEDICAL PROFESSION:

(ii) I SOLEMNLY PLEDGE myself to consecrate my life to the service of humanity.

(iii) I WILL GIVE MY TEACHERS the respect and gratitude which is their due.

(iv) I WILL PRACTICE my profession with conscience and dignity.

(v)   THE HEALTH OF MY PATIENT will be my first consideration.

(vi) I WILL RESPECT the secrets that are confided in me.

(vii) I WILL MAINTAIN by all means in my power, the honor and the noble traditions of the medical profession.

(viii) MY COLLEAGUES will be my brothers.

(ix) I WILL NOT PERMIT considerations of religion, nationality, race, party politics or social standing to intervene between my duty and my patient.

(x)   I WILL MAINTAIN the utmost respect for human life, from the time of conception; even under threat, I will not use my medical knowledge contrary to the laws of humanity.

(xi) I MAKE THESE PROMISES solemnly, freely, and upon my honor.
      (This Declaration has been revised regularly since it was first issued.[44])

The *Declaration of Geneva* was an artifact of the WMA's intention to create a universally acceptable modernized version of the Hippocratic Oath, thereby instilling in future generations of physicians a fundamental commitment to the health of individual patients and thus preventing any future iteration of a medical ethics of *Rassenhygiene*, eugenic sterilization, *euthanasie*, or Aktion T4. To accomplish this, the text of the ancient oath was rescripted and updated to create a declaration that could be pledged by mid-twentieth-century medical men worldwide. Thus, the invocation of gods and goddesses in the ancient Greek oath was replaced by a secular commitment to "THE MEDICAL PROFESSION" (note the capital letters). Line (i) also transformed the oath's original function, which, as is clearly stated in Edelstein's translation, was to commit *entering* students or apprentices to serve as adoptive sons of the Hippocratic family. The Hippocratic's need to adopt nonfamily members as substitute sons has been explained in a *Commentary on the Oath* attributed to Galen, which states that "Hippocrates decided to make instruction available to strangers owing to an insufficient number of family members willing to carry on the medical tradition . . . and [so he] drafted the oath to this effect."[45] Thus, as one of Hippocrates' contemporaries, the Greek philosopher Plato, remarked, "It was possible to study medicine with Hippocrates for a fee,"[46] but—and this part eludes many commentators—inductees also had to be willing to take on the roles once played by sons and nephews who had previously learned medicine by assisting their uncles and fathers in the family's medical practice.

French classicist Jacques Jouanna (1935–) believes that Plato found Hippocrates' decision to allow outsiders into a formerly family trade noteworthy because it constituted "a veritable revolution . . . in the transmission of medical knowledge,"[47] that is, a transition from trade secrets passed down from fathers and brothers to sons and cousins into a form of medical education available to anyone willing to pay a fee—provided they also accepted Hippocratic familial obligations as their own. In effect, the Hippocratics invented a familial version of the medical school–residency–apprentice model of medical education that included formal instruction as "precepts and oral instruction and all the other learning."

As Edelstein's translation makes clear, the oath was to be sworn and signed by students *commencing* their study of medicine, who had agreed to accept the role of adoptive sons to him "who has taught me this art" and to treat him "as equal to my parents . . . and to regard his offspring as equal to my brothers in male lineage and to teach them this art . . . to give a share of precepts and oral instruction and all the other learning to my sons and to the sons of him who has instructed me and to pupils who have signed the covenant and have taken an oath according to the medical law, but no one else."[48] Any approximation to this line would be inappropriate in the Declaration because medical education was no longer a familial affair. Thus, line (iii) of the Declaration reads, "I WILL GIVE MY TEACHERS the respect and gratitude which is their due," and line (viii) states that "MY COLLEAGUES will be my brothers"—a male chauvinist commitment that echoes the original familial construction of the male student/apprentice's role.

There are no counterparts in the ancient oath to line (iv) of the Declaration, "I WILL PRACTICE my profession with conscience and dignity." The concepts of "profession" would first be introduced by the Romans centuries later and would be redefined by Thomas Percival in the nineteenth century. Lines (iv) and (vii), "I WILL MAINTAIN by all means in my power, the honor and the noble traditions of the medical profession," also have no counterparts in the ancient oath. The Hippocratics were inventing formal medical education; they were creating new traditions even as they, somewhat ironically, sought to preserve the Hippocratic familial medical tradition by inducting strangers into their family.

Line (v), "THE HEALTH OF MY PATIENT will be my first consideration," expressly and clearly rejects, in capital letters, any prioritization of the health of a race above care for individual patients. It also preserves the prime moral directive of the ancient oath, that the student/apprentice (and eventually full-fledged Hippocratic physician) is morally obligated to use therapeutic measures "for the benefit of the sick according to [his] ability and judgment; [and to] keep them from harm and injustice." This point was so important in the ancient oath that it is reiterated, stating that when an apprentice shadows physicians, "Whatever houses [the apprentice] may visit, [he] will come for the benefit of the sick, remaining free of all intentional injustice, of all mischief and in particular of sexual relations with both female and male persons, be they free or slaves." As reformulated in the Declaration of Geneva, the Roman word "patient" (the one who endures suffering) is used rather than the Greek concept of "the sick." Prioritizing the health of the individual patient had special resonance in the aftermath of the Nuremberg Trials, at which *Rassenhygiene* was condemned as a "denial of fundamental values on which Medicine is based [because it meant that] the care of the individual patient ceased to be the doctor's primary aim, and the humanitarian purpose of medical science was subordinated to the needs of war."[49]

Line (vi)—"I WILL RESPECT the secrets that are confided in me"— reiterates the apprentice's promise in the ancient oath to keep what they "see or hear in the course of the treatment or even outside of the treatment . . . to themselves, holding such things shameful to be spoken about."[50] Ramm also invokes the Hippocratic Oath as the basis of the *"the physician's duty of silence,"* explaining that "this duty of silence concerning secrets and observations to [the physician] is to be adhered to not only during the life of the patient but even after his death."[51] But, Ramm adds, this *Rassenhygiene* addendum: "if however, a sickness, behavior, or genetic trait conceals in the individual a danger to the *Volk* community then there is a higher viewpoint which transcends the duty of silence leading to the duty to report and thereby serve to protect the community. We are thinking here of giving notice to the health authorities . . . and the reporting of genetic illness for the purpose of sterilization."[52]

Line (ix) of the modern oath, "I WILL NOT PERMIT considerations of religion, nationality, race, party politics or social standing to intervene between my duty and my patient," expressly rejects Ramm's view of transcendent moral obligations to the race. Line (x), "I WILL MAINTAIN the utmost respect for human life, from the time of conception; even under threat, I will not use my medical knowledge contrary to the laws of humanity," also appropriates a central tenet of the ancient oath to contrast it with its betrayal in Nazi medical ethics and medical practices. Edelstein translates these lines in the ancient oath as "I will neither give a deadly drug to anybody who asked for it, nor will I make a suggestion to this effect. Similarly, I will not give to a woman an abortive remedy. In purity and holiness, I will guard my life and my art." This translation renders the oath amenable to the prolife intentions of the authors of the Declaration of Geneva, who viewed it as a means of inoculating future physicians against abortion, *euthanasie*, *Rassenhygiene*, eugenic sterilization, and the extermination of *Lebensunwertes Lebens*. For them, (3. i.) rejects *Rassenhygiene*. Similarly, line (4. i.), "And I will not give a drug that is deadly to anyone if asked [for it], (ii.) nor will I suggest the way to such a counsel. And likewise, I will not give a woman a destructive pessary," prohibits *euthanasie* and all forms of medical assisted killing (as in Aktion T4); extending this prohibition (4.ii) also prohibits any form of abortion or mass sterilization.

The foundational presumption underlying the Declaration of Geneva was that Nazi medical ethics and the consequent abuse of humans were unprecedented historical anomalies. Truth be told, the purity of the pre-Nazi biomedical sciences and medical practices was an idealistic fiction. As Moll pointed out in 1902, academic and hospital physicians were "obsessed by a kind of research mania, have ignored the areas of law and morality in a most problematic manner. For them, the freedom of research goes so far that it destroys any consideration for others. The borderline between human beings and animals is blurred for them. The unfortunate sick person who has entrusted herself to their treatment is shamefully betrayed by them, her trust is betrayed, and the human being is degraded to a guinea pig. . . . There seem to be no national or political borders for this aberration."[53] As Moll also observed, this worldwide phenomenon was associated with developments in biomedical science and with hospital physicians' quest for status and fame—as well as such mundanely immediate rewards as improving salaries.

The deeper point to appreciate is that the National Socialists did not invent *Rassenhygiene*. Race-based biology had been invented well before anyone conceived of Nazism or Nazi medical ethics. During the first half of the twentieth century, "race" was as much a scientific concept/term as "ameba," "bacterium," or "contagion," and the idea of applying biology as social hygiene was as respectable as the germ theory of disease and various public health initiatives. As often as not, however, science validates popular prejudices, including xenophobic prejudices against people who look or speak differently, worship differently, or come from elsewhere. The notion that physicians or scientists, by virtue of their commitments to medicine or science, can somehow rise above the values of their society is more myth than reality. The ancient Greek philosopher Aristotle (384–322 BCE), often characterized as "the first scientist,"[54] famously endorsed the popular "saying of the poets—'Tis meet that Greeks should rule barbarians'—implying that barbarian and Greek are not the same in nature."[55] Such ethnic and racial stereotypes received scientific reification during the Enlightenment, when eighteenth-century Swedish botanist Carl Linnaeus (1707–1778) distinguished four separate human "races" differentiated by geography and skin color: *Africanus niger*, "niger" means black in Latin: hence, "nigger" and "Negro"(black Africans); *Americanus rubescens*, red Americans (i.e., Indigenous Americans); *Asiaticus fuscus*, yellow Asians; and *Europaeus albus*, white Europeans.[56] Influential scientific treatises, such as British scientist Charles Darwin's (1809–1882) watershed 1859 treatise, *On the Origin of Species by Means of Natural Selection, or the Preservation of Favoured Races in the Struggle for Life*, validated the idea of *race*. A quarter century later, in 1883, Darwin's cousin, Sir Francis Galton (1822–1911), coined the term *eugenics*, which he characterized as the science of improving *racial* stock through artificial selection by selective breeding. Race-based eugenicist movements arose thereafter in America, Britain, Scandinavia, and, of course, Germany.[57]

Dr. John Pridham and his fellow veterans, in contrast, envisioned a medicine rising above notions of race. Viewing the future better with his one eye than naysayers could see with two, he and his colleagues took an "old thing" that was passée, revamped it as the Declaration of Geneva, and "behold . . . it had become new." Some version of this Declaration (versions are regularly updated) or some other modernized version of the Oath are currently sworn by medical students at white coat[58] and graduation ceremonies throughout the United States and in many other countries.[59] Moreover, the Oath's moral authority is regularly invoked in campaigns to criminalize abortion, by opponents of physician-assisted suicide and euthanasia, by physicians campaigning against medical complicity in torture,[60] and by clinicians asserting conscience-based refusals to participate in abortions and various forms of medical aid in dying (MAID).[61] Health care professionals also invoked the Oath as a rationale for risking their lives during pandemics. As one quartet of physicians wrote during the

COVID-19 pandemic, "We believe that the modern Hippocratic Oath and the Physician's Pledge are relevant, not only to physicians but also to all of our healthcare colleagues working on the frontlines of this pandemic. We believe that these documents, together with the long tradition of medical ethics, provide a concrete and current basis through which we can navigate the many ethical and moral dilemmas that we may already have faced or will face in the future as this pandemic affects the world."[62]

5

The Declaration of Helsinki: Negotiating Practical Principles for Research on Human Subjects

Principle 1 of the Nuremberg Code would cripple if not eliminate most research in the field of mental disease . . . [as well as] all research on [physical diseases of the] mentally ill or with children . . . [or] unconscious persons.

Most investigators would [conform their research to the Nuremberg Code] if only they knew how to comply.

Until recently the Western World was threatened with the imposition of the Nuremberg Code as a Western Credo. With the wide adoption of the Declaration of Helsinki, this danger is now past!
—Henry K. Beecher[1,2,3]

The judgment rendered by the Nuremberg Military Tribunal concerning human experimentation has never been and probably never will be construed either in Britain or America as legal precedent.
—Maurice Pappworth[4]

The Beecher–Hill Critique

Although the Declaration of Geneva was inspired by the Holocaust, it does not mention the Nazi doctors, Aktion T4, the Holocaust, or abuses of "human material" by Nazi researchers; neither does the Nuremberg Code. Yet their proximity to the Nuremberg medical trial speaks louder than their words: at the moments of their creation, everyone knew that the oath and the code were designed to underscore Nazi medicine's status as an aberration and, not coincidently, to validate respectable physicians' and scientists' research on human subjects as morally permissible. Yet, because Ivy's intent, as encapsulated in his original title, "Permissible Experiments on Humans," was to list some basic criteria for permissible research on humans, the code was never designed to serve as a practical guide directing researchers investigating innovative ways of preventing or curing disease or disability. Yet the impracticality of the Nuremberg principles was evident to a famous whistleblowing research ethics reformer Henry Knowles Beecher (1904–1976). He repeatedly observed that the first principle of the Nuremberg Code, Principle 1, that "The voluntary consent of the human subject is absolutely essential," would prohibit research on children, on the mentally or physically incapacitated, and on unconscious people since none of these people could offer valid consent. In Beecher's own words, Principle 1 of the Nuremberg Code would "cripple if not eliminate most research in the field of mental disease . . . [as well as] all research on [physical diseases of the] mentally ill or with children . . . [or] unconscious persons," because it lacks any provision for surrogate consent.[5] Furthermore, Beecher remarked, the Nuremberg principles requirement of animal experimentation as a prelude to testing on humans would eliminate research on diseases unique to the human species, such as syphilis and yellow fever. Perhaps more worrisome still was Nuremberg Principle 2, which specifically prohibited experiments "random . . . in nature."[6] This became problematic in 1948, the year after the Nuremberg Court published its principles, when epidemiologist Sir Austin Bradford Hill (1897–1991) definitively established a statistical approach involving prospective *randomized* controlled trials (RCTs) as the gold standard for demonstrating the safety and efficacy of new drugs, treatments, and procedures.[7] Unfortunately, Beecher remarked, since the Nuremberg principles "disclaim[ed] 'random' experiments,'" they appear to prohibit the use of randomized controlled trials. What galled Beecher most, however, was the impact on the use of placebos. Proud author of a classic paper "The Powerful Placebo,"[8] Beecher also argued that "use of placebos . . . could hardly be tolerated under [the Nuremberg Code requirements]" because the psychological impact of informing patients that they might receive a placebo might bias a study.[9] Beecher notes wryly that "most investigators would [conform their research to the Nuremberg principles] if only they knew how to comply."[10]

Beecher's comments require a reappraisal of earlier accounts of the relationship between the Nuremberg Code and the WMA's 1964 Declaration of Helsinki. Many commentators regard the Declaration of Helsinki as a morally retrograde document. The authors of the two currently standard histories of bioethics, Jonsen and Rothman, claim that the Nuremberg principles "offered a set of standards on ethical research that . . . might have served as a model . . . for [future] guidelines."[11]

Yet, as is evident in the original title, "Permissible Experiments on Humans," the pressing need motivating Ivy was differentiating the morally permissible experiments on humans as a method of doing good for humankind from the morally impermissible experiments performed by the Nazi researchers. Thus, although Leo Alexander introduced practical refinements (likely drawn from the 1931 German guidelines), it is not clear that either author intended the Nuremberg principles as a practical guide for research using human subjects: their focus was on the need to establish principles differentiating morally permissible from morally impermissible experiments on humans in international humanitarian law. Thus, when active researchers were confronted with plans to implement the Nuremberg principles, they rejected them as so firmly rooted the horrors revealed at the war crimes trials that they could never serve as a practical ethics code for medical research on humans. So, the WMA initiated a decade-long quest to create a new, internationally agreed-on set of workable standards regulating morally permissible research on human subjects.

Principles for Those in Research (1954)

In 1954, when Paul Cibrie (1881–1965), former secretary general of the *Confédération des Syndicats Médicaux Français,* assumed chairmanship of the WMA's Committee on Medical Ethics, he immediately prioritized developing a more functional code of ethics for biomedical research on humans. Shortly thereafter the committee recommended and the WMA approved, *Principles for Those in Research and Experimentation.* The first two of these principles address the responsible conduct of research; the last three principles address research on human subjects. The latter principles require researchers to fully inform research subjects about the nature of an experiment; however, it limited experiments to presumptively therapeutic innovations in "individual and desperate cases" (i.e., to patients)—and then only with the approval of the patient or his next of kin. It also required consent to be documented in writing.[12]

Principles immediately became a paradigm case of an ethics code obsolesced on the moment of its issuance. By limiting experiments to patients, rather than to human subjects, it circumvented the major issues raised by advances in vaccine research and was so minimalist that it did not even redress the major flaws in the Nuremberg principles—except to allow surrogate consent for innovative therapies. Moreover, by restricting experimentation "to individual and desperate cases," vaccine research was ruled out altogether, since it this research had to be conducted on large groups of healthy people; also ruled out were various forms of large-scale prospective random placebo-controlled trials that Bradford Hill established as the gold standard for demonstrating the safety and efficacy of new drugs.

The Search for a Practical Alternative to the Nuremberg Code
Continues (1959)

In 1959, Hugh Anthony Clegg (1900–1983), editor of the *British Medical
Journal* from 1947 to 1965 (*BMJ*, founded 1840), became chairperson of
the WMA's Committee on Ethics. After a review of the Nuremberg principles demonstrated their utter impracticality as a modern code of ethics for
biomedical research on humans, Clegg sought to supplant the 1954 *Principles* with a code directly responsive to then current ethical challenges. He
sought to "draft a code which could serve at least as a guide to doctors
working in different conditions and in different countries."[13] To ensure
the code's relevance to contemporary biomedical research practices, Clegg
asked representatives from the WMA's member associations to comment
on five scenarios.

1. Administration of drugs to medical students to assess their effects.
2. Preventive inoculations using a control group not inoculated against
   whooping cough or tuberculosis.
3. Controlled therapeutic trials of a new drug [Bradford Hill's procedure].
4. Using inmates of prisons, penitentiaries, or mental institutions for controlled prophylactic or therapeutic trials.
5. Investigations on hospital patients, which had no relation to the condition
   that brought them to the hospital.[14]

These scenarios address many of the major issues challenging the
research community in the 1950s. The first scenario probes a conflict of
interest evident in the common medical school practice of recruiting medical students (and departmental and laboratory assistants and secretaries)
to serve as human subjects. The second and third scenarios probe issues
in randomized controlled trials of drugs and vaccines. These issues sur-
faced in discussions of the two-armed (vaccine, placebo) and three-armed
trials (vaccine, placebo, neither vaccine nor placebo) used in developing
the Koprowski, Sabine, and Salk polio vaccines during this period (1950–
1957). To eliminate investigator bias and to control for the positive (placebo) and negative (nocebo, e.g., imaginary negative side effects, e.g., to
sugar pills) and psychological biases, both subjects and the investigators
were often "blinded" (i.e., kept in ignorance of who received a drug or vaccine and who received a placebo). Questions thus arose as to whether to
inform *subjects* that they might receive a placebo as part of the initial consent process, inform them after an experiment ended, or not inform them
at all. Such information could be germane to a subject's health because, if
a vaccine proves effective, subjects receiving a placebo might believe they

were protected by a vaccine when they were not. Note that codes focused on the physician–patient relationship, like the WMA's 1954 *Principles*, would be inappropriate for vaccine research because vaccines are preventive in nature and are given to well persons, not necessarily to patients. Hence, the word "subject" was substituted for "patient."

The fourth and fifth scenarios address issues of medical experimentation on people in total institutions: that is, institutions whose members are cut off from the rest of society and whose conduct is regulated, as in armies, asylums, hospitals, orphanages, prisons, and schools.[15] Total institutions are methodologically ideal for conducting controlled trials, and the initial trials of the Salk vaccine were conducted in children's homes (orphanages) and elementary schools.[16] Yet enrolling subjects from total institutions is morally suspect because their highly controlling context may undermine subjects' freedom to refuse or their surrogates' ability to make voluntary choices—issues that later arose in the Kaimowitz psychosurgery case and the Willowbrook hepatitis experiments. These issues are compounded if hospitalized patients serve as subjects because they are prey to therapeutic misconceptions (subjects' tendency to believe that any experimental intervention is potentially therapeutic)[17] and because medical investigators typically have conflicting interests and loyalties that may undermine their fiduciary responsibilities to their patients.

First Draft of the Declaration of Helsinki (1962)

The initial 1962 draft of the document that would ultimately evolve into the Declaration of Helsinki opens by quoting a line from the WMA's 1949 Declaration of Geneva, "The health of my patient will be my first consideration." There follows a line from the WMA's 1949 International Code of Medical Ethics (ICE): that "any act or advice which could weaken physical or mental resistance of a human being may be used only in his interest." The Declaration then moves from these generic ethics statements to introduce the subject of research, noting that "for scientific progress and the welfare of suffering humanity, it may be essential that the results of laboratory experimentation be verified by human experimentation. . . . For this reason [the WMA has prepared] a code of ethics on human experimentation that will serve as a guide to each doctor, within the framework of his conscience and his national and religious ideologies."[18] Having distinguished innovative therapeutic interventions, "for the benefit of the patient," from those "conducted solely for the acquisition of knowledge," the term "experiment" was defined as "an act whereby the investigator deliberately changes the internal or external environment in order to observe the effects of such a change."[19] Medical experiments are then defined as those supervised by a "qualified medical man."

The first statement of ethical principles for medical experimentation repeats language from the 1954 *Principles*, which require that in any medical experiment, the "nature, the reason, and the risks of the experiment [be] fully explained to the subject," who shall have "complete freedom to decide whether or not to take part in the experiment." Amplifying this principle, subjects also have the right to withdraw from experiments "at any time" (Nuremberg Principle 9), and those conducting the experiment should discontinue it if "it may . . . be harmful to the subject of the experiment" (Nuremberg Principle 10). Since the headline experiment of the era involved mass testing of polio vaccines on children, several principles address the ethics of using children as research subjects. These require parents' or guardians' consent to children's participation in experiments. They also prohibit outright any experiments on "children in institutions [who are] not under the care of relatives," as, for example, would likely be the case with children in orphanages.

The section on "Experiments for the Benefit of the Patient" admonishes "doctors [to] never . . . abuse the trust of the patient." It authorizes surrogate consent and allows physicians the freedom to attempt last-hope experiments to save patients' lives or to alleviate pain or suffering. It also permits "experiments on disease prevention," noting that they "should be based on laboratory and animal experiments or *other scientific data*" (emphasis added)—a provision that remedies the Nuremberg Code's overly stringent Principle 3 requirement of prior animal experimentation. Drawing on the 1949 ICE, the 1962 draft advises special caution in experimenting with drugs that could alter a subject's personality and notes that special ethical rules are needed for controlled trials—however, it does not offer any such rules and thus fails to address the scenarios dealing with controlled trials (Scenarios 2 and 3).

The third and last section, "Experiments Conducted Solely for the Acquisition of Knowledge," responds to Scenario 4, experiments on vulnerable populations in total institutions by forthrightly prohibiting experiments on people hospitalized for mental disease or disability, as well as any experiments using "captive populations" in orphanages, prisons, and reformatories. Harkening back to the Nuremberg Trials, it also protects prisoners of war and "civilians . . . detained [by authorities who] should never be used for human experiment[s]." Finally (responding to Scenario 1), the Declaration recognizes conflicts of interest in asymmetrical power relationships and admonishes that "no doctors should lightly experiment on [those] in a dependent relationship . . . such as a medical student to his teacher, a patient to his doctor, a technician in a laboratory to the head of his department."[20]

Despite its overall stringency and concern to protect vulnerable individuals in total institutions, in one important respect, the 1962 draft was less stringent than the 1954 *Principles*: it did not require written documentation of consent. Perhaps written documentation was assumed, but nowhere in the 1962 draft is there any discussion of written consent or of a consent form. Oral consent would satisfy all the conditions of the draft—leaving no record of the information given to subjects or surrogates about the nature or risks of the experiment and no documentation of consent. Investigators could treat a subject's oral acknowledgment, even a nod or a gesture or the very act of volunteering, as adequate consent. In this respect, *Principles* reflected the ethical and legal views dominant in the pre-bioethics era when eminent authorities like Irving Ladimer, JD, SJD, could write, "I say flatly . . . that consent does not become any better, does not become any more legal, because it is written. . . . For many purposes it may be desirable not to wave papers around and make it look like a real estate transaction or a death notice. There are other ways to get well-documented consent. Whatever is done, however, certainly must be included and incorporated in any record" (Ladimer, November 12, 1964).[21] Sociologist Laura Stark puts this into historical perspective, noting that although "today consent is often associated with forms signed by research participants that have been approved by and IRB . . . informed consent was not synonymous with consent forms in the 1950s and 1960s."[22]

Controversy over the 1962 Draft Declaration

In 1961, Italian physician and WMA president, A. Spinelli, succeeded Clegg as chair of the WMA Committee on Ethics; Clegg, however, continued to assist in an advisory capacity. As historian Susan Lederer documents in detail, the 1962 draft proved highly controversial.[23] The British and French research community found research on people in total institutions morally abhorrent, whereas Americans and Canadians favored the practice, which was in widespread use in both countries. As the debate played out, practical concerns mingled with but were often disguised as matters of principle. The 1950s and 1960s were the apex of the antibiotic revolution during which American and Canadian researchers and pharmaceutical companies often used prisoners and others in total institutions as initial test subjects. Since the 1962 draft prohibited experiments on "captive populations" and others in total institutions, it appeared to threaten researchers' careers and companies' profits.

On the level of principle, senior British commentators, like Sir Austin Bradford Hill, also criticized the 1962 draft. The proposed requirement that experiments should be carried out "under the supervision of a qualified medical man" rankled Bradford Hill, who noted that it would prevent epidemiologist and statisticians, like him, from conducting experiments on human subjects.[24] He also objected to the proposed constraints on experiments with children, observing that it is reasonable to research such questions as "Is gamma globulin more, or less, effective than convalescent serum in the prevention of measles? Was it unethical to find out in the very circumstances in which it was possible to do so (as well as well as important to the subjects)? The [Draft] Guide says Yes [it was unethical]."[25] Bradford Hill believed the correct answer was "No," these were legitimate experiments. Henry Beecher joined with Hill in critiquing the 1962 draft, noting that prohibition on research in mental hospitals and on people with mental disability "would seem . . . automatically to condemn as unethical clinical trials in psychiatry."[26]

The 1964 Declaration of Helsinki

For the next two years, extensive debates over the 1962 draft continued within the WMA's Committee on Medical Ethics, at WMA meetings, and in medical journals worldwide. Throughout, principled objections intertwined with careerist and pocketbook concerns. A compromise was eventually reached in which the controversial verbiage about children and captive populations was deleted and the subject matter of the Declaration was redefined as "Recommendations Guiding Doctors in Clinical Research." The meaning and scope of the expression "clinical research" was left undefined, but it was generally taken to mean research on new ways of treating or preventing disease or disability, ameliorating their symptoms, or coping with their sequelae.[27] The tripartite structure of the 1962 draft was retained in the 1964 Declaration. Five Basic Principles in Section I stipulate that research on humans must (1) be based on "scientifically established facts," (2) be supervised by "a qualified medical man," (3 and 4) involve a prior risk–benefit analysis and not involve risks disproportionate to "the importance of the objective," and (5) special caution must be exercised in research involving personality-altering drugs or procedures. Section II focuses on "last-hope" therapeutic interventions and requires consent of the patient or a "legal guardian" for such interventions.

The more significant section is number III, "Non-Therapeutic Clinical Research." It opens by reminding physicians that when conducting
research, "they remain the protector of the life and health of that person
on whom clinical research is being carried out." The Declaration states
unequivocally that no research on human subjects may be undertaken
without the informed voluntary consent of subjects or their legal guardians and that the investigator should explain to subjects or guardians the
nature, purpose, and risks of the experiment. It also improves on the 1962
draft by requiring documentation of consent, stating, "Consent should, as
a rule, be obtained in writing." However, it is even less direct than the
1962 draft in addressing medical professors' conflicts of interest and loyalties in recruiting their medical students or laboratory assistants as human
subjects: it merely suggests that investigators should respect the potential
subject's "personal integrity . . . if the subject is in a dependent relationship to the investigator." Finally, it accords subjects and their guardians the
right to withdraw permission for research at any time and obligates investigators to discontinue research that, if continued, could prove harmful
to the subject. The revised declaration passed unanimously at the WMA's
1964 Helsinki meeting and, following a precedent set by the Declaration of
Geneva, is called "The Declaration of Helsinki."[28]

Depending on when one starts the clock—1947, 1949, 1954, 1962,
1962 revised, or 1964—it took between two and nineteen years for the
WMA to fashion a consensus around a practical alternative to the Nuremberg Code on the ethics of research on human subjects. Like its precursor,
the Declaration of Geneva, the Declaration of Helsinki promulgated a code
of ethics in an operationally useful form, and again like its precursor, it
quickly gained worldwide acceptance because it answered a worldwide
need. Within two years, the American Society for Clinical Investigation,
the American College of Physicians, the American College of Surgeons,
and the American Medical Association endorsed the Declaration as, in the
course of time, did medical societies and regulatory bodies worldwide.

As to the Declaration itself, unlike the Nuremberg Code, which Beecher
had characterized as "a rigid set of demands," it is revised regularly, commencing with a major revision in 1975. Among the more important changes
in the 1975 revision was the requirement that prior to the initiation of
research, an independent ethics committee (variously referred to as REBs,
RECs, or IRBs) had to review and to approve all proposed research. It was
also recommended that medical journals refuse to publish any research not
in conformity with the Declaration's requirement of documented informed
voluntary consent and a prior review by a research ethics committee. By

2022, the Declaration had been revised seven times. Note that, well before the creation of the field now known as bioethics, the issue facing the medical community was not how to formulate standards for ethical research on human subjects—the standard for that had been set in the 1964 Declaration of Helsinki—but rather how to make recalcitrant researchers accept current standards and how to fine-tune these standards to address issues arising from innovative approaches to medical experimentation and technologies.

6

Kelsey, Pappworth, and Beecher: Moral Awakenings

Hear now this, O foolish people, and without understanding; which have eyes, and see not; which have ears, and hear not.
—Jeremiah 5:21[1]

There are none so blind as those who will not see.
    The most deluded people are those who choose to ignore what they already know.
—John Heywood, 1546

You know, there's none so blind as they that won't see.
—Jonathan Swift, *Polite Conversation*, 1738

Perception without conception is blind.
—Immanuel Kant, 1787, *Critique of Pure Reason*

In the fields of observation chance favors only the prepared mind.
—Louis Pasteur, December 7, 1854

    There is none so blind
    As He who will not see
—Ray Stevens, "Everything Is Beautiful," 1970[2]

On Moralists' Perceptions of Immorality

As noted in chapter 1, Bentham and Moll were moralists. What makes moralists so intriguing is that they perceive as morally questionable, or immoral, what everyone around them does not find morally suspect or immoral. This chapter focuses on moralists in post–World War II America and Britain who viewed as immoral conduct that other biomedical professionals treated as morally acceptable. After a few preliminary remarks about the phenomenon of moral blindness and moral sightedness, the discussion focuses on three moralists, starting with a mature married woman who had the temerity to defy conventional acquiescence to pharmaceutical companies by demanding that physicians who investigate new drugs obtain the informed consent of their subjects. It then turns to two other moralists: one, an uppity Jew with an acerbic personality and a penchant for media attention; the other, a poor boy from a hick town in middle America who rose to success by feigning to be a "Boston Brahmin"—a characterization of upper-class Bostonians at a time when, to quote a well-known limerick, "the Lowells talk only to Cabots, And the Cabots talk only to God."[3] What these moralists have in common is that they perceived, recognized as immoral, and protested the immorality of studies that the editors and reviewers for prestigious medical journals, and their readership, never viewed as immoral—although, a decade or so later, these same studies became paradigms of immorality.

"Woke" is a word that twenty-first-century black protesters use to describe the phenomenon of moral awakening to something as immoral or unjust that one hadn't previously perceived as such. University of California, Santa Barbara linguistics professor Deadre Miles-Hercules traces this use of "woke" to a 1923 collection of writings by Jamaican philosopher, social activist, and black nationalist, Marcus Garvey (1887–1940).[4] Miles-Hercules also found that famous blues singer Huddie (Lead Belly) Ledbetter (1888–1949) popularized Garvey's use of "woke" in a 1938 song about the "Scottsboro Boys," nine black men falsely accused and convicted of raping two white women.[5] White Americans became aware of this use of "woke" after the Black Lives Matter movement used it in a 2014 to protest police brutality and murders of black people. By 2017, the *Oxford English Dictionary* recognized this adjectival use of "woke" as a way of indicating that a person had become sensitive to, had been "awakened," so to speak, to racial or social discrimination or injustice.[6] This usage seems closer to what Jamaican philosopher Marcus Garvey had in mind when he coined the expression in the 1920s (Romano 2020). [7]

As Garvey observed in a short essay, *Negroes Robbed of Their History*, "The white world has always tried to rob and discredit [blacks] of our history" (Garvey 2023, 29).[8] As Garvey might have anticipated, white conservatives soon robbed blacks of the word "woke," appropriating it as a pejorative to derogate "injustices" arising from liberals' reformist zeal. They inverted the meaning of "woke" to stand for something akin to "liberal political correctness," a pejorative that, like "libtard," is intended as both an insult and a parody. According to the *Cambridge Dictionary*, "libtard" is "a combination of 'liberal' and 'retard,' used to indicate a person so completely brainwashed by liberal ideas" that presenting alternative ideas to them is akin to talking to a mentally retarded person.[9] Whatever one's political beliefs, as the series of quotations prefacing this chapter indicates, the "libtard" phenomenon that conservatives experience in trying to talk to liberals (i.e., the phenomenon of "people, . . . which have eyes, and see not, which have ears, and hear not") has been recognized by preachers, philosophers, prophets, and poets, over the ages. Some academic ethicists now characterize this concept as "moral blindness."[10] Yet since recognizing moral blindness presumes an opposite state, it also must be meaningful to talk of the process of becoming morally sighted, that is, the phenomenon of becoming "woke."

Two dictionaries responsive to linguistic innovations, the Urban Dictionary and Wiktionary, now recognize both "woke" and "unwoke."[11] These antonym pairs and their more erudite cousins, "moral sightedness" and "moral blindness," are used in the rest of this book to describe the perspectives of those who "see" or who do not "see" as immoral something that is later recognized as immoral by people in the same community. Wokeness has consequences; insofar as a community is morally awakened to conduct that it formerly deemed moral, its members are bound to ask philosopher Kwame Anthony Appiah's (1954–) question, "What were we thinking? How did we do that for all these years?"[12] An important correlative of accepting the phenomenon of moral blindness and moral sightedness is the epistemological status of judgments of moral progress and regress as inherently perspectival. Thus, Brandt and Ramm were "woke," or morally sighted, with respect to *Rassenhygiene*, seeing its implementation as moral progress and decrying bourgeois medical morality focused on individuals as morally blind because it put the interests of individual patients ahead of those of the *Volkskörper*. Conversely, Alexander, Ivy, and Pridham condemned followers of *Rassenhygiene* as morally blind, and their condemnation led to one of the foundational documents of modern medical ethics, the document now called "The Nuremberg Code." What one can or cannot recognize as moral progress or moral regress depends on one's vantagepoint—which is why conservatives complain of "libtardness" in response to liberals' "wokeness."

Frances Kathleen Oldham Kelsey: The Female Conscience of the FDA

Moral sightedness is often the prerogative of outsiders accepted as insiders. A paradigm case is the Canadian American, Frances Kathleen Oldham Kelsey (1914–2015). Born and raised in Cobble Hill, a small town on Vancouver Island, Frances Oldham earned her Master of Science degree in pharmacology from McGill University. In 1936, just after the peak of the

Great Depression (1928–1933), Frances applied to a graduate program at the University of Chicago. By dint of excellent grades and a masculine-appearing first name, "Frances," she bypassed the barriers of sexist gender discrimination and was admitted as a graduate student and research assistant.[13] She earned her PhD in pharmacology from the university and later participated in the research that became the basis for the 1938 US Federal Food, Drug, and Cosmetic Act (FD&C), the legislation that empowered the US Food and Drug Administration (FDA) to regulate the safety of food, drugs, and cosmetics. Frances Oldham added "Kelsey" to her last name in 1943 by marrying another physician, Freemont Ellis Kelsey (1912–1966). After joining the University of Chicago faculty, Frances earned an MD degree in 1950 and then served a two-year stint as associate editor of the AMA's journal, *JAMA*.

She and her husband started a family in South Dakota, where they worked as general practitioners and where she also had an appointment as a university professor. Frances was later hired by the FDA as "one of only seven full-time and four young part-time physicians reviewing drugs" for the agency.[14] One month into her new job, she received her first assignment: an application to market a drug developed in 1954 that had been approved for sale in West Germany as a safe sleeping pill for pregnant women. Since the drug was also approved in Britain, Canada, parts of continental Europe, and South America, its approval seemed to be an ideal easy-peasy walk-in-the-park for a novice regulator. Yet, "The European data [submitted to the FDA] left [Kelsey] 'very unimpressed.' She had lived through cycles before in which a drug was acclaimed for a year or two—until harmful side effects became known."[15] So Kelsey requested more data, either from existing studies or from new studies, on the drug's impact on the fetus. In response, the company (Merrell Pharmaceutical, 1950–1996) ignored her request for additional data and instead blitzed her with testimonials from satisfied physicians. When Kelsey persisted in demanding more data, the company responded with a pressure campaign: "it made repeated phone calls and personal visits to Kelsey and complained to her superiors that she was unreasonable and nit-picking, and that she was delaying the drug's approval unnecessarily."[16]

Despite the pressure and personal harassment, Kelsey refused to approve the drug. Her concerns deepened in December 1960 when the *British Medical Journal* published a case report of neurological side effects from the drug. So, Kelsey again pressed the pharmaceutical company for more clinical data. As she explained in a May 5, 1961, letter, the FDA's standard for approving new drugs should be that "the burden of proof that the drug is safe . . . lies with the applicant." "In this connection," she continued, "we are much concerned that apparently evidence [of] peripheral neuritis [tingling in feet or hands] in England was known to you but not forthrightly disclosed."[17] Angered by the accusation that Merrell had suppressed information, an executive from the company telephoned Kelsey's supervisor threatening a lawsuit on the grounds that "he considered [the letter] somewhat libelous."[18] Her supervisor supported Kelsey.[19]

About the same time physicians worldwide began to report a spike in miscarriages, neonatal deaths, and birth defects ranging from malformed internal organs, to eye or ear defects, to phocomelia (abnormally short limbs or flipper-like arms, often attached close to a newborn's trunk). In November 1961, German pediatrician Widukind Lenz (1919–1995) determined that the cause of this outbreak of fetal congenital deformations was thalidomide, the very drug the American pharmaceutical company was pressing Kelsey to approve.[20] In an odd twist of history, Widukind was the son of a Nazi party member Fritz Lenz (1887–1976), one of the foremost theorists of *Rassenhygiene* and coauthor of the standard textbook on the subject.[21] Following in his father's footsteps, Widukind had been a leader in the Hitler Youth and had joined Hitler's "Brown Shirt" paramilitary organization (SA). After the war, both father and son were "rehabilitated." Nonetheless, Fritz continued to espouse eugenicist ideas about the need to prevent the propagation of *Lebensunwertes Lebens*. His son, Widukind, inherited this familial obsession but became a researcher focused on preventing fetal anomalies. This concern motivated his identification of thalidomide as the cause of phocomelia and other congenital deformities. Alerting the world, he prevented disability and deaths in thousands of unborn children of all races. Unfortunately, an estimated 8,000 infants, worldwide, had already been born with thalidomide-induced missing or malformed limbs; another 5,000 to 7,000 are estimated to have died in utero. The United States was a notable exception. Thanks to Widukind Lenz's research and to Kelsey's timely refusal to approve thalidomide, only seventeen American-born children are known to have suffered congenital defects from the 2.5 million "investigational" samples of thalidomide tablets that Merrell Pharmaceutical supplied to 1,267 American physicians.[22]

Fortuitously, Kelsey's actions caught the eyes of the staff of Democratic Senator Estes Kefauver of Tennessee (1903–1963, senator 1949–1963). The senator had been investigating pharmaceutical companies' pricing practices since 1959 but he could not get a bill through Congress. Seeking to spark support for the senator's flagging legislative initiatives, his staff leaked information about Kelsey's refusal to approve thalidomide to the

Washington Post. The July 15, 1962, edition of the *Post* headlined on its front page, "'Heroine' of FDA Keeps Bad Drug Off Market."[23] The resulting near-scandal reenergized Kefauver's mordant efforts to reform the pharmaceutical companies, culminating in the passage of 1962 Kefauver–Harris Amendment to the Federal Food, Drug, and Cosmetic Act. This amendment supported as official FDA policy Kelsey's view that pharmaceutical companies had the burden of proving to the FDA that novel drugs (and vaccines) were safe and effective. Moreover, at Kelsey's urging, for the first time in US history, the law also required that investigators seeking FDA approval for new drugs obtain the informed consent of their human subjects.

In some autobiographical remarks, Kelsey recollects her struggles with what bioethicists would later characterize as "scientistic medical paternalism," that is, the parental-like authority over all aspects of medicine accorded to physicians due to their scientific knowledge and medical experience. As Kelsey explains to an interviewer,

> The 1962 Kefauver-Harris Amendments and the 1963 investigational drug regulations introduced a number of new procedures which led to the strengthening of the control of drugs entering the market in the United States. . . . One very dramatic last-minute addition to the 1962 amendments was by Senator Jacob Javits of New York. He had raised the question, "Do people know they are getting investigational drugs?" It was very clear from our survey of the 1,000 doctors in the thalidomide case that many of the mothers and patients had not been told this, and the doctors themselves did not quite understand the status of the drug. So, a very important amendment to the law, not a regulation, was that patient consent must be obtained before a new drug, an unapproved drug, was given in a clinical trial.
> . . . the statements in the 1962 law and the [1964] Food and Drug regulations . . . are exactly the same. They used the same words, because frankly this was a new concept for the Food and Drug Administration. We never imagined we could have gotten away with anything, however much we thought the doctor should do [it], because at that time the doctors felt they were the Lord Almighty. That the patient should take what the doctor gives them because doctor knows best. And if the doctor thinks it is important that this drug be studied in a fashion that the patient does not know he is getting an unproven drug—not to worry. Big Daddy will take care of you.[24]

More specifically, in responding to Senator Jacob Javits (1904–1986, senator 1957–1981), Kelsey suggested language from a 1962 draft of the Declaration of Helsinki that required that investigators obtain the informed consent of research subjects receiving experimental drugs. In the words of the 1962 statutes, repeated in the 1963/1964 FDA regulations,

> Experts using such drugs for investigational purposes certify to such manufacturer or sponsor that they will inform any human beings to whom such drugs,

or any controls used in connection therewith, are being administered, or their representatives, that such drugs are being used for investigational purposes and will obtain the consent of such human beings or their representatives, *except where they deem it not feasible or, in their professional judgment, contrary to the best interests of such human beings.* (Italics added)[25]

Although the statute restates the consent requirement in the 1962 draft of the Declaration of Helsinki in legislative language, the italicized words created a loophole big enough to drive the proverbial truck through. At this moment, Kelsey was also urging officials at the National Institutes of Health (NIH) to promulgate a similar regulatory requirement since "there is no generally accepted professional code [of ethics] relating to the conduct of clinical research [this is] a mounting concern . . . over the possibility of untoward events . . . [because] highly consequential risks are being taken by individuals and institutions as well as NIH" (the National Institutes of Health, founded 1887).[26] Nonetheless, scientism triumphing over ethics, the NIH declined to require informed consent on the grounds that "whatever the NIH might do by designing a code or stipulating standards for acceptable clinical research would be likely to inhibit, delay, or distort the carrying out of clinical research rendering any such effort unacceptable,"[27] that is, scientistic visions of medical progress overrode ethical concerns.

In 1962, President John F. Kennedy (1917–1963) awarded Kelsey the Medal of Freedom, and the popular press knighted her the "Feminine Conscience of the FDA" and "Guardian of the Drug Market."[28] One issue that still vexed Kelsey, however, was the loophole in the law that its opponents had created to allow investigators to waive the requirements for obtaining and documenting informed consent. In 1963, she insisted that the grounds for waiver were very narrow. In a memo, she

> made it clear that such exceptions to the obtaining of patient consent might include unconscious patients; a child in an emergency if the parents cannot be reached; mentally incompetent patients with no known representatives; or patients suffering from an incurable disease when the doctor feels knowledge of the nature of such disease would be detrimental to the welfare of the patient. There is no known basis for concluding that such exceptions would include circumstances in which the investigator feels that informed consent would interfere with the design of the experiment or would disturb the "doctor–patient" relationship.[29]

Thus, it was Kelsey's official opinion that the only exceptions to the consent requirement were (1) the infeasibility of obtaining valid consent, as in emergencies, or (2) therapeutic experiments in which requesting informed consent would be detrimental to the welfare of a patient.

In 1964, Henry Beecher, who championed professional self-regulation, but abhorred the notion of bureaucratically imposed rigid rules, and seemed fearful that informing patients might lead to disruptive selection (i.e., by discouraging some types of potential subjects from participating), wrote to the FDA commissioner "contesting Frances Kelsey's authority."[30] As these criticisms mounted, Kelsey "lost formal control over investigational drugs and suffered what one reporter would later describe as a 'humiliating *bare desk* treatment, she was generally ignored and given little to do of consequence.'"[31] Nonetheless, some of the specialist literature on the regulation of research on human subjects recognized Kelsey's pivotal role in fighting to require that investigators for pharmaceutical companies document having the informed consent of the people on whom they tested experimental drugs.[32]

On August 30, 1966—that is, after the publication of Beecher's 1966 whistleblowing *New England Journal of Medicine* article, "Ethics and Clinical Research"—the FDA published "A Statement on Policy Concerning Consent for the Use of Investigational New Drugs on Humans," initiating Kelsey's proposal that exceptions to the consent requirement were very narrow. Furthermore, in the wake of Beecher's whistleblowing article, the NIH joined the FDA in officially recognizing a potential conflict of loyalties and interests between physician-investigator roles as researchers pursuing science (and their own career interests) and as physicians acting in the interests of their patients. Thus, in the mid-1960s, well before the conception or birth of bioethics, both the FDA and the NIH held that in the clinical "setting in which the patient is involved in an experimental effort, the judgment of the investigator is not sufficient as a basis for reaching a conclusion concerning the ethical and moral set of questions in that relationship."[33]

In chapter 11, I will reflect on why Kelsey's role as a bioethics reformer is barely acknowledged in standard histories of bioethics. I will contend that, among other things, acknowledging Kelsey's role would have highlighted the role of the WMA's 1962–1964 Declaration of Helsinki, thereby undercutting the presumption of American exceptionalism that runs through Jonsen's and Rothman's standard histories of bioethics. Moreover, it is also clear that governmental regulation of research on human subjects by the FDA and NIH was prior to and thus not a product of the creation of bioethics. Kelsey is the mother of these regulations, and her successful efforts to require documented informed consent in law and regulations in 1962–1964 came to fruition prior to the formation of the Society for Health and Human Values (SHHV, founded 1969) and first bioethics institutes (founded 1969 and 1971), prior to the Library of Congress's formal acknowledgment of bioethics as a separate field (1974), and prior to the formulation of bioethical principles for human subjects research in the 1978 *Belmont Report*. Perhaps Kelsey's contributions were ignored because she was female, and acknowledging her femininity, or her role as the martyred mother of informed consent, might clash with masculine conceptions of bioethics' founders as "fathers" of the field. Females saving babies can be accepted as a properly feminine role; a female as a heroine of research ethics reform would place her outside of a woman's submissive feminine role. Discussion continued in chapter 11.

In 2001, forty years after the *Washington Post*'s thalidomide story, a surprising new voice, the AMA's *Virtual Mentor* (founded 1999, since 2015 renamed the *AMA Journal of Ethics*) crowned Kelsey a heroine not of bioethics but of professional medical ethics. In an article lauding Kelsey for her "commitment to the professional ideal of patient health and safety," she was styled "a role model in medicine."[34] Yet this laudatory narrative left untold the dark side of Kelsey's encounter with Lords Almighty physicians and the pharmaceutical industry. It failed to mention Beecher's campaign, which, with the help of the business press, sidelined Kelsey, demoting her to a "bare desk bureaucrat." Nonetheless, it is impressive that a female fellow of the AMA's Ethics Standards Group, Karen Geraghty, recognized Frances Oldham Kelsey's well-deserved status as heroine of biomedical ethics by crowning her a role model for professional medical ethics.

Maurice Pappworth, Whistleblowing Moralist

Coincidentally, in the same year that Kelsey received her medal, 1962, a British physician, Maurice Pappworth, published an article revealing researchers' abusive use of patients as "human guinea pigs," in the British National Health Service and in the United States. Pappworth was a veteran who served in the Royal Army Medical Corps (RAMC) in Greece, Italy, and North Africa during World War II and who oversaw a military hospital in India during the immediate postwar period. Upon returning to Britain, however, Pappworth, was unable to get an appointment at a London teaching hospital. So, he became a Harley Street private practitioner who gave lectures and tutorials to postgraduate students. It was these students who made him "woke." As Pappworth tells the tale,

> Several of [his] postgraduate students ... told [him] about unethical experiments that they had personally observed in British hospitals in which they were either junior staff or attending courses. Some told me of their dilemma: Whether or not to take an active role, or even a passive one, in persuading a patient

to volunteer, knowing that noncooperation might jeopardize their careers. For many years as background to my tutorials I had spent hours in the Royal Society of Medicine library scanning journals in which experiments in humans were described that seemed to be unethical and sometimes illegal. A further concern was that promotion in teaching hospitals depended primarily not on clinical or teaching ability but on published work. . . . So, whenever I read an account of an unethical experiment, I wrote a letter to the journal protesting, often as not to have it rejected. Medical research had become sacrosanct, based on the dubious dogma that its continuation must be the prime concern of teaching hospitals.[35]

Several points are worth noting about Pappworth's account of becoming woke: first, the students, as novices, having yet to become acculturated into the ethos of hospital practice, could still see that their professors were asking them to be complicit in morally questionable activities. Like the child in the tale of the unclothed emperor, the unacculturated students could see naked immorality, whereas their acculturated professors, morally blinded by the culture of academic medicine, could not. Second, when the students reported what they had seen to Pappworth, he took their concerns seriously enough to investigate and confirm their observations. Pappworth also acted on their concerns by writing letters critical of these experiments to the editors of the journals in which they were printed.[36]

Papworth's letters were typically rejected, and he came to believe that editors were rejecting his letters because "medical research had become sacrosanct, based on the dubious dogma that its continuation must be the prime concern of teaching hospitals."[37] Yet another factor in these rejections was Pappworth's ungentlemanly blunt language: his tendency to call experimenters "dastardly," or to describe British teaching hospitals as "dominated by ghoulish physiologists masquerading as clinicians," or to call for researchers to be replaced by "true physicians whose main interest is the welfare of their patients and not the publication of papers."[38] As one editor wrote to Pappworth, "We cannot accept [this letter] in its present form," explaining, "I know that there are times when good comes of speaking strongly and by giving maximum publicity to what appear to be public scandals; but you haven't persuaded us here that this is one of those occasions."[39] Note that the editor seems more affronted by Pappworth's strong language than by his reports of an ethical transgression.

As Pappworth's daughter and biographer, Lady Joanna Seldon, PhD (1954–2016), wrote, "Many British doctors during the 1950s and 1960s, though aware of the unethical research taking place in this country, nevertheless turned a blind eye to it."[40] Frustrated by journal editors' refusal to publish his letters, Pappworth writes that when in 1962 I was approached by the editor of *Twentieth Century* (a now defunct literary magazine) to contribute to a special number entitled "Doctors in the Sixties." [He published an] article, called "Human Guinea Pigs: A Warning" [that] appeared in the autumn of 1962, describing 14 experiments in lay language; no names were named, and no journal references were given. This was an early event in the debate in Britain about human experimentation. The debate excited much interest in both the lay and medical press and led eventually to the establishment of local medical research ethics committees.[41]

In this initial article, Pappworth cited fourteen cases, ten British, four American.[42] And, like Moll, his long-forgotten German precursor, Pappworth chose not to disclose the names of the researchers who performed these unethical experiments, nor their institutions nor the names of the journals that had published them. Nonetheless, again like Moll, Pappworth was soon vilified by his medical peers—in this case, the British medical establishment, which wrote him off as a noisome disloyal crank. Unlike Moll, however, Pappworth's whistleblowing article goaded other physicians to action. The first of these was an eminent physician, Sir Hugh Anthony Clegg, editor of the *British Medical Journal*. In the October 1962 issue of the journal, Clegg, noting that Pappworth's warnings and Kelsey's actions had "recently received a lot of attention in the press in this country [Britain] and in other countries [the US], it is thought desirable that the medical profession be made aware of what progress [the WMA Committee on Medical Ethics] has made in this admittedly difficult subject."[43] This said, Clegg published, without authorization, a revised working draft of the 1962 version of the document that would, after further revision, become the WMA's 1964 "Declaration of Helsinki."[44] By his unauthorized disclosure, Clegg made the scandal-to-code connection by relating the need for an ethics code to address the scandals revealed by Kelsey and Pappworth—invoking an aura of notoriety to pressure the WMA to get on with the unfinished business of finalizing the world's first operational code of ethics for morally permissible experimentation on human subjects.[45]

Henry K. Beecher, Whistleblowing Moralist

On the other side of the Atlantic, Pappworth's publication, "Human Guinea Pigs—A Warning," attracted the attention of Andrew Ivy, coparent of the Nuremberg Code. Ivy initiated correspondence in which he supported Pappworth's efforts to draw attention to unethical research.[46] So too did another American physician, Harry Unangst (1904–1976), of

Peck, Kansas, a small town fifteen miles from Wichita. Like Pappworth, Harry was a veteran who had served in field hospitals in North Africa and Europe during World War II. And, again like Pappworth, he was not born with the proverbial silver spoon in his mouth. His father, Henry Eugene Unangst, had trouble keeping a job and had a weakness for gambling and alcohol.[47] Henry's grandson, Jonathan, reports that his fraternal grandfather was "a 'scoundrel'—[who spent] much of his time, and the family's money, drinking, gambling, and philandering. Dad, I think, never forgave his father for these things and especially for cheating on his mother."[48] By default, many familial responsibilities fell on the shoulders of young Harry, a bookworm who managed to amuse his friends by playing ragtime and popular tunes on the piano as he worked to pay his way through high school and college—his ne'er-do-well dad, Henry, having refused to support him.

Against the odds, Harry overcame these obstacles to become the first member of his family to graduate college, earning BA and MA degrees from the University of Kansas (1926, 1927). At college, one biographer observes, Harry

> developed a foppish side—[Johnathan] recalls a "rather dapper photo of him, tennis racket in hand, from about that time" . . . evidence, perhaps, of a protective veneer, an overdone trust in appearances, an unsettling sense of vanity. Or perhaps it reflects a wish to distance himself from his father. . . . He was remaking himself, fashioning a new self-image, not for the first time and certainly not for the last. Just a couple of years later—as early, perhaps, as 1924, his junior year—he began experimenting with changes to his name, the formulation *Harry K.U. Beecher* arising from the erasure marks over "*Harry K. Unangst.*"[49]

Harry's experiments with nominal self-reinvention led him to seek to change his name in the summer of 1928, when, on the verge of applying to Harvard Medical School, he legally adopted as his family name the maiden name of his maternal grandmother, Mary Julian Kerley née "Beecher" (1807–1885). In September 1928, a court granted his request and Harry applied to Harvard Medical School using the pared-down version of his new name, "Henry Knowles Beecher."[50] Harry had nominally rebirthed himself as an apparent descendant of Presbyterian minister Lyman Beecher (1775–1863), whose progeny included the famous abolitionist Henry Ward Beecher (1813–1887) and his even more famous sister, Harriet Beecher Stowe (1811–1896), author of *Uncle Tom's Cabin*. Having undone his plebian birthright by rejecting the name of his philandering wastrel alcoholic father, Harry embraced his new persona and his fabricated "Beecher" heritage for the rest of his life—which forced him, as one biographer noted, to cultivate a "penchant for deflection and evasiveness in discussions of his family history."[51]

As Henry K. Beecher, Harry became "a fixture of Boston society, fond of ballroom dancing, lobster, and fine Scotch, the latter often consumed in generous quantities at the Somerset Club, an exclusive WASP redoubt on Beacon Hill": that is, he became a Boston Brahmin.[52] Combined with his ambition, talent, charm, and hard work, Harry's new persona transformed a life once foreordained for obscurity into one bound for the limelight. After graduating from Harvard Medical School, Harry, now known as "Henry Knowles Beecher"—but still "Harry" to his friends—climbed the professional ranks to become the Henry Isaiah Dorr Professor of Anesthesiology at Harvard Medical School. In 1934, he married an obstetrician's daughter, Margaret Swain (1906–1973), an elegantly slim young woman who belonged to a fashionable charity society, the Boston Junior League. When his country called during World War II, Harry served honorably, receiving five battle stars.

Beecher was honorably discharged at the rank of lieutenant colonel, and he had been awarded a silver star medal for his heroism tending to American and allied troops during the bloody battle of Anzio, Italy, where about one-third of the US and Allied troops were killed or wounded in the battle (circa 40,000 casualties). He was called back into service by the US Army and the CIA, which tasked him with evaluating data from the Nazi doctors' experiments and then with advising on and performing secret experiments on performance-enhancing drugs and with evaluating mescaline and LSD as potential truth serums. The context for these requests was Prime Minister Winston Churchill's 1946 "Iron Curtin" speech, followed by a battle of wills in the 1948–1949 Berlin airlift, exacerbated by the Soviets' explosion of a nuclear device (a version of an atomic bomb) in Semipalatinsk on August 29, 1949. The Cold War threatened to turn kinetic as the Soviet Union undermined and replaced governments throughout Eastern Europe and as a communist army took over mainland China. These events led to the formation of the NATO alliance and set the backdrop against which the CIA and US military exploited Beecher's military experience and neuroscientific fascination with mind–body relationships (as expressed, for example, in his 1955 paper, "The Powerful Placebo."[53,54]) to recruit him as a scientific investigator.

As required of all US government-sponsored researchers in the pre-bioethical era. Beecher would have secured his subjects' signatures on the "waiver and release" forms. These forms were designed to protect researchers and their government sponsors and were often vague about the actual purpose or details of an experiment. More to the point, frontline field hospital physicians, like Beecher, had been accustomed to making such tough decisions as whether to sacrifice some wounded men to use scarce resources to restore health to those who could better serve military objectives by returning to the frontlines more rapidly.[55] As a world-renowned scholar on the ethics of military medicine observed, "The hallmark principles that drive bioethical decisionmaking in ordinary clinical settings are largely absent in military medicine. Military personnel do not enjoy a right to life, personal autonomy, or a right of self-determination to any degree approaching that of ordinary patients.... Military necessity grants paramount authority to reason of state, proportionality limits but does not eliminate excessive harm, and the doctrine of double effect permits unintentional harm to non-combatants."[56] Insofar as Cold War researchers, like Beecher and his colleague Louis Lasagna, MD (1923–2003), believed the excuse of war applied to those working on the frontlines of the Cold War, they could, and did, assume the prerogatives afforded to frontline physicians during World War II.

Yet, troubled by his own conduct, Beecher began to discuss and to write about the ethics of research on human subjects. In February 1958, he sent a memo to Harvard's Committee on Research indicating that neither Harvard nor "other similar institutions [have] faced up to the problems surrounding experimentation in man," although he also believed that "these matters are too complex ... to permit the establishment of rigid rules."[57] Like many researchers in America during the Cold War period (1947–1991), Beecher believed that nobility of purpose framed in terms of some ideal, in this case, patriotism, sufficed to guarantee the morality of a well-constructed scientific experiment: a view that founding bioethicist Robert Veatch (1939–2020) characterizes as "social utilitarianism."[58] On this view, if a well-constructed experiment holds the promise of yielding socially or militarily useful results, ethically questionable means of obtaining these results would be excusable. Nonetheless, to his credit, although he had accepted contracts from the US Army and the CIA to explore psychoactive drugs as possible truth serums, Beecher began to have moral qualms about his research. As neuro-anesthesiologist George Mashour observed, "It may appear paradoxical that Beecher, who advocated the ethical treatment of human subjects, had also engaged in potentially unethical work on hallucinogens for the government. A more compelling hypothesis, however, is that Beecher advocated ethical treatment of human subjects largely because of such work."[59]

As Beecher's questions about the ethical conduct of experiments on human subjects intensified, he sought answers from a variety of sources. Lacking training in moral philosophy, Beecher, like other autodidacts, mixed and matched sources as they appealed to him, often with little rhyme or reason. Thus, as late as 1970, he quotes the following line from the English philosopher George E. Moore's (1873–1958) ideal utilitarian treatise, *Principia Ethica* (1903), as cited by situational ethicist Joseph Fletcher (1905–1991): "'Right' does and can mean nothing but cause of a good result' and is thus identified with 'useful': whence it follows that the end always will justify the means, and that no action that is not justified by its results can be right."[60] This quotation is cited as a prefix to a section subtitled "Situation Ethics," in which Beecher describes an experience in North Africa when penicillin was in short supply and

> hospitals were overflowing with wounded men. Many had been wounded in battles; many also wounded in brothels. Which would get the penicillin? By all that is just it would go to the heroes . . . who were still in jeopardy, some of whom were dying. They did not receive it; nor should they have; it was given to those in brothels . . . there were desperate shortages of manpower at the front . . . those with [serious wounds] would not swiftly be restored to the battle lines even with penicillin, whereas those with venereal disease on being treated with penicillin would in a matter of day . . . return to the front. . . . I believe that the course chosen was the proper one.[61]

Thus, Beecher believed that accomplishing socially useful ends may excuse a morally suspect means of achieving them. Beecher also cites with approval Fletcher's view that "*not only means but ends too are relative*, only extrinsically justifiable. They are good only if they happen to contribute to some other good than themselves."[62] David Rothman notes that these ethical ruminations were designed to justify Beecher's own Cold War experiments, in which he had related to potential colleagues that he was "asked by the [US] Army to study compounds that . . . give access to the subconscious. The Army has a further interest as well: . . . Can one individual obtain from another, with the aid of these drugs, willfully suppressed information? If we undertake the study this latter question will not be mentioned in the contract application. We request that it is not referred to outside of this room"[63] (i.e., this was secret military research). Rothman notes further that Beecher expressed a belief common among World War II and Cold War researchers that, in Beecher's own words, "In time of war, at least, the importance of the . . . [a military] purpose hardly appears debatable."[64]

As Beecher's colleague Louis Lasagna testified to President William Clinton's Advisory Committee on the Human Radiation Experiments

(1994–1995) from 1952 to 1954, Beecher and he had carried out US Army–sponsored hallucinogen experiments on uninformed Harvard students.

> The idea was that we were supposed to give hallucinogens to healthy volunteers and see if we could worm out of them secret information. And it went like this: a volunteer would be told, "Now we are going to ask you a lot of questions, but under no circumstances tell us your mother's maiden name or your social security number." ... We'd give them a drug and ask them a number of questions and sure enough one of the questions was "What was your mother's maiden name?" ... The subjects were not informed about anything.... [We felt that] if we ask for consent, we lose our subjects ... we were so ethically insensitive that it never dawned on us that you ought to level with people that they were in an experiment.[65]

By sheer happenstance, one of the Harvard students who served as an underinformed subject in another Beecher–Lasagna experiment was none other than Daniel Callahan, cofounder of the first bioethics thinktank, the Hastings Center. As Callahan tells the tale, when he was

> a grad student at Harvard in 1960 or so, a notice appeared in the paper that a research project was beginning that called for the participation of former swimmers and runners. I had been a swimmer in college. We were not told the goal of the research or who was running it. ...
>
> We were not told what the research was about, nor did I know the name of the person directing the swims; and I don't think there was any informed consent. [The published paper states that "The subjects were told that 'pep pills' and placebos were being used."]
>
> Then, around 1971 or so, I read Beecher's CV since he was part of [the Hastings Center's] project on brain death. I noticed in his list of publications a project he had run on amphetamines that involved swimmers and runners in the early 1960s—and that's how I first learned what the research was and who had run it! Sydney [Dan Callahan's wife] recalls it well because I came home after the trials with the faster second dose [of an amphetamine] high, happy, and agitated—and then in a few hours became very nasty.[66]

Despite this discovery, Callahan continued working with Beecher and, in recognition of Beecher's role as a whistleblower and medical ethics reformer, established the Henry Knowles Beecher Award in medical ethics in 1976, making Beecher himself its first awardee.

As to Beecher, after conducting these experiments, he began writing articles in which he attempted to reconcile them with his conscience. The moral dicta that Beecher typically quoted seemed to justify serving one's country by giving amphetamines or LSD to underinformed Harvard students in secret military-financed experiments. Quotations from Fletcher's situational ethics appealed to Beecher because they emphasized that an operative morality had to be flexible, because moral "matters are too complex . . . to permit the establishment of rigid rules."[67] Yet, in a consummate example of trying to have one's conceptual cake after having eaten it, Beecher follows these observations by invoking Immanuel Kant's (1724–1804) dictum that "people must always be treated as ends, never as means alone."[68] As anyone who has ever taken an introductory ethics course could have informed Beecher, this meant that one may not use uninformed and unconsented people as a mere means to find out about the military utility of psychoactive drugs. Yet Beecher reiterated his Kantian leanings by quoting a later statement of neo-Kantian ethics by British-Polish-Jewish mathematician Jacob Bronowski (1908–1974), "that the end for which we work exists and is judged only by the means we use to reach it."[69] The fact that these statements about ends and means are inconsistent with each other does not seem apparent to Beecher. Perhaps, at some level, he was conscious of his inconsistencies because he ends this chapter by quoting a line Rainer Maria Rilke's *Letter to the Young Poet*: "That we must seek to do the difficult is a certainty that may never leave us."[70]

Quite inadvertently, Beecher seems to be embracing the views of Ralph Waldo Emerson (1803–1882), the Sage of Concord, who wrote, "A foolish consistency is the hobgoblin of little minds, adored by little statesmen and philosophers and divines. With consistency a great soul has simply nothing to do. He may as well concern himself with his shadow on the wall. Speak what you think now in hard words, and to-morrow speak what to-morrow thinks in hard words again, though it contradicts everything you said to-day."[71] One virtue of a self-contradictory ethics is that it enables one to endorse actions inconsistent with some of its precepts. Beecher's inconsistencies troubled Robert Veatch, a founding bioethicist well versed in moral philosophy. Veatch observed that although Beecher seems to have embraced some form of situationist utilitarianism (hence the quotations from Fletcher and Moore and the story about rationing penicillin), he inconsistently dismisses as "a pernicious myth [the] view that ends justify means" in research, claiming that "a study is ethical or not at its inception. It does not become ethical merely because it turned up some valuable data."[72] Veatch concludes it "sounds like [Beecher] has some criterion of ethical rightness, other than social utilitarianism," but what this was remains a puzzle.[73]

I think Veatch gives Beecher's amateur philosophizing too much credit. In fact, Beecher's publications and his private letters of the 1960s and 1970s reveal a man perpetually ambivalent. Sometimes he writes like the scientific investigator who surreptitiously gave psychoactive drugs to unsuspecting people on the grounds that it could serve country, medicine, science, or society—and/or satisfy his curiosity about mind–brain interactions. At other times he is the observant bible-reading full-immersion Congregational Methodist who takes to heart St. Paul's admonition not to do evil that good may come of it (Romans 3:8). He seems to have lived a life of situational ethics in which one adjusts one's name and persona— and one's ethics—to the needs of the moment, finding supportive quotations to justify whatever ideas or actions he has in mind, even if they are inconsistent with each other. Although Beecher is unlikely to have read the preface to English philosopher F. H. Bradley's 1893 book *Appearance and Reality*, he, like Bradley, may believe that philosophy "is the finding of bad reasons for what we believe upon instinct."[74] Yet, throughout his life, it was instinctively important to Beecher to do, and to be seen as having done, the right thing: to stand by his family, his faith (he never abandoned the Congregationalist Methodist church), his profession, the ideal of professional self-regulation—and to serve his country.

Beecher and Pappworth's Correspondence

Andrew Ivy was not the only physician on the other side of the Atlantic to notice the notoriety surrounding Pappworth's 1962 article in the popular quarterly literary magazine, *Twentieth Century*. Henry Beecher also contacted Pappworth, and although the two were physically, religiously, and professionally worlds apart, for a time they became intercontinental pen pals. It was an odd correspondence. On one end was an orthodox Jew who had assumed the role of Old Testament prophet chastising the sinful for their sins by publicly condemning researchers for performing morally dubious experiments on unsuspecting patients without their consent; on the other was a Congregationalist Methodist struggling with guilt feelings about having committed unethical experiments in the interests of national security. Yet, like Pappworth, Beecher came from a strict religious tradition disdained by their fellow physicians (a form of Congregationalist Methodism that involved baptism by full emersion in bodies of running water like rivers and streams). Both had done exceptionally well in school: Pappworth earned degrees in both medicine and surgery from a newly founded red brick state school, the University of Liverpool; Beecher earned a bachelor's degree and master's from the University of Kansas (1926, 1927). But the two men had more in common than plebian origins and a disdained religious background. Both had undergone nomological reinvention. Pappworth's family of Polish-Jewish immigrants had given him the

Hebrew name, "Moshe [Moses] Elkanan ben [son of] Yitzakh Yaakov ve [and] Miriam Devorah." To pacify the dictates of Russifying governmental authorities in Poland, the family had added a Russian-sounding name "Papperovitch." When the family moved to Britain it Anglicized this name for reasons of acculturation, and so "Moishe Papperovitch" was renamed "Maurice Pappworth."

Pappworth, like Beecher, had served in the brutal North Africa campaign during World War II and was honorably discharged at the rank of lieutenant colonel. Unlike Beecher, however, when Pappworth returned to London after the war, he was not welcomed into an academic medical post. Instead, he was refused appointments at London teaching hospitals, apparently because of anti-Semitic prejudice. Pappworth's path to career success had been blocked when he first moved to metropolitan London in 1939 because "deficiencies" in his background and education were made evident to him and he was barred from appointments to London teaching hospitals because, he was informed, such positions were reserved for gentlemen and "no Jew could ever be a gentleman."[75] Forced into private practice, Pappworth supplemented his income with lectures that prepared medical students to take qualifying exams. As it happened, Pappworth's outsider perch gave him the perspective to see what gentlemanly insiders could not: that some British physicians were using patients as unconsenting "human guinea pigs."

Beecher first reached out to Pappworth in a January 7, 1965, letter on his official Harvard Medical School–Massachusetts General Hospital Henry Isaiah Dorr Professor of Research in Anesthesiology stationery.

Dear Dr. Pappworth,

I read with great interest, or perhaps I had better say I studied with great care, your interesting article, "Human Guinea Pigs: A Warning."
. . . . I heartily agree with your thesis and indeed have often spoken along the same lines myself.
As in most hospitals like the Massachusetts General Hospital we have had an explosion (I think no other term describes this situation) in research in man. As you have so well pointed out, this entails heavy responsibilities which are not always recognized. I am doing my best to see that they are recognized in this institution and for this reason would be especially grateful for your help.
. . . I would like a reprint [of your article] if you have one available.
Very sincerely
Henry K. Beecher M.D.[76]

Once Beecher and Pappworth discovered each other, Beecher began to look to Pappworth as a role model and soon adopted Pappworth's method of compiling lists of unethical experiments published in respectable medical journals. At this point, the mutual support team's correspondence focused on exchanging lists and recounting details of unethical experiments. It took a different turn after Beecher's January 25, 1965, letter to Pappworth. In that letter, Beecher explains that he is preparing a paper, "Ethics and the Explosion of Human Experimentation," in which he would report that at Massachusetts General Hospital, "there has been a 17-fold increase in funding [for research] in 20 years." Beecher would then suggest that this has led to a generational change leading to an exponential increase in unethical experiments: that is, an explosion of funding had attracted a younger generation of investigators who were more interested in publication than in patients' welfare. He thus concurred with Pappworth's publicly stated observations on generational change and a quest for publication as the source of the increase in unethical experiments.

Switching topics, Beecher exclaimed that in the previous year, he discovered "the most shocking example [of unethical experiments] has been described in great detail in SCIENCE magazine February 7, 1964, page 551, under the title 'Human Experimentation: Cancer Studies at Sloan-Kettering Stir Debate on Medical Ethics.'"[77] In a pivotal remark, Beecher observes that

> certain elements in this country have long held the view that these matters must be swept under the rug. I do not at all agree with this point of view and I think with the "explosion" just referred to, that these matters must be faced. I have been invited to participate in a symposium for senior medical writers from the best newspapers and journals in this country and I expect to speak on the title mentioned above. . . .
>
> While the individual who does it, well may be severely criticized, as I fully expect to be for presenting in the Middle West the paper I mentioned, I think that someone has got to call attention to the problems as they exist. It seems to me that a man must be prepared to stand behind the work he publishes; he must expect it will be referred to. Indeed, that is the purpose of publication.[78]

Beecher's letter ends with a discussion of journal editors' responsibility for monitoring the ethics of the research published in their pages. He notes that a recent survey of journal editors about their responsibility to monitor the moral status of the research reported in their journals found that while "fifty-eight per cent said 'yes, of course' [they had such a responsibility], an astonishing 28 per cent said 'no.' In the latter group one individual made the extraordinary comment '. . . the implications of effectively monitoring the standards of conducting experiments either with humans or animals would be highly antagonistic to the perpetration and expansion of the research process.'" In closing, Beecher reiterates the idea that "it does seem that some of us need to call attention to the not uncommon excesses."[79]

What is evident from this letter is that, apparently scandalized by an experiment described in *Science*, Beecher was going to use his speaking slot at a midwestern conference attended by medical journalists to publicly call out unethical experiments on humans. The *Science* article that inspired Beecher to publicly denounce unethical research describes experiments conducted in 1963 by Dr. Chester M. Southam (1919–2002) of the Sloan Kettering Cancer Center and Cornell University Medical College. Seeking to challenge the then current view of cancer as a disease caused by external agents, like viruses, Southam conducted a series of experiments designed to test the hypothesis that cancers spread in people's bodies because their immune system is unable to adequately reject them. As evidence for his idea that healthy people can reject cancer, Southam recruited fourteen healthy subjects from an Ohio Penitentiary who were "fully informed about the experiment and its possible risks and nonetheless eager to take part in the experiment in which they would receive injections of cancer cells."[80]

At this point, according to the *Science* article, "the doctors faced a choice that has confronted researchers since the beginning of experimental medicine: Should they use themselves as subjects?" Some researchers, including Andrew Ivy, held that self-experimentation was a prerequisite for legitimizing potentially harmful experiments on other people, provided that potential subjects gave their informed voluntary consent. Words to this effect were included in his second draft of the Nuremberg Code and ultimately became Article 5 of the code, which states that "no experiment shall be conducted where there is an *a priori* reason to believe that death or disabling injury will occur; except, perhaps, in those experiments where the experimental physicians also serve as subjects."

Post-Nuremberg, precedents for serving as a subject in one's own experiments had been set in the mid-1950s when each of three polio vaccine researchers—Hilary Kroprowski (1916–2013), Albert Sabine (née Abram Saperstejn, 1906–1993), and Jonas Salk (1914–1995)—declared that they had taken their own polio vaccines as proof of their confidence in its safety.[81] Mindful of this precedent,

> Sloan Kettering . . . issued a press release stating that the researchers did inject themselves with cancer cells and established the safety of the procedure before trying out larger-scale experiments at Ohio State Penitentiary. Southam, however, had been unwilling to inject himself or his colleagues when there was a group of normal volunteers . . . fully informed about the experiment . . . eager

to take part in it. "I would not have hesitated," Southam said, "if it had served some useful purpose. But ... to me it seemed like false heroism. ... I do not regard myself as indispensable ... and I did not regard the experiment as dangerous. But, let's face it, there are relatively few skilled cancer researchers, and it seemed stupid to take even the little risk."

"Somewhere along the line," the *Science* article continues,

> the practice of fully explaining the experiment to the patients and obtaining their informed consent was replaced by the practice of obtaining oral assent only to a vague description of the procedures, in which the word "cancer" was entirely omitted and patients were merely told that they would be receiving "some cells." ... "We stopped telling them they were getting living cancer cells when it was well established that there was no risk," Southam said last week. ". . . All I can say is that within any reasonable definition of the words 'no risk' there was no risk." ... To inform them more explicitly about the experiment, ... To what purpose? I told them that they would be getting some cells, and I described what would happen, but—since I believed that there was no risk to them under the circumstances—to tell them the nature of the thing injected seemed irrelevant.[82]

By the summer of 1963, Southam had arranged with the research director of the Jewish Chronic Disease Hospital (founded 1925, Brooklyn, New York City) to conduct the experiment on twenty-two diseased noncancer patients. The diseased patients rejected the cancer implants just as promptly as healthy subjects did, providing further evidence supporting Southam's hypothesis that the immune system was implicated in the human body's susceptibility to cancer.

In August 1963, three physicians resigned from Jewish Chronic Disease Hospital in protest demanding an investigation by a medical committee. A committee was formed but, finding no ethical issues, commended the research instead.[83] The protesting physicians then contacted William A. Hyman (1893–1966), a New York lawyer who took the hospital to court, informing the New York newspapers of his lawsuit.[84] The *Science* article reports further that "it is now established that the Brooklyn hospital did not tell the patients that they were receiving cancer cell injections, and that they were not asked for written consent. [The] Hospital director ... asserts that the patients were told they were receiving 'some cells,' and that they gave oral consent."[85]

What was it about the case described in this article that led Beecher to describe it as "the most shocking example!" of unethical experiments on humans? What led to his exclamation point "(!)" or his statement that "some of us need to call attention to the not uncommon excesses?" No doubt Beecher, a faithful Methodist who read his bible regularly, was familiar with 1 John 19, "If we confess our sins, he is faithful and just to forgive us our sins, and to cleanse us from all unrighteousness." Given Beecher's religious beliefs, some commentators suspect that, recalling that he himself had sinned, Beecher was seeking absolution. One typescript of his conference speech suggests that he contemplated playing the role of repentant sinner. In this typescript, Beecher wrote that he "was obliged to say that 17 years ago a group in [his] laboratory prolonged anesthesia 23 to 74 minutes beyond the necessary in order to study kidney function under ether as opposed to cyclopropane. The anesthesias were uneventful and no unusual strains were placed on the fifteen patients [crossed out: Some of those involved were minors.] . . . no adequate explanation was given to the subjects, nor was their consent adequately obtained. . . . All of this was quite wrong and I would not today, 17 years later, participate in such a study."[86] Yet, as historian Susan Lederer observes, "in the version delivered at the conference Beecher . . . omitted specific reference to the anesthesia comparison, merely stating: 'Lest I seem to stand aside from these matters I am obliged to say that in years gone by work in my laboratory could have been criticized:' words that fall short of the full-throated confession of a contritely repentant sinner."[87]

One can speculate that Beecher was reluctant to play the repentant sinner role because, when he shared a copy of his ethics speech with Louis Lasagna, his former research fellow, Lasagna responded, "Gee, Harry, when I was with you from 1952 to 1954, we never got consent on anybody." Referencing the then standard waiver and release contracts required by government-supported research, Lasagna added, "We got releases from some of our healthy volunteers because we had an Army contract and the Army wanted to say if anything bad happened to them, they wouldn't sue the Army, but that was it." So, "[Lasagna] said, 'How about . . . Let him who is without sin cast the first stone?'"[88] Beecher certainly had no wish to have his secret Army and CIA experiments with psychopharmaceuticals thrown back at him, so, perhaps he concluded that playing the penitent's role would be counterproductive.

Then again, Beecher did not believe that every experiment on human subjects required the informed consent of the subject or a surrogate. Instead, Beecher believed as most researchers did prior to the bioethical paradigm shift, that the moral issue was informing potential subjects of risk of possible harms. In his view, the extent of information conveyed to subjects should be titrated to correlate with the potential risk of harm to which subjects might be exposed.[89] If the risk was slight, fully informed consent was unnecessary. Hence, in an era in which most psychedelic drugs were considered relatively harmless but potentially useful, perhaps Beecher believed that informed consent was optional—and counterproductive if it discouraged people from volunteering as subjects. A similar line of argument was, in fact, Southam's rationale for not asking for fully informed consent: that is, since, in his view, the cancer implants posed no risk whatsoever, there was no need for fully informed consent. As Southam put this point, "All I can say is that within any reasonable definition of the words 'no risk' there was no risk"—thus a full description mentioning the word "cancer" was unnecessary.

The catch, as became evident from the account in *Science*, was that Southam did not really believe that there was no risk. He explicitly forbade members of his research team to be injected with cancer cells, and he refused to have himself injected with them. As he put it, "Let's face it, there are relatively few skilled cancer researchers, and it seemed stupid to take even the little risk." "Little risk," however, is not the same as "no risk." More to the point, from Beecher's Methodist perspective, Southam was violating the Golden Rule: he was not "doing unto others as he would wish others to do unto him" (Leviticus 19:18, Luke 6:31, Matthew 7:12). Worse yet, the real reason Southam avoided using the word "cancer" was, as Southam himself put this point, patients' "bizarre, defensive reaction" to the word "cancer." Given prospective subjects' defensive reaction, using the scary words "cancer cells" would virtually guarantee their refusal. So, Southam committed what, in Beecher's view, was a sinful action: telling an ignoble lie by omitting words indicative of risk of harm because he knew that potential subjects would not willingly subject themselves to this risk and then covering up this sin by seeking oral consent based on misleading descriptions of the risks involved (i.e., by deceiving patients). As Beecher, an avid and daily reader of the Bible would recall from Proverbs 6:16–17, among the things that "the Lord hates as an abomination unto him are, a lying tongue."

Although something akin to these thoughts may have crossed Beecher's mind, what appears to have motivated him most was the line in the *Science* article reporting that an "ad hoc medical grievance committee, convened by the hospital to investigate their charges, found no irregularities and instead commended the research."[90] This clearly troubled Beecher, and he says as much to Pappworth.

> Certain elements in this country have long held the view that these matters must be swept under the rug. I do not at all agree with this point of view . . . I have been invited to participate in a symposium for senior medical writers from the best newspapers and journals in this country and I expect to speak on the title mentioned above. . . . While the individual who does it, well may be severely criticized, as I fully expect to be for presenting in the Middle West

the paper I mentioned, I think that someone has got to call attention to the problems as they exist.[91]

Any one of these factors might have rendered the experiment scandalously immoral from Beecher's perspective. In combination, they overdetermined his description of Southam's cancer experiments as "the most shocking example" of an unethical experiment! Worse yet, in the Southam cancer case, a medical investigative committee had found no irregularities and instead commended the research. In the words of Jeremiah 5:21, they "have eyes and see not." To make sure that the public and their profession had eyes and would *see*, Beecher decided to follow Pappworth's example by speaking out in a forum that ensured media attention.

Pappworth had a similar experience. He too had been provoked into publishing "Human Guinea Pigs—A Warning" by journal editors' persistent refusal to publish his letters. It is important to underline the fact that *none* of the cases deemed "unethical" by Beecher and Pappworth were hidden from the medical community. They were published in such leading medical journals as the *Journal of the American Medical Association* (*JAMA*), *The Lancet, New England Journal of Medicine*, and, even more publicly, the *New York Times*.[92] Beecher was, to reiterate his words in the January 25 letter, angry that "certain elements in this country have long held the view that these matters must be swept under the rug. I do not at all agree with this point of view . . . these matters must be faced. . . . I think that someone has got to call attention to the problems as they exist."[93] In actuality, of course, his task in "calling attention" was to make the medical community morally sighted to abuses in experiments that they were aware of, but to which they were morally blind, "unwoke," so to speak.

In a file at the Countway Library at Harvard Medical School, historian Susan Lederer discovered on "a yellowed sheet of paper in Beecher's handwriting . . . he had written 'Why me?'"[94] Beecher "offered three reasons. He was at the end of his career rather than the beginning. He could incur the risk of rocking the research establishment with his critique of medical research gone awry, because, at age 62, he was near retirement . . . and he was he had worked as both a clinician and an investigator on the wards of Massachusetts General Hospital [and was] qualified to offer his critique because he had worked as both a clinician and an investigator on the wards of the Massachusetts General Hospital. He had a well-established record of research."[95] So at the March 22, 1965, symposium on the "Problems and Complexities of Clinical Research," at Brook Lodge in Augusta, Michigan, Beecher intended to "call attention to the problems as they exist."

The conference setting, Brook Lodge, was a former summer home of the Upjohn family (owner of Upjohn pharmaceutical, now part of Pfizer) that had been repurposed for use as a corporate retreat. It had forty-eight guest rooms, a dining hall, and a conference center that offered guests a view of an 82-acre pastoral paradise. Beecher intended to disrupt this bucolic tranquility by calling out sinners. It was not his first attempt. He had been trying to alert fellow medical professionals to unethical experiments on humans since the late 1950s. Yet none paid attention. He is said to have exclaimed at one point that his efforts had "all the impact of a soap bubble." So, emulating Pappworth, Beecher intended to publicize a Pappworth-style list of unethical experiments, focusing on the unethical cancer experiments described in the *Science* article.

Beecher had no training in moral philosophy, and not surprisingly, his ethical critique was an inconsistent jumble of ideas; nonetheless, he repeatedly stressed several points. First, he believed (concurring with Pappworth) that unethical experiments had become more commonplace because an abundance of government and pharmaceutical funding for research had drawn a new generation into the research enterprise that was more focused on finding publishable results than in protecting the welfare of their patients or subjects. Second, this new generation of investigators was "thoughtlessness and carelessness" but "not willful[ly] disregard[ing] . . . patients' rights" in the manner condemned at the Nuremberg Trials.[96] Third, because patients had limited ability to understand or appreciate the possible risks and benefits of experiments, the best moral safeguard on experiments involving human subjects was prior ethics review by professionals like himself, "skillful, informed, intelligent, honest, responsible, compassionate, physician[s]" [97] who can understand and balance the beneficial ends that the experiment was designed to serve, against the dangers inherent in the means the experimenters had contrived to reach these ends—not rigid rules!

The wording Beecher, Kelsey, and Pappworth used to condemn unethical experiments indicates that their conception of the function of informed consent was what might be called "proto-bioethical." In a context in which "lack of informed consent was a routine, though not universal feature of clinical research with patient-subjects"[98] these reformers sought to reshape subjects' "waiver and release" consent forms designed to protect researchers, sponsors, and funders from lawsuits into something different: a shield against researchers' attempts to exploit their subjects. The impetus for formalizing researchers' moral responsibilities toward patients traces back to the Nuremberg Code, which made the subject's informed voluntary consent its first principle. But this ideal received greater attention after 1952, when it became the subject of Pope Pius XII's September 14 address to a medical audience in which he construed a patient's right to informed consent as a moral issue, not a legal matter. As the pope put this point, "As a private person, the doctor can take no measure or try no course of action without the consent of the patient."[99] About five months later, on February 26, 1953, Charles E. Wilson (1890–1961), secretary of defense (1953–1957) for President Eisenhower (US president 1953–1961), issued a memorandum, "Use of Human Volunteers in Experimental Research," requiring all US military-sponsored or military-funded research to conform to a version of some provisions originally stipulated in the Nuremberg statement of "Permissible Experiments on Humans" (later known as "The Nuremberg Code").

Beecher, however, objected on the grounds that "such a vast code would be restricting and crippling to experimentation in man."[100] He pressed his view that the "best protection" for human subjects was ultimately the medical researcher's "character, wisdom, experience, honesty, imaginativeness and sense of responsibility."[101] Ultimately, Beecher and his Harvard colleagues forced the US Army to downgrade its demands for compliance with the Nuremberg principles to the status of mere "guidelines." Indicative of Harvard's prestige and of Beecher's deployment of that prestige, a US military that had demanded and received the unconditional surrender of the German and Japanese armed forces retreated in the face of a handful of Harvard faculty led by the Henry K. Beecher, the Henry Isaiah Dorr Professor of Anesthesia Research. Beecher's victory, however, would be short-lived since he himself would soon become woke to moral issues where he had once seen none.

Becoming Woke: Beecher and Pappworth as Moral Reformers

In his study of scientific revolutions, Thomas Kuhn observes that revolutionaries are typically inspired by anomalies in conventional treatments of a subject. Had Kuhn written about moral revolutions in medical ethics, he no doubt would have noted that the two most effective whistleblowing moralists, Henry Beecher and Peter Buxtun (1937–), were also inspired by anomalies. Beecher says as much in the opening lines of "Ethics and Clinical Research," when he observes that "it must be apparent that [research subjects] would not have been available [for these experiments] if they had been truly aware of the uses that would be made of them. Evidence is at hand that many of the patients in the examples to follow never had the risk satisfactorily explained to them, and it seems obvious that further hundreds have not known that they were the subjects of an experiment although grave consequences have been suffered as a direct result of experiments described here."[102] This anomaly is evident in Beecher's index case, Southam's cancer implantation experiments. As reported in *Science*, Southam admitted that he never told his subjects that they were being injected with cancer cells because, "by any reasonable definition of the words 'no risk,' there was no risk." And yet, according to the same article, Southam himself refused to be injected with cancer cells because, in his own words, "there are relatively few skilled cancer researchers and it seemed stupid to take even the little risk."[103] As noted previously, "a little risk" that leads a fully informed person, like Southam, to reject an intervention is not "no risk." So, Beecher concluded, as Southam himself no doubt did, that Southam's subjects would not have participated in experiments had they been told that they were being injected with cancer cells. Of course, if Lasagna's account of why Beecher and he failed to fully inform the subjects of their own psychoactive drug experiments is accurate, Beecher would have been familiar with this motive, since he had once disinformed his own subjects for the same reason.

Buxtun, like Beecher, was induced to act by an anomaly. He had been employed by the US Public Health Service (USPHS) to prevent the spread of sexually transmitted infections (STIs). His daily routine involved tracking down people who may have contracted syphilis or other STIs. So, imagine his shock when he learned that the USPHS itself was knowingly allowing hundreds of men with documented cases of syphilis, an STI, to spread the disease untreated and was actively preventing these men from receiving treatment. The USPHS's Syphilis Study seemed self-evidently at cross-purposes with the USPHS's basic obligation of preventing the spread of STIs.

Yet neither Beecher nor Buxtun was a moral revolutionary. Buxtun quit health care and did not foment transformational change after his interview with Jean Heller of the Associated Press and his testimony at the Kennedy hearings. Beecher sought to initiate reforms (informed consent, IRBs), but he believed in professional self-regulation and engaged in whistleblowing to spur fellow professionals to vigilance and to prevent jurispathic bureaucratic interventions that would undermine professional self-regulation. More personally, Beecher had no intention of undermining the scientistic paternalist paradigm that gave him the academic status and the social prerogatives he enjoyed throughout his adult life.

Nonetheless, sometime in the 1960s, Beecher and Pappworth each became woke to the fact that the editors and readers of leading medical journals were oblivious to unethical research routinely published on their pages. So, each undertook the challenge of awakening their colleagues to the unpalatably routine immorality of everyday research practices. Some people might characterize Beecher and Pappworth not as moral reform‐ers but as whistleblowers. But, as used in this book, this characterization doesn't fit because the term "whistleblowers" designates people who alert the public to recognized wrongdoing: that is, to actions that the public understands to be morally wrong, such as frauds, or kickbacks. In the 1960s, however, neither the medical world, nor the public, nor the media recognized Southam's experiments—often described in the *New York Times*—as morally problematic. Thus, Beecher and Pappworth are best understood as moralists, like Bentham, since they too recognized and decried as immoral conduct that their colleagues did not recognize as immoral. Consequently, their initial attempts to awaken their fellow professionals had, in words attributed to Beecher, "little more impact than a soap bubble." Facing indifference or rejection, each of them decided to awaken their colleagues by turning to popular media.

Pappworth led the way with his article in *Twentieth Century*. Beecher was determined to do the same at the Brook Lodge Conference for science journalists. He circulated advance copies of his presentation to those planning to attend the conference and peppered his case descriptions with colorful remarks about experimenters who "evidently believe that they had the right to choose martyrs for science." His remarks were soon quoted in newspapers like the *Los Angeles Times* and in popular magazines like *Good Housekeeping*.[104] Blowback swiftly followed. Beecher had "the most humiliating experience": two Harvard Medical School faculty members, David Rutstein (1909–1986), a prominent physician and fellow veteran military physician (soon to head the Veterans Administration's research and education division) and host of the groundbreaking television series *Facts of Medicine*, joined with Thomas Chalmers (1917–1995) to call "a press conference to refute what [Beecher] had said without finding out whether or not [Beecher] could be present."[105] These eminent physicians publicly dismissed Beecher's cases as unrepresentative and noted that just because the published articles did not state that patients had given informed consent, it did not follow that the researchers had failed to obtain informed voluntary consent. Their press conference made national news. But not everyone turned against Beecher; someone wrote a letter to Beecher saying, "I want you to know how much I enjoyed the article [about you] and I know there are many more people who feel the same way. We need more, a lot more Doctors like you who will tell the people the honest truth. But try and find them today. It is really sad. Thank you, Dr. Beecher, and continue to speak the truth. God Bless you always."[106]

Following their efforts to use mass media to awaken the public to the scandal of the unethical abuse of patients as human guinea pigs, Beecher and Pappworth each sought to impart their message in some less ephemeral format: a journal article or a book. Since Beecher's primary audience was his fellow physicians, he sought publication in a medical journal. Pappworth, having been denied professional platforms, sought to write a book for a lay audience. To achieve these objectives the two moralists would not only have to cope with reviewers' unwoke hostility, they would also have to deal with the challenge of transforming their repetitive laundry lists of the unethical experiments into an engaging narrative.

Moreover, although each had a lifetime's experience reading, teaching, and writing about medicine, neither had the literary skills, command of moral philosophy, nor the ethics lexicon needed for the task they were undertaking. Nonetheless, both moralists pursued publication. Beecher's first thought was to send his manuscript to *Science*, the journal that published the article that catalyzed his decision to go public. Finding no encouragement in that quarter, Beecher next turned to *JAMA*, the journal that published his earlier reflections on the ethics of experimenting on humans—and the journal with the largest circulation. So, he sent a copy of his talk to *JAMA*'s editor, John Talbott (1902–1990), with a letter explaining that the manuscript summarized "about ten years of as careful thought as [he, Beecher] was capable of doing. [Moreover] many individuals . . . though appalled by the information, agree that it should be published, the sooner the better. . . . It is rather long. I do not believe it can be shortened significantly and carry the same message, which so urgently needs to be disseminated." Talbott was receptive and, following standard procedure, informed Beecher that he had submitted his manuscript to several external reviewers. Alert to the possibility of blowback from unwoke reviewers, Beecher sent a note to Talbott indicating that "reverberations" from his Brook Lodge presentation "are still continuing. . . . Unquestionably the shoe pinched a lot of feet."[107]

Beecher was right to feel apprehensive. None of the external reviewers favored publication. One wrote that the manuscript was so "poorly organized [that] frankly, I was surprised that a thoughtful physician of Doctor Beecher's stature expected you to review this manuscript."[108] The reviewers also rejected Beecher's review of fifty cases of unethical experiments on the grounds that it exceeded the journal's page limits and violated the journal's editorial standards by not identifying the publications cited. Upon receiving the negative reviews, Talbott, no doubt mindful of the controversies surrounding Beecher's Brook Lodge presentation and Beecher's avowed refusal to pare down the number of cases, rejected the article outright (i.e., without encouraging revision and resubmission).

Beecher next approached Joseph Garland (1893–1973), editor of the oldest and, in the eyes of many, the most prestigious American medical journal, the *New England Journal of Medicine* (founded 1812). Garland sent the article to seven reviewers. Six recommended against publication of the article on the grounds that there were too many cases, that Beecher had not allowed investigators to tell their side of the story, that many medical readers would recognize the "anonymous" cases, and that Beecher's critique had already received extensive media coverage. Only one reviewer supported publication of the article as submitted, and then only on the condition that the *Journal* obtain a legal opinion "regarding any possible problems."[109] Yet all but one reviewer believed the article merited publication in some revised form.

Some editors are transactional, submitting manuscripts for review and accepting the results. Great editors, like Maxwell Perkins (1884–1947) of Scribner Publishers, are transformational: discovering, mentoring, and standing by their authors even in the face of reviewers' objections. (Among Perkins's authors were F. Scott Fitzgerald, Ernest Hemingway, and Thomas Wolfe.) Garland was, by all accounts, a transformational editor; so, when his editorial board voted to reject Beecher's manuscript, Garland overruled it.[110] Still, a laundry list of fifty examples of unethical experiments would be a dull read with all the deficiencies cited by the reviewers. So, Garland wrote to Beecher requesting that he cut the number of cases in half. He also requested a list of references for all the experiments described so that they could be fact-checked against Beecher's accounts.[111] Ultimately, the paper was revised, and on March 3, 1966, Garland wrote to Beecher summarizing his editing of the revised paper:

> The result of my editorial labor in which I have attempted to reduce your important data to a relatively unemotional statement of factual material. I have tried to omit anything accusatory or especially critical, since what we want is not an indictment but a sober and undramatic presentation of what has been done and is being done in violation of basic ethics.
>
> To my mind this makes the message all the more impressive; res ips[sic] loquitur; ["*res ipsa loquitur*," the thing speaks for itself.].[112]

Some commentators believe that in the editing process, Garland blurred the line between editor and coauthor, but excision is editing, not coauthoring.[113] More to the point, the content of the trimmed-down arti-cle is a faithful compendium of Beecher's thoughts during the decade in which he gradually became morally sighted.[114] The article opens with a statement from Beecher's January 7, 1965, letter to Pappworth, "Human experimentation since World War II has created some difficult problems with the increasing employment of patients as experimental subjects." It then continues,

> In many of these experiments it must be apparent that patients would not have been available if they had been truly aware of the uses that would be made of them. Evidence is at hand that many of the patients in the examples to follow never had the risk satisfactorily explained to them, and it seems obvious that further hundreds have not known that they were the subjects of an experiment although grave consequences have been suffered as a direct result of experiments described here.[115]

Note the absence of the word "consent." For Beecher, the moral issue was not whether experimenters managed to get patients to sign a waiver and release document or signal oral consent or assent; he did not charge any researcher with the issue that his Harvard colleagues Chalmers and Rut-stein harangued him with in their press conference. For Beecher, the moral issue turned on the question of whether patients had the risk satisfacto-rily explained so that they understood and appreciated the risks of harm to which they would be exposed if they agreed to become experimental subjects. Did they understand that they would be given "pep pills and a placebo," for example?

Beecher next paraphrases a line from his January 25, 1965, letter to Pappworth, "There is a belief prevalent in some sophisticated circles that attention to these matters would 'block progress.'" Beecher forthrightly rejects this view, citing a single moral authority, Pope Pius XII (the 1952 Encyclical), "who stated that . . . science is not the highest value to which all other orders of values . . . should be subordinated."[116] It is not clear why Beecher believed his medical readership—mostly non-Catholics, like Beecher himself[117]—would consider the pronouncements of a Roman Catholic pope morally authoritative, but asserting papal moral authority was apparently the best moral argument Beecher could think to offer.

Seeking to head off the sort of post–Brook Lodge anti-whistleblower blowback that he had received from his Harvard colleagues, Beecher reas-sures fellow medical professionals "that American medicine is sound, and most progress in it soundly attained. There is, however, a reason for con-cern in certain areas, and I believe the type of activities to be mentioned [i.e., unethical experiments] will do great harm to medicine unless soon corrected. It will certainly be charged that any mention of these matters does a disservice to medicine, but not one so great, I believe, as a continuation of the practices to be cited."[118] Thus, Beecher asserts here—as he did in his letters to Pappworth—that he is revealing unethical research not to blame or shame miscreant physicians or the journals that published their research; to the contrary, he sought to do the opposite, to protect the medical profession and medical research from "a great harm" by revealing a problem that requires correction.

In the next section, "Reasons for Urgency of Study," Beecher provides data on the expansion of medical research, using the figures he had shared with Pappworth in his January 7, 1965, letter. He then details his generational change explanation of the increase in unethical experimentation observing that because "medical schools and university hospitals are increasingly dominated by investigators; Every *young man* knows that he will never be promoted to a tenure post, to a professorship in a major medical school, unless he has proved himself as an investigator. If the ready availability of money for conducting research is added to this fact, one can see how great the pressures are on ambitious *young physicians*."[119]

Focusing on "the problem of consent," Beecher next cites his *JAMA* article on "consent . . . myth and reality"[120] as the source of his claim that "consent in any fully informed sense may not be obtainable [from patients although it should] remain a goal toward which one must strive for sociologic, ethical, and clear-cut legal reasons." The reason why consent should be viewed as a symbolic act and legal formality, rather than an effective moral safeguard, is that "if suitably approached, patients will accede, on the basis of trust, to about any request their physician may make." Given the inadequacy of consent as an effective safeguard against unethical experiments, Beecher concludes, "A far more dependable safeguard than consent is the presence of a truly *responsible* investigator."[121] Beecher then reemphasizes that the twenty-two cases he presents are not cited to condemn the investigators conducting the research: "they are record[ed] to call attention to a variety of ethical problems found in experimental medicine, for it is hoped that calling attention to them will help to correct abuses present." Beecher then reiterates his standard refrain "that thoughtlessness and carelessness, not a willful disregard of the patient's rights, account for most of the cases encountered. Nonetheless, it is evident that in many of the examples presented, the investigators have risked the health or the life of their subjects." Thus, he claims, although the investigators may be guilty of thoughtlessly or carelessly endangering patients' lives or health, they are not guilty of "willfully disregarding patients' rights." Consequently, "references to the examples presented are not given, for there is no intention of pointing to individuals, but rather, a wish to call attention to widespread practices."[122]

After summarizing the twenty-two cases, Beecher concludes by reprising his view that, of the two standard "safeguards" against the abuse of patients as human guinea pigs, "the more reliable safeguard is provided by the presence of an intelligent, informed conscientious, compassionate, responsible investigator," rather than through the process of consenting the patient. He follows this claim by reiterating his utilitarian constraint on the ethics of research, stating that no experiment is ethical unless "the gain anticipated from an experiment [is] commensurate with the risk involved." He then reinforces this utilitarian constraint with a somewhat neo-Kantian provision that "an experiment is ethical or not at its inception; it does not become ethical post hoc—ends do not justify means"; he finally concludes with a bit of Fletcher's situation ethics, "There is no ethical distinction between ends and means."[123]

Since Beecher stated these ideas in the decade before he submitted his paper to Garland, this hodgepodge of ideas are clearly his, not his editor's. The paper ends with a comment about a professional initiative that he had mentioned in his January 25 letter to Pappworth: "In the publication of experimental results it must be made unmistakably clear that the proprieties have been observed. It is debatable whether data obtained unethically should be published even with stern editorial comment."[124] This idea flows naturally from Beecher's belief that the quest for publication is the source of researcher immorality, but it is tacked onto the paper with no analysis. It is just part of the potpourri of ideas running through Beecher's mind inspired by his newly woke awareness of unethical research. Consistency may be the hobgoblin of little minds, but inconsistency is no bar to influence. Beecher's article remains the most frequently cited journal article in the bioethics literature.[125]

Among the editorial challenges that impeded publication of Beecher's article was the fact that lists make dull reading. Garland addressed this by more than halving Beecher's list. Any publishing house contemplating publication of Pappworth's manuscript would confront the exponentially greater challenge of transforming a list of 500 cases into a persuasive narrative. The first publishing house attempting to publish the manuscript was Victor Gollancz Ltd., which, in October 1962, issued Pappworth a contract for a manuscript of between 60,000 and 70,000 words due March 1963. As is not uncommon, Pappworth submitted the manuscript well after the deadline. Following standard practice, the manuscript was submitted to reviewers. Their comments were uniformly negative. One wrote that "Dr. Pappworth's . . . argument [is] extremely disjointed. The writing is often very muddled. . . . On the case histories . . . the verdict is that that they become terribly monotonous in their sameness, and so readability vanishes." Another reviewer came to the same conclusion: the list of cases was "virtually unreadable" and would be "so to the great majority of laymen." This report concludes, "It is not at all a *persuasive* book, and it becomes dull as a result of repetition." After receiving the readers' reports, in July 1963, Victor Gollancz wrote to Pappworth, "I am sorry to have to tell that we cannot publish 'Human Guinea Pigs' in anything like its present form. The real point is, to put it bluntly, it just isn't in its present form, a book."[126] After this, Pappworth sent his manuscript to three more publishers, who rejected it for the same reasons.

It was during this frustrating period, when his manuscript was receiving repeated rejections, that Pappworth wrote to Beecher about naming names and publicly blaming researchers and institutions, apparently to make his manuscript more newsworthy. "What would your reaction be to a suggestion," Pappworth wrote, if "summaries of some of the most offensive experiments be published together with the journal references and names?"[127] In a later, February 20, 1965, letter, Pappworth informed Beecher that "after much thought and great hesitation I have decided that I shall publish a book on Human Guinea Pigs with quotations, names and references to about 150 of the 500-odd questionable experiments of which I have details."[128] Beecher did not reply. Pappworth, however, reacted to Beecher's June 1966 paper, "Ethics and Clinical Research," by demanding to know why Beecher protected "the guilty" by not revealing their names. At the same time, Pappworth requested Beecher's references for the twenty-two cases of unethical research cited in that article. Beecher replied that he and his editor had agreed that anonymity was the preferable format since the intent of the paper was to reveal a widespread practice, not to subject fellow physicians to potential criminal prosecution.[129] He also explained, "My attitude about what you call the 'guilty' is not one of protection for them alone. I did not want to detract from my object of pointing out a widespread practice, not individuals."[130] Beecher later told Pappworth, "I am afraid I cannot send you further references. Everybody and his brother have been after me to divulge these names . . . I decided that the only sensible way was to decline in all cases to give information not contained in the paper itself."[131] After this letter, collaboration between the two moralists came to a halt.

Around the same time, the summer of 1966, *Human Guinea Pigs* found a willing publisher, Routledge and Kegan Paul. Recognizing the challenges presented by Pappworth's manuscript, the publishing house commissioned a freelance editor, Kathleen Orr, to work with Pappworth on editing the book. When the book came out in 1967, it drew the attention of such internationally renowned reviewers as the moral philosopher Sir Bernard Williams (1929–2003) and well-known author Arthur Koestler, CBE (1905–1983). It also commanded headlines in newspapers from tabloids to broadsheets. The *Daily Express* screamed, "What are your chances of being a human guinea pig?"; the *Daily Mail* stated more sedately, "Children, pregnant women, mental defectives, and old people were being used in National Health hospitals as 'guinea pigs' without consent ... [Pappworth] likens some of the experiments to research carried out in the Nazi concentration camps."[132] A review in the more staid and scholarly *Economist* reported that "on the evidence scrupulously documented in this book, some practitioners of experimental medicine are wolves in white coats and are prepared to do hair-raising things to people admitted to a hospital's care ... [Pappworth] is out to give a much-needed shock and will not be popular in medical high places; but he is right to insist that scientific enthusiasm should be tempered by humanity and prudence, and that the patient's interests should at all times be paramount."[133] On the other side of the Atlantic, noted cardiologist and author Michael Halberstam (1932–1980) wrote in the *New York Times* that Pappworth documented that "too often research physicians lose sight of the particular patient. ... No physician who reads this book will ever make that mistake again and we shall all be better for it."[134]

Taking the opposite position, British gastroenterologist and award-winning researcher, Sir Christopher Booth (1924–2012), a doyen of the British medical establishment who later became president of the British Medical Association (BMA) and the Royal College of Physicians (RCP founded 1518), challenged Pappworth's claims. "It is palpably untrue to claim that the majority of doctors are either genuinely ignorant of the immensity and complexity of the problem [of unethical experiments on patients] or wish purposely to ignore the whole matter by sweeping it under the carpet ... [Pappworth's] work treats the serious subject of the involvement of human beings in research without sufficient understanding, personal experience, or knowledge." Booth concludes that Pappworth "is to be condemned for his continued attacks on his professional colleagues who, by their research, are doing all they can to improve the treatment they give to sick people who come to them for help."[135] In this ad hominem critique, Booth denies Pappworth's standing as an expert, claiming that he lacks "sufficient understanding, personal experience, or knowledge" and goes on to condemn Pappworth as a traitor to fellow medical professionals for "his continued attacks on his professional colleagues."

The blowback against Pappworth differs markedly from that directed against Beecher. In part, this was because Pappworth, smarting from a lifetime of anti-Semitic rejections, was "not known as a 'man to mince words.'"[136] Pappworth's blunt, often acerbic style was paid back in kind by reviewers. In *The Lancet*, Pappworth's cases were dismissed as a "vitriolic amplification of some of the charges [that Pappworth has been] firing at the medical profession for years, especially in his cannonade of a book." "Shorn of such excesses and some of its haughtiness his book would have a greater impact on those he is presumably trying to influence."[137] Yet Pappworth believed that professional status should not shield unethical experimenters from public accountability, so he named and publicly shamed prominent professionals, powerful medical institutions, and distinguished journals. When eminent, prominent, and powerful professionals or institutions attempted to shush him, he replied, "Those who dirty the linen and not those who wash it should be criticized. Some do not wash dirty linen in public or private and the dirt is left to accumulate until it stinks."[138] As physician-historian Paul J. Edelson observes,

> Pappworth was personally attacked—he was called "shrill," his work was described as "slanted," he was said to lack the "restraint" necessary to write a "more effective" book—in ways that Beecher, always referred to, in print, as "Professor Beecher" or as *Time Magazine* put it "Harvard's Dr. Beecher," as Pappworth never was. And, unlike Beecher, whose article has been widely cited and reprinted at a crucial moment in the debate over medical experimentation, [by 2004] Pappworth's book had been cited in the medical literature fewer than a dozen times since its appearance in 1967. . . . Publicly [Pappworth] was repeatedly dismissed as a troublemaker and, it may be as a result, he was not considered an acceptable person to be referred to publicly at meetings or professional discussions regardless of the substance of his remarks.[139]

It should be noted that just over a half-century later, at the end of 2021, Pappworth's effort as a pioneering bioethicist has been more widely recognized, and *Human Guinea Pigs* has been cited over 500 times.[140]

"At the core of [Pappworth's] story," observed Professor Martin Gore, CBE, FRCP (1951–2019), "is the clash between a little Jew brought up in poor circumstances, who came from [provincial] Liverpool rather than [cosmopolitan] London, had none of the requisite charms or behaviors apparently required, was revered by his students but dismissed by a powerful establishment of academics who know everything but understood nothing." "Pappworth's life story [was also about] the anti-Semitism Pappworth undoubtedly encountered as a medical student, during the war, and in the 1950s and 60s."[141] In contrast, as Edelson observes, "Beecher . . . was generally treated respectfully in public, which, perhaps, allowed people to use him both substantively and symbolically as a sponsor of medical reforms in which they were interested."[142] Another factor was that Pappworth, having little faith in the profession that rejected him as an uppity Jew, publicly identified, named, and shamed miscreant researchers, institutions, and journals, whereas Beecher declined to do so. Just as importantly, whereas Pappworth looked to external bodies outside the profession to prevent unethical research (that is, to regulation and regulatory bodies); Beecher, in contrast, believed that the profession should reform itself by acting without recourse to externally imposed regulations or laws. Thus, he recommended that professional journals require that researchers document how they had safeguarded their patients interests and rights as a prerequisite to publication. He also urged journal editors refuse to publish morally dubious experiments. Beecher's focus on professional self-reform made his proposals less threatening to fellow professionals. This may be why his collaboration with Pappworth ceased after Pappworth asked for the names and journal references for the twenty-two cases that Beecher described in "Ethics and Clinical Research." It was more than a difference in methodology whether to focus on moral reform or on punishing immoral researchers; another factor was that Beecher was a former outsider who now proudly held a named professorship at Harvard. Having become an eminent insider, he sought to avoid returning to his outsider status. Pappworth, forever an outsider, had nothing to lose—and notoriety and free publicity to gain—by naming and shaming.

Curiously, with a few minor changes, Pappworth's story might have been Beecher's as well. At the core of Harry Unangst's story is the clash between his youth as a bible-reading born-again full-immersion Congregational Methodist who, like Pappworth, was brought up in poor circumstances in a provincial town, Peck, Kansas, rather than in a cosmopolitan city, like Boston or London. Unlike Pappworth, however, Beecher, having the requisite charms and behaviors, acculturated successfully by assuming the persona of the Boston Brahmin, Henry Knowles Beecher. He also married up in social status and became a doyen of the Boston and Harvard establishments. As Edelson observes, status accounts in large measure for how differently Beecher and Pappworth were treated by their professions, and it helps to explain why Beecher's paper achieved its now iconic status as a classic of medical ethics and why both Harvard and the prestigious bioethics thinktank, the Hastings Center, offered Henry K. Beecher awards in medical ethics.[143] Pappworth's book, in contrast, has the secondary status of a whistleblowing tract and, as of 2022, no professional society or ethics center offers a Pappworth award. To the contrary, the profession treated Pappworth as a pariah, denying him the title of Fellow of the Royal College of Physicians for an unheard-of fifty-seven years. He finally received the honor a few months before his death and then only after a fellow East European Jew from Manchester, Professor Leslie Turnberg (later Barron Turnberg), became the president of the Royal College of Physicians. At age of eighty-three, Pappworth, the brash whistleblowing Manchester Jew, was, for a few months before his death, accepted as a Fellow of the Royal College of Physicians (FRCP).

7

Protesting the USPHS's Syphilis Study: Gibson, Schatz, Buxtun, and Jenkins

There is no implication that the subjects of this study were aware that treatment was being deliberately withheld. . . . It seems to me that the continued observation of an ignorant individual suffering with a chronic disease for which therapeutic measures are available, cannot be justified on the basis of any accepted moral standard: pagan (Hippocratic Oath), religious (Maimonides, Golden Rule) or professional (AMA Code of Ethics).
—Count Gibson, 1955[1]

I am utterly astounded by the fact that physicians allow patients with a potentially fatal disease to remain untreated when effective therapy is available.
—Peter Schatz, 1964[2]

The group is 100% Negro. This in itself is political dynamite and subject to wild journalistic misinterpretation. It also follows the thinking of Negro militants that Negroes have long been used for "medical experiments" and "teaching cases" in emergency wards of county hospitals.
—Peter Buxtun, 1965[3]

It took us weeks to realize, "Oh my God. There's something wrong." It's always amazing to me that young people when you say, "the Tuskegee Study," and people immediately say they knew it was unethical. That wasn't the way it was in the beginning. We had to struggle with understanding.
—William Carter Jenkins, 1969[4]

In common life, to retract an error even in the beginning, is no easy task . . . but in a public station, to have been in an error, and to have persisted in it, when it is detected, ruins both reputation and fortune. To this we may add that disappointment and opposition inflame the minds of men, and attach them, still more, to their mistakes.
—Alexander Hamilton, 1774[5]

Birth is a natural lottery. Some of us are born in peaceful times and have silver spoons fashioned for us; others enter the world in dangerous places and unfavorable times. In 1937, Peter Buxtun was born in Czechoslovakia (now the Czech Republic) to a Jewish father and a Roman Catholic mother. Before he reached his first birthday, his life was permanently altered by an event in Munich, Germany. There, two heads of state, British Prime Minister Arthur Neville Chamberlain (1869–1940, prime minister, 1937–1940) and German *Führer und Reichskanzler* Adolf Hitler (1889–1945, leader and imperial chancellor, 1935–1945), agreed that the Sudetenland, the German-speaking part of Czechoslovakia, would be ceded to its German mother-land. Reading the writing on the wall, Buxtun's father immediately began the process of immigrating to the United States with his family. After arriving in America, the family settled on a farm in Oregon where, like many young men growing up on American farms, Peter fell in love with guns, joined the National Rifle Association, and became a Republican. Unlike most American youths, Peter also grew up listening to tales of the Holocaust told by his uncle, a tank officer in the German army, who described his futile attempts at preventing Nazi soldiers from murdering civilians.[6]

Endowed with curiosity about the Holocaust by his family's history, Buxtun took several courses on the subject as a student at the University of Oregon. Among other things, they "reminded [him] of what could have been his fate if his family had not left Czechoslovakia in time."[7] After his service as a psychiatric social worker and combat medic in the US Army during the Vietnam War, Buxtun answered an advertisement recruiting people to work in venereal disease control in San Francisco. As historian Susan Reverby (1946–) recounts the story, some months into his new job as a contract tracer for sexually transmitted infections (STIs) for the US Public Health Service (PHS or USPHS), Buxtun

wandered into a coffee room of his clinic to hear an older PHS officer finishing a tale to a workmate . . . "the patient was insane . . . secondary to long term syphilitic infection. The doctor of course treated this man with penicillin and shortly thereafter had the county medical society and the CDC jump down his throat. 'See here doctor so and so . . . you have spoiled one of our subjects. You have *treated* someone who was not to be treated.'"

Buxtun was shocked by the story . . . he recalled, "working five days a week . . . to track down men who had been named as contacts and dragging them in for treatment. I believed in it . . . and these men who would otherwise have been quite ill thanked us."

Buxtun pressed his PHS colleague for more of the story, and out poured the details of the [USPHS's Tuskegee Syphilis] Study as it had been handed down within the PHS and in the published reports. "I thought," Buxton recounted, "we can't be doing this." The contrast between what he did every day and his knowledge of German history and of the Nuremberg Trials caught Buxtun up. He got on the phone to the CDC and asked the public relations person for any reprints ... Thus, he tracked down syphilitics during the day and the published Study reports during the night. The contrast could not have been starker.[8]

In search of more information about the ethics of the study, Buxtun soon went to a public library to look up German war crimes proceedings. There he discovered the Nuremberg Doctors Trial and the Nuremberg Principles, whose first principle states that morally permissible experiments on human subjects requires their informed voluntary consent. He also discovered that the principles forbade research that could lead to the injury, disability, or death of a subject. Buxtun recalled, "It was toward the end of the evening in that library downtown, and I thought: I've got to do something."[9] He did. In November 1965, he wrote a report to the CDC/USPHS comparing the USPHS's Tuskegee Syphilis Study to the Nazi medical experiments condemned at Nuremberg. Before mailing the report, Buxtun shared it with coworkers. "It caused a stir where I worked [Buxtun reports]. People (including my boss) were coming up to me saying, 'Well, when they fire you, for God's sake don't mention my name.' One doctor came back, fire in his eyes, marched up to me, threw [the report] at my chest and said, 'These people are all volunteers,' wrongly implying the Tuskegee participants had provided informed consent."[10]

Without support from his peers, and expecting to be fired, Buxtun mailed the report to the director of the Public Health Service's Division of Venereal Diseases. There was no response. However, Reverby observes, "unlike Gibson and Schatz, Buxtun persisted. ... He wrote [directly] to William J. Brown, then the chief of the Venereal Disease Division of the CDC" in November 1966. Still, "Nothing happened [for months]. And then suddenly [in March 1967, he] got orders to 'Come to a meeting in Atlanta.' In Atlanta, after waiting all day [at the USPHS's headquarters] Buxtun was led to a room with a large table. [As he describes it] Over here was the American flag. Over there was the flag of the CDC. All the bureaucrats, of course, sitting at the head of the table in front of the flags. ... The door was shut, and we all sat down."[11]

Buxtun was next given "a proper tongue lashing" by Assistant Surgeon General John C. Cutler (1915–2003), director of the USPHS's Venereal Disease Division and lead researcher on the USPHS's Tuskegee Study. "See here, young man [Cutler said] We are getting a lot of valuable information from the Tuskegee Study. This is something that is going to be of great help to the Black race here in the United States."[12] He then informed Buxtun that "medical judgment" had determined that the study should continue. Some six months later, in November 1967, Buxtun resigned. He enrolled that fall in the University of California's Hastings College of Law, a public law school located in San Francisco (founded 1878—not to be confused with the Hastings Center, a prestigious New York State bioethics institute).

The following year, 1968, was the height of the civil rights movement: Reverend Martin Luther King Jr. (1929–1968) was assassinated in April, followed in June of that year by the assassination of civil rights advocate and presidential candidate, former US Attorney General Robert Kennedy (1925–1968). Throughout that summer, African Americans across the United States rioted in protest. Responding to this context, "Buxtun worried that 'if people found out about [black] people with [untreated] syphilis who were allowed to suffer, that would be blown out all over the country . . . and damaged other things the CDC was doing.' I was going around saying 'We don't do this sort of thing.' Or 'we shouldn't do this sort of thing.'" So, in November 1968, Buxtun wrote another letter to Director Brown, pointing out that "the group is 100% Negro. This in itself is political dynamite and subject to wild journalistic misinterpretation. It also follows the thinking of Negro militants that Negroes have long been used for 'medical experiments' and 'teaching cases' in emergency wards of county hospitals."[13]

Before responding to Buxtun, on February 6, 1969, Director Brown convened a committee of the USPHS and outside experts to review the situation. James H. Jones (1943–), the first historian to publish a monograph on the USPHS's Tuskegee Study, observes that "no one with training in medical ethics was invited to the meeting, none of the participants was black, and at no point during the discussion that followed did anyone mention the PHS's own guidelines on human experimentation."[14] The USPHS's 1966 guidelines required prior review of a proposal by an independent committee (later known as an institutional review board or IRB) to protect "the rights and welfare of the [subjects and to assess], the appropriateness of the methods used to secure informed consent, and of the risks and potential medical benefits of the investigation."[15] Had the 1966 USPHS guidelines been applied to the study, it should not have proceeded. Yet, presuming the moral purity of their own actions, the USPHS only applied its guidelines to applications from investigators external to the USPHS. Consequently, since the USPHS's Syphilis Study was conducted by USPHS's own investigators, the guidelines were presumed irrelevant, and African Americans enrolled in the study were kept uninformed of their diagnosis so that their diseases would remain untreated and their corpses could serve to document the effects of untreated syphilis.

Still, Buxtun's reports, and his letters, raised questions about the propriety of continuing the study. So, on February 6, 1969, a meeting was held to discuss whether to ameliorate the situation by discontinuing the study and offering antibiotic treatment to the men enrolled in it. During the meeting, Buxtun's concerns that the "racial issue" might tarnish the reputation of the USPHS were mentioned briefly. It was decided that they "will not affect the study. Any questions can be handled by saying these people were at the point that therapy would no longer help them. They are getting better medical care than they would under any other circumstances."[16] In the end, the "racial issue" was dismissed with a note that, should questions arise, they were to be handled as a public relations matter—not as an ethical issue.[17]

Only one person at the meeting voiced objections to continuing the study, a Jewish American World War II veteran, Dr. Gene Stollerman (1921–2014), chair of the Department of Medicine at the University of Tennessee. Dr. Stollerman had been visiting the USPHS to attend a different meeting, the Advisory Committee for Immunization Practices (ACIP), when Brown invited him to attend the USPHS's Tuskegee Study meeting. According to a memo in Brown's files, "Stollerman told him that this appears . . . to be a hot potato from many standpoints—racial, public relations, etc. . . . He thinks 'we should go all out to get this worked out as soon as possible.'"[18] Dr. Stollerman himself recalled being "astounded that the [USPHS] had tolerated the study for so long."[19] During the meeting, Stollerman repeatedly contended that the USPHS had an obligation to discontinue the study and to offer antibiotic treatment to men enrolled in it. Rather than responding directly, those attending deflected Stollerman's objections by debating whether antibiotic treatments would do the syphilitic men in the study more harm than good. Scientism was so deeply embedded in the thought processes of the USPHS that, having concluded it made no medico-scientific sense to administer antibiotics, and having devised a seeming solution to the "public relations problem," the committee members dismissed Stollerman's objections as irrelevant. As was typical prior to the bioethics moral revolution of the 1970s, faith in scientism obviated questions of morality.

Worse yet, acting on their assumption that antibiotics would be harmful to the men enrolled in the study, the USPHS contacted local black physicians and got them to agree not to prescribe antibiotics to anyone enrolled in the study. Historian James Jones remarks that "apparently no one thought to question the morality of withholding [antibiotic] treatment [since it] was not specifically limited to syphilis. Antibiotics . . . are given for a wide variety of infections" [such as pneumonia];[20] it is also likely that many in the study group had received antibiotics for other ailments. Sadly, the net result of this racist saga of deception, disability, needless disease, and unnecessary deaths was irreparable damage not only to the health of the African American subjects and their sexual partners but also to the ethical and scientific reputation of the USPHS and white doctors among African American and other minority communities. Moreover, since the study group had likely been contaminated by treatment with antibiotics, there were no offsetting advances in medical knowledge. The sole impact was an enduring legacy of distrust of medicine in the African American and other minority communities.[21]

Perhaps the best account of why people, committees, and institutions persist in continuing erroneous policies was penned by Alexander Hamilton (circa 1756–1804). He wrote that "in a public station, to have been in an error, and to have persisted in it, when it is detected, ruins both reputation and fortune. To this we may add that disappointment and opposition inflame the minds of men, and attach them, still more, to their mistakes."[22] So, as Hamilton might have predicted, when Buxtun confronted the USPHS with its ethical failings, and later, when Stollerman raised objections, the scientists at the USPHS attached themselves still more stubbornly to their mistakes.

As it turns out, Buxtun's were not the first letters written to the CDC or the USPHS protesting the Tuskegee Study. A young Jewish Canadian cardiologist, Irwin Schatz (1931–2015), wrote a letter expressing his shock on reading an article in the December 1964 issue of the *Archives of Internal Medicine*. The article had opened with the statement, "The year 1963 marks the 30th year of the long-term evaluation of the effect of untreated syphilis in the male Negro conducted by the Venereal Disease Branch, Communicable Disease Center, United States Public Health Service. This paper summarizes the information obtained in this study, known as the 'Tuskegee Study.'"[23] Included in the article is a summary that states that "the [untreated] syphilitic group continues to have higher mortality and morbidity than the uninfected controls, with the cardiovascular system most commonly involved. As of December 1, 1963, approximately 59% of the syphilitic group and 45% of the control group were known to be dead."

Schatz recalls having "reread the article several times to make sure he understood it. 'I could not believe my eyes,' he recalled discuss[ing] the article with several colleagues and was met with shrugged shoulders and apathy . . . well aware of the debates over the Holocaust and Nazi doctors, in a letter to the article's senior author Donald H Rockwell (1931–2013) of the USPHS Schatz made clear . . . what he thought."[24]

> I am utterly astounded by the fact that physicians allow patients with a potentially fatal disease to remain untreated when effective therapy is available . . . I assume you feel that the information which is extracted from observation of this untreated group is worth their sacrifice. If this is the case, then I suggest the

United States Public Health Service and those physicians associated with it in this study need to re-evaluate their moral judgments in this regard.[25]

His letter went unanswered. However, a 1972 Freedom of Information Act request uncovered Schatz's letter. Stapled to it was a memo from Dr. Anne Yobs, coauthor of the article in the *Annals*, proclaiming, "This is the first letter of this type we have received. I do not plan to answer this letter."[26]

In 1955, a forty-four-year-old associate professor at the Medical College of Virginia, Count Dillon Gibson (1921–2002), had written an even earlier letter querying the ethics of the study. After hearing a presentation on the USPHS's Tuskegee Study by a former colleague, Sidney Olansky (1914–2007), Gibson wrote to him explaining that "I am gravely concerned about the ethics of the entire project. . . . There is no implication that the subjects of this study were aware that treatment was being deliberately withheld. . . . It seems to me that the continued observation of an ignorant individual suffering with a chronic disease for which therapeutic measures are available, cannot be justified on the basis of any accepted moral standard: pagan (Hippocratic Oath), religious (Maimonides, Golden Rule) or professional (AMA Code of Ethics)."[27]

Olansky wrote back, "I know exactly how you feel & what must be going through your mind," he assured Gibson, "Come join the 'next roundup [i.e., examinations of subjects],'" he suggested, or come to Washington so that "Dr. John C. Cutler . . . can fill you in with more detail than I can in a letter."[28] Seeking counsel from others on how to respond, Gibson was told "that if [he] wanted to get along, succeed and thrive in his medical career he'd better shut up about this and stop raising questions. He was going up against very senior and powerful men." Bullied by powerful senior physicians, Gibson kept his thoughts to himself—but he never forgot the incident and was later involved in Martin Luther King Jr.'s nonviolent civil rights movement and became a leading civil rights advocate.[29]

William (Bill) Carter Jenkins (1945–2019), a Georgia-born African American biostatistician and epidemiologist, was another stymied whistleblowing moralist. When he came across the study, he consulted his mentor, Geraldine A. Gleeson (1911–), who, it turned out, had worked on the study. She discouraged Jenkins from inquiring further, claiming that "you [Jenkins] just don't understand."[30] According to Jenkins, after being told, "Don't worry about it," he

> decided to try to read something about the study myself that was al[ready] published. One of the reasons why the study is so well known is because it was so well documented. That's when I found out that my advisor was actually a statistician on the study. Then that gave me a bit of a 'hmm' kind of reaction.

I decided to talk to some friends, read more articles and try to make a decision. It took us weeks to realize, "Oh my God. There's something wrong." It's
always amazing to me that young people when you say, "the Tuskegee Study,"
and people immediately say they knew it was unethical. That wasn't the way it
was in the beginning. We had to struggle with understanding.[31]

Eventually, Jenkins understood all too well. He and fellow African American professionals working with the Student Nonviolent Coordinating
Committee (SNCC, 1960–1976) bookstore, Drum and Spear (1968–1974),
produced a newsletter, *Drum* (circa 1968). In an issue of the newsletter, they
denounced the USPHS's Tuskegee Syphilis Study as racist and sent copies to
the *New York Times* and the *Washington Post*—which ignored them.

There appears to be no currently available evidence that anyone other
than these four plus one—Gibson, Schatz, Buxtun, and Jenkins—and, in
a different context, Stollerman, objected to the Venereal Disease Division
of the US Public Health Service's Study of Negroes with Untreated Syphilis. Yet the USPHS repeatedly published reports in medical journals that it
was allowing about 400 Negro men to go untreated for syphilis. The earliest known objection, Count Gibson's 1955 letter to Olansky, was written
when Gibson was a young faculty member and was penned just shy of a
quarter century after the study was initiated in 1932. Two factors appear
to have predisposed this near-universal moral blindness: pervasive scientism
and systemic racism. As Kelsey, a physician herself, observed, doctors' scientism led others to treat physicians as "Lords Almighty" whenever they
invoked the ideal of sacrifices needed to advance medical science. Proponents of the study invoked this idea in response to the few occasions that
someone question them. Race essentialism and systemic racism, both rife in
the medical science of the interwar period, further abetted moral blindness
to the USPHS investigators' exploitive immorality. One leading authority
championed the study's "immense value [because] syphilis in the negro is
in many respects almost a different disease from syphilis in the white."[32]
Underlying the USPHS's Syphilis Study was racial essentialist presumption
that black people, by virtue of being black, differ from white people in their
physiological responses to an infection by the bacterium, *Treponema pal-
lidum*, the causative agent in syphilis, and that these differences were genetically encoded in the two races. Moreover, the idea that this was, as initially
claimed, "an experiment in nature" flies in the face of the sociopolitical context of American apartheid—the blatantly racist legal segregation of African
Americans in the Jim Crow American South—that made a test of untreated
syphilis in black men seem "natural" because the USPHS presumed that, by
virtue of America's legalized apartheid, these men would never have their

syphilis treated. Ironically, as historian Alan Brandt caustically remarked, the researchers directing the USPHS's "Tuskegee Study made that a self-fulfilling prophecy"[33] by proactively working to prevent them from receiving treatment.

One could envision a morally unproblematic version of the study as a study in the effects of racial apartheid (i.e., legally enforced racial segregation) on the health of African American men. The "Tuskegee Study" was first proposed after funds *to provide treatment* to the infected men. After hopes for funding faded, serendipitously, or so it seemed, someone discovered that "in Macon County . . . [there was] an unusually high rate [of syphilis] and, what is more remarkable, the fact that 99 per cent of this group was entirely without previous treatment." This discovery seemed to offer "an unparalleled opportunity for carrying on this piece of scientific research which probably cannot be duplicated anywhere else in the world."[34] Yet any study leaving syphilis untreated would conflict with the USPHS's fundamental mission of preventing sexually transmitted infections because, as Brandt observes, "every patient with latent syphilis may be, and perhaps is, infectious for others."[35]

> Moreover, every major textbook treatment of syphilis at the time of the [USPHS's] Tuskegee Study's inception strongly advocated treating syphilis even in its latent stages. . . . If a complete cure could not be effected, at least the most devastating effects of the disease could be avoided. [As a leading expert] wrote in his 1933 textbook: "Though it imposes a slight though measurable risk of its own, treatment markedly diminishes the risk from syphilis. In latent syphilis . . . the probability of progression, relapse, or death is reduced from a probable 25–30 percent without treatment to about 5 percent with it." Furthermore, in 1932, the year in which the [USPHS's] Tuskegee Study commenced, the USPHS sponsored and published a "paper . . . that strongly argued for treating latent syphilis."[36]

Yet the lead author of that paper still favored the study because, by virtue of his unempirical presumption of racial essentialism, the disease had to be different in Negroes.

The basic problem with conducting the study as a "study in nature," or even as a study in the medical effects of racial apartheid, was that, unlike monitoring the natural flight paths of migrating birds, gathering data on the progress of syphilis in humans requires interventions to monitor the course of the disease. Subjects would have to agree to premortem medical tests and postmortem dissection. Yet to fully inform potential subjects virtually guaranteed their refusal to participate. Why? Because it is too much to ask hundreds of people suffering from a medically treatable, serious, and deadly infectious disease that they voluntarily forgo treatment, periodically undergo blood draws and spinal taps, and then donate their corpse for autopsy. No statistically significant number of infected men would have volunteered for such a study had they known what was involved. Presuming that informing subjects virtually guaranteed their nonparticipation, the investigators told subjects they would be *treated* for "bad blood"—a colloquial term for syphilis[37]—even though they would actually be preventing their subjects from receiving treatment for their disease. To put it bluntly, the researchers lied.

Truth-telling is a social institution. Cultures have elaborate conventions about who is, and who is not, entitled to truthful information. In our culture, parents are encouraged to collude in fabricating falsehoods about the existence of Santa Claus and the tooth fairy (i.e., to lie to their children). People in authority, or of higher social status, including white physicians acting as "Lords Almighty," often accord themselves the paternalistic prerogative of lying, sometimes in the name of softening bad news (e.g., a terminal prognosis) or, as happened in this case, to advance science. Asserting their white paternalistic prerogatives as physician-investigators, these respectably middle-class well-educated USPHS investigators justified their systematic lies to hundreds of disenfranchised poor black men in the segregated South by invoking the idea of "advancing medical science."

Among their deceitful acts was circulating a recruitment letter rife with purposeful disinformation and falsehoods. The letter offered free "treatment" for "bad blood" (i.e., syphilis), purposefully falsifying the study's actual purpose: to observe the natural course of *untreated* syphilis. The "treatment" referenced in the letter was a diagnostic lumbar puncture (spinal tap) to collect subjects' cerebrospinal fluid as a baseline against which researchers would measure the progress of the untreated disease. The repeated offers of "treatment" and "FREE TREATMENT"—notice the capitalization of the lie—systematically disinformed the letter's recipients to disguise the fact that they were not patients being "treated" but subjects in a study designed to leave them monitored and untreated so that their corpses would provide publishable data. Here, for example, is the text of a letter that about 399 African American men with untreated or undertreated syphilis received after an initial screening exam.

> Some time ago you were given a thorough examination and since that time we hope you have gotten a great deal of treatment for bad blood. You will now be given your last chance to get a second examination. The examination is a very special one and after it is finished you will be given a special treatment if it is believed you are in a condition to stand it. REMEMBER THIS IS YOUR LAST CHANCE FOR SPECIAL FREE TREATMENT. BE SURE TO MEET THE NURSE.[38]

Internal correspondence between researchers illustrates both the under-lying racial biases of the researchers and their willful duplicity in tricking their African American subjects into "volunteering" for things that they would have rejected had they been accurately informed about them. Here is an example: "If the colored population becomes aware that accepting free hospital care [by becoming a subject in the USPHS's syphilis study] means a post-mortem, every darky will leave Macon county . . . however, if the doctors are . . . requested to be very careful not to let the objective of the plan be known [it should work]."[39] "Darky" is a pejorative term that defines people in terms of their skin color. The racial pejorative serves to legitimize a plan to exploit and deceive African American men who, quite unknowingly, were participating in an experiment in which their black corpses became more valuable than their living black bodies. In this com-munication, the racial pejorative prefaces a plan to systematically deceive hundreds of African American men into accepting postmortem autopsies against their presumed wishes. As one USPHS researcher put this point, other than to document the course of the disease, investigators had "no further interest in these patients *until they die*."[40] Medical historian Susan Lederer underscores the grisliness of these communications, commenting that "the investigators who staffed the study over four decades regarded their African American subjects neither as patients, nor as experimen-tal subjects, but as cadavers, who had been identified while still alive."[41] Lederer comments further that in Alabama, "surviving friends and family members [could] claim the body [of the deceased] so long as they could provide the money for the burial. . . . Intent on getting postmortem evi-dence for the syphilis study USPHS investigators scrupulously complied with the legal requirement for family permission for autopsy [by providing families with burial funds]. For . . . the Public Health Service the dead took precedence over the living," and so, ironically, informed surrogate consent for autopsies was secured from families on behalf of the dead, although it had been denied to the deceased while they had been alive.[42]

In 1945, thirteen years after the study began, Sir Alexander Fleming (1881–1955), Ernst Boris Chain (1906–1979), and Sir Howard Walter Flo-rey (1898–1968) received the Nobel Prize in Medicine for the discovery of penicillin. Within a few years, supplies of penicillin became plentiful, and the USPHS initiated a series of campaigns to treat syphilis in the American South, but no offer of penicillin was made to the subjects in the USPHS's Tuskegee Study. Just five months after the passage of civil rights laws ended legalized segregation in the South, USPHS researchers published a cele-bratory article on the study in a leading medical journal, the *Archives of*

Internal Medicine. Its title, "The Tuskegee Study of Untreated Syphilis: The 30th Year of Observation,"[43] proudly proclaims that the USPHS had left about 400 African American men untreated for syphilis for thirty years. Yet with the singular exception of the Jewish Canadian Irwin Schatz, the medical community reading this title or perusing the accompanying article seems to have been oblivious to the fact that, for almost two decades after penicillin had become widely available, the USPHS was allowing hundreds of African American men with syphilis to go untreated.

Moreover, the USPHS continued to leave them untreated for seven more years until forced to discontinue the experiment shortly after Associated Press reporter Jean Heller used Buxtun's information to write an article that was published as the front-page headline in the *New York Times* and the *Washington Post.*[44] On entering law school, Buxtun brought the USPHS's study to the attention of several professors, who brushed him off in various ways. One mentioned the statute of limitations; another suggested he contact the American Civil Liberties Union. At that point, Buxtun, then a first-year law student, was "up to [his] rear in alligators" and had neither the time nor the focus to drain the USPHS's Tuskegee Study swamp.[45]

Relentlessly seeking to call out injustice, Buxtun's persistence paid off when he told his story to Jean Heller of the Associated Press. On August 28, 1972, about one month after Heller's article was published, the secretary of the Department of Health Education of Welfare (DHEW 1953–1979), alert to the possibility of a public relations nightmare in the wake of the Heller article, convened a USPHS Syphilis Study Ad Hoc Advisory Panel to "to fulfill the public pledge . . . to investigate the circumstances surrounding the Tuskegee, Alabama, study of untreated syphilis in the male Negro initiated by the United States Public Health Service in 1932."[46] Unlike the 1969 committee, which consisted entirely of white men associated with the USPHS, the majority of the 1972–1973 committee were African Americans unaffiliated with the USPHS. The committee's chair, Broadus N. Butler, PhD (1923–1996), was president of Dillard University, a historically black college founded in 1930. Butler had been a member of the legendary 332nd Fighter Group and the 477th Bombardment Group, the first African American aviators in the US Army Air Force. Popularly known as "The Tuskegee Airmen," throughout their terms of service the Airmen fought a two-front battle: fighting Nazis in North Africa, Sicily, and continental Europe in the air as they grappled with Jim Crow laws and endemic anti-black discrimination in the military and in the racially segregated American South. Among the other African American committee members were Ronald H. Brown (1941–1996), general counsel for the National Urban

League (founded 1910), an African American civil rights organization, and Vernal Cave, MD (1919–1997), an alumnus of the City College of New York (CCNY founded 1847, now part of the City University of New York) who received his medical degree from Howard Medical College, a historically black medical college founded in Washington, DC in 1868 (i.e., during the post–Civil War reconstruction period). After earning his medical degree, Dr. Cave spent five years as an Army Air Force physician. He then practiced in Brooklyn, New York, eventually becoming director of New York City's Bureau of Venereal Disease Control.

The two African American women on the committee were Jean L. Harris, MD (1931–2001), who represented the historically black National Medical Association's Foundation and was the first black woman to graduate from the Medical College of Virginia (founded in 1888). The other was Jeanne C. Sinkford, DDS (1933–), associate dean of Howard University's College of Dentistry, the first woman to become the dean of an American dental school. Several white men also served on the committee: Alabama Labor Council President (AFL-CIO, 1957–1983) and civil rights advocate Barney Weeks (1913–2003) and anti–death penalty activist-lawyer Fred Speaker (1930–1996). Representing ethics at a time when Americans treated "ethics" as a religious or legal matter were Seward Hiltner (1910–1984), a professor at Princeton Theological Seminary, and Jay Katz, MD (1922–2008), a Jewish American USAF veteran, Yale medical school psychiatrist, and adjunct law school professor (who became a founding bioethicist).

The committee was charged with addressing two questions: Was the study justified in 1932? Should the study have been discontinued twenty years later, in 1953, when penicillin became generally available? The committee found that

> in retrospect, the Public Health Service Study of Untreated Syphilis in the Male Negro in Macon County, Alabama, was ethically unjustified in 1932. . . . [Because] one fundamental ethical rule is that a person should not be subjected to avoidable risk of death or physical harm unless he freely and intelligently consents. There is no evidence that such consent was obtained from the participants in this study.
>
> . . . [Moreover], the conduct of the longitudinal study as initially reported in 1936 and through the years is judged to be scientifically unsound and its results are disproportionately meager compared with known risks to human subjects involved. Outstanding weaknesses of this study, supported by the lack of written protocol, include lack of validity and reliability assurances; lack of calibration of investigator responses; uncertain quality of clinical judgments between various investigators; questionable data base validity and questionable value of the experimental design for a long-term study of this nature.[47]

Notice that this pre-bioethical moral condemnation rests on the "fundamental ethical rule that a person should not be subjected to avoidable risk of death or physical harm," Thus, the moral issue arises solely because of unconsented exposure to "avoidable risk of death or physical harm."

In response to the second question (Should the study have been discontinued twenty years later, in 1953, when penicillin became generally available?), the committee concluded

> that penicillin therapy should have been made available to the participants in this study, especially as of 1953 when penicillin became generally available.
> Withholding of penicillin, after it became generally available, amplified the injustice to which this group of human beings had already been subjected. The scientific merits of the Tuskegee Study are vastly overshadowed by the violation of basic ethical principles pertaining to human dignity and human life imposed on the experimental subjects.[48]

The committee also describes the exploitation of vulnerable populations.

> History has shown that certain people under psychological, social, or economic duress are particularly acquiescent. These are the young, the mentally impaired, the institutionalized, the poor, and persons of racial minority and other disadvantaged groups. These are the people who may be selected for human experimentation and who, because of their station in life, may not have an equal chance to withhold consent.[49]

Notably missing are the words "Negro" and "racism." Although the committee does not use these words, it is not unreasonable to suspect that its largely African American membership might have been cautious about using these words because of their lifelong confrontation with the antiblack racial bias, endemic white supremacism, and legalized racial apartheid in the American military, in the American South, and structurally embedded elsewhere in America.

Both the all-white 1964 committee and the majority-black 1972 committees made their decisions after African American insurrections of the 1960s. However, the 1972 committee differs from the 1969 committee in more than its racial composition: the 1969 committee's decision to continue the study was made sub rosa, behind closed doors and out of the public eye, and was disguised with a false narrative designed to counter bad publicity. In contrast, the 1973 USPHS review issued a public report condemning both the initial 1932 study and the later 1969 decision on the clear moral grounds that the subjects were exposed to risk of harm and death without being informed of these risks or consenting to them. The absence of transparency in the 1969 decision permitted questionable morality, just as the publicity of the 1973 committee's report attested to its unquestioned morality.

In an important addendum to the 1973 Report, Jay Katz questioned the DHEW's formulation of the questions it put to the committee. He noted that unless one succumbs to race essentialism, there was no need to conduct the study because

> (1) It was already known before the [USPHS's] Tuskegee Syphilis Study was begun . . . , that persons with untreated syphilis have a higher death rate than those who have been treated. The life expectancy of at least forty subjects in the study was markedly decreased for lack of treatment.

On a deeper level Katz points out that

> (2) . . . the investigators seem to have confused the study with an "experiment in nature." But syphilis was not a condition for which no beneficial treatment was available, calling for experimentation to learn more about the condition in the hope of finding a remedy. The persistence of the syphilitic disease from which the victims of the USPHS's Study suffered resulted from the unwillingness or incapacity of society to mobilize the necessary resources for treatment. The investigators, the USPHS, and the private foundations who gave support to this study should not have exploited this situation in the fashion they did.

As Katz underlines: there was nothing natural about the study. Treatment unavailability was solely because an uncaring society practicing legal racial apartheid failed to make it available to these African American men and persisted in doing so in the name of science. Moreover, Katz argued,

> (3) It has long been a principle of medical and surgical morality (never to perform) "on man an experiment which might be harmful to him to any extent, even though the result might be highly advantageous to science" . . . without the knowledgeable consent of the subject (Claude Bernard 1865).[50] This was one basis on which the German physicians who had conducted medical experiments in concentration camps were tried by the Nuremberg Military Tribunal for crimes against humanity. Testimony at their trial by official representatives of the American Medical Association [Andrew Ivy] clearly suggested that research like the USPHS's Syphilis Study would have been intolerable in this country or anywhere in the civilized world. Yet the USPHS's study was continued after the Nuremberg findings and the Nuremberg Code had been widely disseminated to the medical community. Moreover, the study was not reviewed in 1966 after the Surgeon General of the USPHS promulgated his guidelines for the ethical conduct of research, [which require the informed voluntary consent of subjects] even though this study was carried on within the purview of his department. . . .
>
> In conclusion, I note sadly that the medical profession, through its national association, its many individual societies, and its journals, has on the whole

not reacted to this study, except by ignoring it. One lengthy editorial appeared in the October 1972 issue of the *Southern Medical Journal* which exonerated the study and chastised the "irresponsible press" for bringing it to public attention.[51] When will we take seriously our responsibilities, particularly to the disadvantaged in our midst who so consistently throughout history have been the first to be selected for human research?

Jay Katz M.D.[52]

Here again, Katz, like other Jewish moralists (Buxtun, Hyman), compares an American medical study with those condemned at the Nuremberg Doctors Trial and, like Hyman, invokes the Nuremberg Code by that name.

Quotidian Immorality That No One Notices: Transparency without Accountability

In his addendum, Katz invokes (but misquotes) the French physiologist Claude Bernard (1813–1878) as a moral authority; however, since bioethics had yet to be invented, he appeals to no moral authority other than the Nuremberg Trials and the Nuremberg Code, which served him as it had Buxtun and Hyman (and likely Schatz and Stollerman) as a touchstone of immorality. Another point Katz raised is that, although these experiments were evidently immoral on their face, they were not secret, and yet no one seemed to notice their obvious immorality. Jeremy Bentham once observed, "Publicity is the very soul of justice. It is . . . the surest of all guards against impropriety."[53] The USPHS's reaction to Buxtun's outing of the USPHS's Syphilis Study by way of Heller's Associated Press article is an excellent example of the efficacy of publicity as an instrument of justice. Yet, as Katz notes, leading medical journals had routinely published papers detailing the results of the study, yet no one (except for the four currently known protestors) seemed to notice the study's palpable immorality. The point to appreciate is that for decades, major medical journals had been reporting on the USPH's Syphilis Study. Yet, except for the morally sighted foursome—Buxtun, Gibson, Jenkins, and Schatz—reviewers, medical editors, and readers treated these publications as a quotidian affair, an everyday research report. No one seemed to notice that hundreds of African American men were allowed to suffer from—and likely spread—a treatable disease so that the USPHS, the very agency responsible for providing treatment for this disease, could study its progress in black bodies. But, to purloin Beecher's opening sentence in "Ethics and Clinical Research," had anyone paused to reflect for a moment, it should have been evident to them that the USPHS's Syphilis Study involved the "employment of patients as experimental subjects when it must be apparent that they would not have been available if they had been truly aware of the uses that would be made of them."[54]

The UPS's syphilis study is a paradigm case of highly visible quotidian immorality. "Quotidian" is a fancy word for "every day." It is derived from the Latin (*quot*), how many, and (*diēs*) days or daily, as in the line in the Lord's Prayer "give us each day our daily bread." Which raises the question: Why didn't public exposure prompt an outcry? Why didn't everyone reading these articles recognize that the study was unethical and speak out? And why did just these four recognize the study as immoral? And why did those who did call the study out as immoral meet with such adamant opposition and veiled threats? One factor, the endemic and pervasive racism and judgmental moralism of that era likely blinded the medical community to the immorality of exploiting syphilitics with black bodies for the (presumed) greater good of scientific knowledge. As to the second question, as an African American and civil rights activist, Jensen could and did recognize discrimination when he saw it on the pages of a medical journal. Similarly, as Jews functioning in a predominantly Christian society during a time when racist anti-Semitism was commonplace, Buxtun, Schatz, and, later, Katz and Stollerman could readily recognize racism when they encountered it. Buxtun and Katz expressly compared the USPHS's Syphilis Study to the actions of Nazi physicians during the Holocaust.

Gibson may have been sensitive to quotidian immorality because he belonged to several minority Christian religions by choice rather than heritage. In 1947, he joined Dorothy Day's (1897–1980) nonviolent Catholic Worker Christian social justice movement (founded 1933) and later converted to another minority religion, the Ruthenian (Byzantine) Greek Catholic Church. He was sensitive to racism and, in the 1960s, joined the Reverend Martin Luther King Jr.'s nonviolent civil rights movement and on "Bloody Sunday" (March 7, 1965) marched with Reverend Martin Luther King Jr. and other peaceful marchers across the Edmund Pettus Bridge, where they were brutally assaulted by Alabama state troopers.[55] More to the point, like the Jewish protestors, Gibson was aware of and had attended the Nuremberg Doctors Trial, so he knew about the Holocaust and the Nazi medical experiments. Knowledge of these trials prepared him to recognize racism when he saw it, whether directed against Jews or African Americans.

The most prominent of the moral whistleblowers, Henry Beecher, was not motivated by the USPHS's study. As discussed in the previous chapter, his index case of medical immorality was Southam's cancer cell studies at the Jewish Chronic Disease Hospital in Brooklyn, New York. If we date Beecher's concerns about the ethics of experiments on humans from his

February 1958 letter to Dean Berry, by the time he came across the article in *Science*, Beecher had spent at least seven years reflecting on the subject. In a letter he submitted with the first draft of "Ethics and Clinical Research," he remarks that he had been thinking about the subject for about a decade, and we know from early drafts of the Brook Lodge paper that he was mulling over his own complicity in unethical experiments as well. Ironically, we also know from his opposition to implementing the Nuremberg Principles and his references to the Declaration of Helsinki that Beecher had been keeping track of proposed research ethics reforms for over a decade. Furthermore, when he first came across the article describing Southam's experiments in *Science*, he was not initially motivated to act. He resolved to take a stand only after exchanging cases of unethical experimentation with Pappworth. Another factor may be that memories of Harry Unangst still resided beneath the cosmopolitan Boston Brahmin façade that Harvard's Henry K. Beecher had invented for himself. Harry never abandoned his Unangst family. He gave them financial support, and he seems never to have forgotten that he was once an impoverished young man from Peck, Kansas.

White Christian whistleblower, Count Dillon Gibson Jr., had a lot in common with Harry Unangst. He too was raised by religious parents in a small town, Covington, Georgia, and remained religious throughout his life. Like Unangst/Beecher and Pappworth, Gibson had served in an army medical corps in the 1940s and for a time headed a laboratory on a military base in Vienna, Austria. Moreover, like Unangst/Beecher and the Jewish moralist whistleblowers, he was familiar with the Nazi medical experiments because he had attended the medical war crime trials of the Nazi doctors at Nuremberg. While attending, he discovered the possibility of simultaneous translation, which he later introduced to American medical centers. After discharge from the military, Gibson, like Unangst/Beecher, experienced life-changing events: one conducting secret military experiments that later troubled his conscience, the other converting to Catholicism and becoming involved with Dorothy Day and her nonviolent Christian social justice Catholic worker movement. Both men married spouses of different social and religious backgrounds: Beecher married the fashionable Margaret Swain in 1934; Gibson, now a New Yorker, married Katherine Vislocky (1919–2002), daughter of a Greek Orthodox priest in 1950. He then converted to her religion, the Ruthenian (Byzantine) Greek Catholic Church. The thread common to all five of the white whistleblowing moralists—Beecher, Buxtun, Gibson, Pappworth, and Schatz—was an awareness of Nazi medicine, knowledge of the Nuremberg Doctors Trial, or of the principles later called the Nuremberg Code. Nazi medical ethics and practices became their touchstone of immorality; their knowledge of the trials of Nazi doctors and the Nuremberg Code enabled them to recognize similarities when they heard them in a lecture or read them on the pages of a medical or scientific journal.

Moreover, each of these six whistleblowing moralists—Beecher, Buxtun, Gibson, Jenkins, Pappworth, and Schatz—was in some significant way an outsider who circumstances empowered to give voice from an insider's perspective. By virtue of their unusual backgrounds, particularly their personal experience with anti-Semitism or antiblack racism, or with war crimes, each became morally sighted ("woke") when he read or heard about morally questionable experiments, and each asked the same question: did the subjects understand and appreciate the nature and risks involved before they volunteered for the experiment? Answering, "No," their prior history allowed each one to view medical experiments differently than other health care practitioners and to challenge the ethical justification in various ways. All save one were aware of the Nuremberg Trials and the Nuremberg Code. So, when these moralists compared the experiments that they were reading or hearing about to the paradigmatic examples of medical immorality condemned at Nuremberg, they were shocked to discover that the enemy was us: that civilized American and British researchers were performing experiments comparable to those that the Nazi researchers had performed—and they had the courage to protest.

Beecher, Buxtun, and Pappworth succeeded in publicizing unethical experiments where Gibson, Jensen, and Schatz did not because, as historian Susan Reverby sums it up, "Buxtun persisted."[56] So too did the other successful whistleblowers, Beecher and Pappworth. Yet of the six whistleblowing moralists considered in this chapter, standard historical accounts usually credit only one for initiating reforms in research ethics, Henry K. Beecher. This raises the question of why Beecher gets the credit for research ethics reform, whereas Buxtun, Pappworth, and Kelsey tend to get little or none. As bioethicist Carl Elliott observes,

> Buxtun's revelations triggered Senate hearings, a federal inquiry, a class action lawsuit, and in concert with several other research scandals of the period, a lasting set of federal guidelines and institutional structures intended to protect the subject of medical research.
>
> It would be difficult to name a figure in the history of American medical ethics whose actions have been more consequential than Buxtun's. Yet most ethicists have never heard of him, and many accounts of the Tuskegee scandal do not even mention his name. [It] did not appear in Jean Hellers 1972 exposé, nor was he mentioned in the first scholarly article written about the scandal written in 1978 by [a] Harvard historian [Alan Brandt]. In the well-known play (and later

film) based on the scandal, *Miss Evers' Boys*, Buxton is completely absent. . . . If the role played by Buxtun in exposing the scandal is at all familiar it is largely because of James H. Jones influential 1981 history of the Tuskegee scandal, *Bad Blood*, . . . and Susan Reverby's 2009 book, *Examining Tuskegee*.[57]

One reason why Buxtun was ignored whereas Beecher was remembered (as either hero or a hypocrite) was that Buxtun was not a member of the club. Although Buxtun had been a psychiatric social worker, a combat medic, and a STI contract tracer, he abandoned the health care professions to become a lawyer. Since he could and did abandon his role in health care, he was insusceptible to peer pressure or intimidation by senior physicians. He had already "moved on" to become a lawyer when he contacted Jean Heller.[58] Perhaps also, as a gun-collecting, conservative, Republican, NRA member, Buxtun may have been too conservative and libertarian to fit liberal bioethicists' conceptions of the origins of their field. Finally, Buxtun, like Kelsey, never claimed the role of bioethics icon: he just wanted to protect people and, ironically, the USPHS/CDC, from the scandal of researchers who acted like Nazi doctors.

With respect to Pappworth, historian Duncan Wilson points out that "doctors largely ignored Pappworth's recommendation" to have people experienced in ethics (i.e., bioethicists) and laypeople (minorities, people with disabilities, research subjects) on research ethics committees (RECs). They preferred "a group of doctors including those experienced in clinical investigation. . . . Clinical investigation should be free to proceed without unnecessary interference and delay. Imposition of rigid or central bureaucratic controls would likely deter doctors from undertaking investigations . . . medical knowledge would inevitably diminish with resultant delay in advanced in medical care."[59] Wilson argues that the British medical profession rejected Pappworth's "calls for outside involvement [because they] conflicted with the longstanding and continued support for [self] regulation," without external oversight by disability advocates, ethicists, patients, or public bodies.[60] Communities create their heroes, heretics, and villains. The physicians' club was unlikely to credit a rude, uppity Jew who outed and shamed fellow physicians. Central casting would never select Pappworth for the hero's role—neither have most historians of bioethics.

In contrast, Edelson observes, "Beecher . . . was generally treated respectfully in public, which, perhaps, allowed people to use him both substantively and symbolically as a sponsor of medical reforms in which they were interested."[61] Pappworth, the uppity Jew with an acerbic personality, preferred to publicly identify, name, and shame fellow physicians. Beecher declined to do so. He never turned against fellow members of the club. Also, in striking contrast to Pappworth, who wanted outsiders (disability advocates, ethicists, patients) and external regulators to oversee research, Beecher believed that the profession should reform itself without recourse to government regulation (e.g., by having fellow researchers review research proposals and having editors review research submitted to professional journals). Beecher sought reforms to prevent unethical experiments in large measure because these experiments threated two things he valued highly: the medical profession and biomedical research. In return, his fellow researchers, his profession, and—for a time—bioethicists knighted him a hero of research ethics reform.

8

Scientistic Medical Paternalism
and Its Discontents

The "obedience of a patient to the prescriptions of his physician should be prompt and implicit. He should never permit his own crude opinions as to their fitness, to influence his attention to them."
—American Medical Association, Code of Ethics, 1847

The physician is a technically competent person whose competence and specific judgments and measures cannot be competently judged by the layman. The latter must therefore take these judgments and measures "on authority." But [since] there is no system of coercive sanctions to back up this authority. All the physician can say to the patient who refuses to heed his advice is, "well, it's your own funeral"—which it may be literally.
—Talcott Parsons, 1951[1]

When modern medical ethics emerged in the post–World War II era, the veteran military physicians who drafted the Nuremberg Principles and the Declaration of Geneva inverted the values of *Rassenhygiene* by emphasizing physicians' commitment to the medical needs of individual patients. Whereas *Rassenhygiene* focused primarily on benefiting the *Volkskörper* and its gene pool and was eliminative to the point of exterminating peoples with improper genetic or racial profiles, modern medical ethics was principledly inclusive from its earliest expression. Its oaths and codes expressly forbade discrimination based on ethnicity, gender, race, or religion. This list of protected groups expanded in the aftermath of the bioethical turn of the 1970s, so that by 2017, the Declaration of Geneva forbade discrimination on the basis of age, creed, disability, disease, ethnic origin, gender, nationality, political affiliation, race, religion, sexual orientation, or social class.[2] Yet, despite its idealistic objectives, prior to the Great Society initiatives of the 1960s and the bioethics turn of the 1970s, modern medical ethics still reflected the ableism, classism, ethnocentrism, male chauvinism, racism, and scientistic paternalism of the Western medical world: that is, physicians' belief that, in Kelsey's words, white medical men were entitled to act as "Lords Almighty."

A Brief History of Anglophone Scientistic Paternalism

In Anglophone medical culture claims to lordly or noble status trace to the eighteenth century when physicians seeking higher social status emulated the dress and conduct of their patrons, the nobility and gentry, by wearing wigs, donning coats with gold buttons, and sporting gold-handled canes. To appear as good Christians they also offered pro bono charity care to the deserving poor, typically on the day prior to the sabbath or certain other Christian celebrations.[3] By the end of that century, medical treatment of the deserving poor in industrializing areas of Britain began to be systemized through two comparatively novel institutions: dispensaries providing medications to the deserving sick poor and charity hospitals providing the deserving sick poor with housing, meals, beds, and medical care. Although nominally supervised by the affluent gentry and nobility that funded them, in practice, hospitals were total institutions controlled by physicians and surgeons. Having dressed and assumed the role of lords and gentlemen, physicians and surgeons assumed the prerogatives of this elevated status in hospitals where they could and did lord it over the sick working-class men, abandoned women, and other poor folks who became their patients.

Physicians and surgeons' presumptions of gentlemanly status also led to outbreaks of flyting: a form of verbal dueling in defense of personal, familial, or professional honor that took the form of oral or written invectives demeaning one's opponent's ungentlemanly character or conduct by insulting it as "tradesman-like," "illiberal," or, in various ways, ungenteel or incompetent. Written diatribes were often publicized as pamphlets, the social media of the period, sparking "pamphlet wars" in which physicians and surgeons publicly exchanged insults with each other. Occasionally, these exchanges led to duels with pistols.

Flyting was particularly problematic in charity hospitals where formerly solo practitioners were forced to work side by side. Reacting to a series of flytes and pamphlet wars that threatened a major charity hospital, the Manchester Infirmary (founded 1752),[4] an eminent, philosophically inclined physician, Thomas Percival, penned a seminal 1803 work, *Medical Ethics; or, a Code of Institutes and Principles, Adapted to the Professional Conduct of Physicians and Surgeons.* His eponymous book introduced the expression "medical ethics" into the English-language medical lexicon and, with it, the concept of medicine as something more than a paid occupation:

it was an occupation that, unlike trades, should be self-regulating and dedicated to serving such communal goods as treating the sick poor and preventing communal illnesses. By accepting a medical "office," Percival contended, physicians and surgeons implicitly committed themselves to common standards of ethical conduct—and did so simply by virtue of their occupation, their "office," or "profession," irrespective of whether they had signed an oath. Percival's code formalized the moral commitments health care professionals had toward each other, toward hospitalized and private patients, and toward society.

Although Percival's *Medical Ethics* was designed to prevent or resolve flytes and other intrapractitioner disputes, as indicated in the opening lines of the code, practitioners' first and primary duties were toward their hospitalized patients. Thus,

> Hospital PHYSICIANS and SURGEONS should minister to the sick, with due impressions of the importance of their office; reflecting that the ease, the health, and the lives of those committed to their charge depend on their skill, attention, and fidelity. They should study, also, in their deportment, so to unite *tenderness* with *steadiness*, and *condescension* with *authority*, so as to inspire the minds of their patients with gratitude, respect, and confidence.[5]

It is noteworthy that this foundational document of the ethics of scientistic paternalism recognizes the asymmetrical relationship between literate, educated, middle-class physicians and surgeons and their lower-class, typically uneducated and often illiterate, hospitalized patients, "the deserving poor." These health care practitioners held an "office" (from the Latin *officium*, for duty) that requires them to attend to their patients skillfully, attentively, and faithfully so that their patients would be thankful, respectful, and confident in their treatment. To accomplish this, Percival encourages practitioners to unite "tenderness with steadiness, and condescension with authority."

Percival's conceptions of tenderness, steadiness, and authority are similar to what we understand them to be today. Then as now, physicians and surgeons' authority has been earned by virtue of their superior scientific knowledge, medical training (by attending medical college), and experience (gained through their apprenticeship). Percival's eighteenth-century understanding of "condescension," however, differs from our current understanding because, in our presumptively egalitarian Anglophone culture, presumptions of superior status are considered illicitly inegalitarian. This, however, inverts the meaning of "condescension" as Percival would have understood it in 1794 or 1803. As anyone who has ever read a Jane Austen (1775–1817) novel will appreciate, eighteenth- and nineteenth-century Britain was a class-conscious Christian society. In this cultural context, to "condescend" means "to descend with," as described in Exodus 19:20, which describes the deity graciously condescending to reveal itself to Moses on Mount Sinai. Moses was not yet ready to ascend to the rank of divinity, so the divine descended with him on a human level, that is, it condescended—literally descended to equal level with him—at the top of a high mountain. In urging "condescension" toward hospital patients, Percival, a devout Unitarian Christian, is exhorting his nineteenth-century medical readership to follow the example set by divinity by descending from the privileges of their higher status to communicate with their lower-class charity hospital patients in ways that these patients can understand and appreciate (i.e., they must *not* act as "Lords Almighty"). Following the precedents set by Divinity on Mount Sinai, and by Divinity's humanizing itself as Jesus Christ, physicians and surgeons should suspend the privileges of their superior lordly status to communicate with, and care for, their patients effectively and tenderly.

Percival also urges practitioners to attend to the "*feelings* and *emotions* of their patients" and not to condemn or harshly oppose their prejudices.[6] In addition, he believes that clinicians have an ethical commitment "not to be restrained, by parsimonious considerations, from prescribing . . . *drugs*, even of *high price* in diseases of extraordinary malignant and dangerous . . . on principles, of . . . beneficence."[7] Note that, as described by Percival, beneficence is expressly identified as a characteristic of scientistic paternalism, thereby preserving the Hippocratic ideal that physicians should act to benefit the sick person. In the opening section of his second chapter, which deals with physicians and surgeons' conduct toward private (i.e., non-hospitalized and therefore higher-status paying patients), Percival restates his view that "every patient committed to the charge of a physician or surgeon, should be treated with attention, steadiness and humanity."[8] Absences are always worth noting: Percival had no need to bring up the subject of "condescension" in this context because private practice patients would be of the same or a higher status as their health care practitioners. What is clear in both chapters is that Percival is urging that, insofar as practically possible, every patient, including hospitalized charity patients, should be treated with attention, steadiness, and humanity. In this respect, Thomas Percival was the George Wiley of eighteenth- and early nineteenth-century medicine; that is, he insisted that charity patients receive respectful and considerate care.

Through a complex turn of events, in 1847, at its founding meeting as a national medical association, the AMA adopted an Americanized version of Percival's *Medical Ethics*, creating, in the process, the world's first national code of professional medical ethics.[9] To adapt Percival's code, which addressed moral issues in hospitals, to an American environment where hospitals were a rarity, drafters of the AMA's code had to revise physicians' first duty. Their revision obligates physicians to be "ever ready to obey the calls of the sick"—that is, to promptly attend to sick or injured people when called to their houses, workplaces, or other locations. In doing so, they were to treat patients attentively, skillfully, and faithfully, just as Percival recommended. Toward the end of the AMA code, its authors added a duty not stated in Percival's code: that "when pestilence prevails, it is [physicians'] duty to face the danger, and to continue their labors for the alleviation of suffering, even at jeopardy of their lives."[10] Note the absence of the word "cure": there were no known effective cures for pestilent diseases at that time, but doctors could use opiates to alleviate suffering, and so the AMA stipulated that doctors had a duty to provide this care "even at jeopardy of their [own] lives."

The AMA's code was formulated as a social compact, stipulating reciprocal duties binding on patients as well as physicians.[11] Consequently, just as physicians were required "to obey the calls of the sick," patients were obligated to reciprocate and so the "obedience of a patient to the prescriptions of his physician should [also] be prompt and implicit." Patients, moreover, should accept physicians' superior knowledge and "should never permit [the patient's] own crude opinions as to their fitness, to influence his attention to them." Patients should also prioritize physicians' convenience and so "should always, when practicable, send for their physician in the morning before this usual hour of going out. . . . Patients should also avoid calling on their medical adviser unnecessarily during hours devoted to his meals or sleep. They should always be in readiness to receive the visits of their physician as the detention of a few minutes is often of inconvenience to him."[12]

Save for two votes, the code was unanimously approved by the founding delegates to the AMA's 1847 convention. This laid the historic foundation for "Lords Almighty" scientistic paternalism that Kelsey found objectionable 115 years later. In the AMA code, patients' opinions were dismissed as "crude," whereas the physician's educated and scientific opinions were to command the patient's prompt and implicit obedience. Moreover, patients owed it to their physicians to arrange their requests for medical attention to suit the convenience of their physicians by, for example, not disturbing physicians during mealtimes or when they are sleeping.

A century later, as the locus of medical care shifted from patients' homes to physicians' workplaces (i.e., clinics and hospitals), the role AMA physicians had wistfully scripted for patients in 1847 became a reality. In a

1951 book, *The Social System*, American sociologist Talcott Parsons (1902–1979) wrote a classic description of mid-twentieth-century American hospital physicians' conception of their role. Parsons characterizes physicians and patients' roles as a "trust relationship" in which patients are expected to trust their physicians by entering "the sick role . . . [an] exemption from normal social role responsibilities"[13] that excuses the sick from their daily responsibilities, enabling them to take to the sickbed provided they recognize that illness is "itself undesirable." By virtue of this recognition, American society imputes upon the sick person a correlative "obligation . . . to seek technically competent help, namely, . . . that of a physician and to cooperate with him in the process of trying to get well."[14] "This relationship is expected to be one of mutual 'trust,' in which the patient puts his faith in a physician who is trying his best to help the patient and in which, conversely, the patient is 'cooperating' with him to the best of his ability. . . . It makes the patient, in a special sense, responsible to his physician. . . . [Thus] the doctor-patient relationship has to be one involving an element of authority—we often speak of 'doctor's orders.' . . . the patient's obligation [is] faithfully to accept the implications of the fact that he is 'Dr. X's patient' and so long as he remains in that status must 'do his part' in the common enterprise."[15]

Parsons makes the further point that "the nature of [medical] help imposes a further disability or handicap upon [the patient]. He is not only not in a position to do what needs to be done, but he does not 'know' what needs to be done or how to do it. . . . Only a technically trained person has that qualification. . . . If [the sick person] were fully rational he would have to rely on professional authority, on the advice of professionally qualified."[16] Parsons concludes that in American medicine, "the situation of the patient is such as to make a high level of rationality of judgment peculiarly difficult. He is therefore open to, and is peculiarly liable to, a whole series of irr- and non-rational beliefs and practices."[17] Thus, Parsons contends, "the only rational choice a patient can make is to obey her or his doctor's orders."[18] Finally, Parson reiterates, that since

> the sick person is . . . peculiarly handicapped in arriving at a rationally objective appraisal of his situation. [Whereas] the physician is a technically competent person whose competence and specific judgments and measures cannot be competently judged by the layman. The latter must therefore take these judgments and measures 'on authority.' But [since] there is no system of coercive sanctions to back up this authority, all the physician can say to the patient who refuses to heed his advice is, "well, it's your own funeral"—which it may be literally.[19]

Parsons's description of the physician–patient relationship in mid-twentieth-century America is that, by virtue of their training and experience and emotional equipoise, physicians alone are the sole rational decision makers, and they alone should determine what is best for a patient's care. Thus they, and they alone, should be vested with the prerogative of making health care decisions on behalf of the patient. No one else—not patients, not patients' families or friends, not the public, not bureaucrats in the federal government, no one other than physicians themselves are competent to make decisions about a patient's health care. Should any patient presume to intrude on physicians' decision-making prerogatives, Parsons claims that, both metaphorically and literally, "would be [the patient's] funeral." This authoritative sociological reprisal is a less eloquent update of the medical paternalist conception of the physician–patient position propounded in the 1847 Code of Ethics—that is, that the view that the "obedience of a patient to the prescriptions of his physician should be prompt and implicit."[20]

Lord Almighty Physicians Run Amok

Parson's portrayal of the physician's role presumes that patients and their families are so irrational or nonrational that, to use the language of the 1847 AMA Code of Ethics, "they should never permit [their] own crude opinions as to their fitness, to influence [their] attention to [their doctor's orders]."[21] Parsons also assumes that physicians endorse an American version of Moll's dictum, *Salus aegroti suprema lex*, the health of the patient is the supreme law and so they are trying their best to help their patients. Yet the structural sexism and racism of American health care institutions in the pre-1960s, compounded by the structural apartheid of the North and the legalized segregation of the Jim Crow American South and the institutionalized classism endemic everywhere in a country in which the deserving poor were forced to stand in lines for "interminable lengths of time for the most meager and inadequate and most stingily given out medical treatment,"[22] meant that in reality, clinics and hospitals were run for convenience of their staff and did not prioritize the needs of those they nominally served.

As founding bioethicist Robert Veatch summed up the situation, in the 1960s and 1970s, when ordinary people

> learned what decisions physicians were making about laypeople's health, they were often appalled . . . they discovered that physicians . . . were making controversial moral moves, choices that . . . some laypeople considered morally indefensible. Physicians intentionally withheld grave diagnoses from patients; they did research on them without informing them; they sterilized some patients who they thought were not worthy of being parents; they routinely held critically and terminally ill patients alive against the wishes of those patients or their families;

they refused to perform sterilizations, abortions, and provide contraceptives if they thought that patients shouldn't have them. . . . The more laypeople learned about the ethic that had become embedded in the medical profession, the more they protested.[23]

Thus it became increasingly apparent that medically paternal relationships with patients and research subjects were rife with conflicted loyalties, values, and interests that were often resolved in ways inimical to the well-being of patients and research subjects. To address these conflicts during this period, reformers sought to deconflict the issues by creating an alphabet soup of new institutions—HECs, IACUCs, and IRBs—and by providing explicit regulatory guidance on the treatment of patients (e.g., bills of rights, research guidelines).

George Wiley and the Patients' Rights Movement

America's Great Society initiatives of the 1960s, including the new Medicaid and Medicare programs, were designed to be implemented in an egalitarian manner as prescribed by the new civil rights laws. Yet, old habits dying hard, health care institutions were still inclined to treat newly funded Medicaid and Medicare patients as they had treated them when they were charity patients. Civil rights advocates and feminists had other ideas. They argued that one does not lose one's civil or human rights by entering a clinic, a doctor's office, a hospital, or any other health care facility. One well-known civil rights activist championing the rights of newly entitled Medicaid and Medicare patients was former Syracuse University chemistry professor, George A. Wiley, PhD (1931–1973), an African American military veteran who had done postdoctoral work in chemistry and taught at University of California, Berkeley before becoming a faculty member at Syracuse University. In the 1960s Wiley left academia to join the Congress on Racial Equality (CORE, founded 1942), an African American civil rights group that used nonviolent direct-action tactics to advance the cause of racial equality (e.g., they occupied institutional and corporate offices to draw media attention to racial inequities).

Wiley's biographers, Nick Kotz (1932–2020) and Mary Lynn Kotz, believe that "George Wiley saw himself as a connector—a social chemist who was combining unknown elements."[24] One of the unknown elements that Wiley sought to connect to the civil rights movement was the 8.3 million Americans receiving welfare in the 1960s: 2.1 million receiving "Old Age Assistance," 85,000 receiving "Aid to the Blind," 583,00 receiving "Aid to the Permanently and Totally Disabled," 649,000 receiving "General

Assistance," and, the largest cohort, 4.6 million women receiving "Aid to Dependent Children (ADC)."[25] To give these people a potent public voice, Wiley formed the National Welfare Rights Organization (NWRO, 1966–1975), whose mission was to demand the rights for these people, particularly African Americans and women with children, in the new Medicaid and Medicare programs.[26]

The NWRO soon found that to address these issues, it had to simultaneously campaign against, and yet at the same time, negotiate with, the US Department of Health Education and Welfare (DHEW or HEW, 1953–1979) and three formidable professional medical organizations: the American Medical Association, the American Hospital Association (AHA, founded 1898), and the Joint Commission for Accrediting Hospitals (JCAH, founded 1951, "the Joint Commission"). The Joint Commission grew out of an "end result" (data driven) quality improvement initiative pioneered by a surgeon, Ernest Amory Codman (1869–1940). In 1951, the Codman-style evidence-based medical improvement programs run by the American College of Physicians (ACP, founded 1915), the AHA, the AMA, and the Canadian Medical Association (Association Médicale Canadienne, CMA/AMC, founded 1867) merged to form the Joint Commission.

To ensure high-quality health care for Medicare and Medicaid recipients, US government agencies adopted the Joint Commission's standards as a criterion of accreditation for institutions seeking to receive Medicaid or Medicare funding. Joint Commission accreditation thus became a prerequisite for any health care institution seeking to receive Medicare or Medicaid funding. In the fall of 1969 the Joint Commission called for suggestions regarding accreditation standards from a consumer viewpoint. Hearing opportunity knock, Wiley responded. "Characteristic of [his] direct action style, Wiley insisted on delivering the suggestions in person and directly to the Board of Commissioners . . . threaten[ing a] sit-in by [the NWRO's] organization of women and children, if they refused his request."[27] "Exactly what George Wiley had to say [to the Board] . . . cannot be ascertained, but the themes that run through the NWRO programs and their demands on the [Joint Commission] may be glimpsed in the address [Wiley] presented before the National Health Council at its annual meeting":

> For too long . . . have [poor people] been required to wait interminable lengths of time for the most meager and inadequate and most stingily given out medical treatment . . . [for too long] has there been the inconvenience of scattered and inadequate facilities [for] inoculations or TB shots . . . or vaccinations . . . or chest X ray or . . . family planning or other kinds of things that any citizen would need in the way of health care . . . [for too long] have the poor been

[kept] on the short end of the medical facilities . . . have [they] had access to sophisticated miracle treatment only as the guinea pigs.

. . . [Professional medical organizations] have dominated the medical establishment [and] have run a colonial empire on the poor, on the black, on the minorities in this country . . . [and] thwarted every serious effort of the Congress of the United States from providing significant, comprehensive health insurance . . . so that every person can have access [to health care]. For too long . . . have [they] protected their membership and controlled the numbers of needed health personnel for economic gain and defined the limits of access to health care by the people of the ghettos and barrios of our country.[28]

In this speech, Wiley channeled the frustration of the African American, Asian American, and Hispanic American (Latinx) former charity recipients, particularly unmarried or widowed women with dependent children. As a delegate to a 1973 NWRO conference explained,

We have *financial* problems. We can't pick and choose our doctors; we don't have the option not to pay [for bad service]. Most private M.D.'s refuse to take us. We get whatever doctors are at the clinic and the clinic is crowded, and we are waiting in line a long time, and then we are pushed through. There's no time to ask questions. WE need money and we need services. That's what my group expects me to tell them about [when I report what I learned at this NWRO conference].[29]

After extensive negotiations with the Joint Commission on behalf of the NWRO, they agreed on a list of demands for hospital patients' rights on June 18, 1969. The first and primary right demanded was that "the patient has a right to considerate and respectful care." Given that hospitals had forced their "charity cases" to use "whatever doctors [are] at a crowded clinic," where they stood "waiting in line a long time" only to be "pushed through"—or, as Wiley phrased this point, waiting "interminable lengths of time for the most meager and inadequate and most stingily given out medical treatment . . . the poor [have] been [kept] on the short end of the medical facilities"[30]—it makes sense that the first demand agreed on between Wiley and the Joint Commission was the right to considerate and respectful care. The second right involves two rudimentary courtesies typically denied charity patients: the right to know the name of health care practitioners responsible for their care and the right to have complete current information about their diagnosis, treatment, and prognosis explained in terms *they* can understand (i.e., not to be "pushed through . . . or denied time to ask questions"). The third right addressed the practice of not seeking charity patients' consent. Practitioners would now be required to give their patients the information necessary to understand alternative treatment options and to have treatment contingent on patients giving informed consent prior to its initiation. Patients should also be informed whether the recommended treatment was experimental.

The fourth and fifth demands involve health care facilities' obligation to post a public notice of their accreditation status. It also required them to have community members on their boards of governance. A sixth right required health care facilities to make patient advocates/ombudsmen available to patients.[31] The "patient ombudsman/patient advocate" role was "to resolve patient complaints about billing, quality of care or conditions under which that care is rendered.... The hospital must provide for an ombudsman or patient advocate to coordinate an administrative hearing mechanism. This ombudsman or patient advocate must not be an employee of the hospital but may be contracted from community organizations."[32]

Absences, as always, speak loudly. Given the ascendency of bioethics in the next decade, it is important to note the absence of the words "ethics" and "morality." The rights Wiley demanded and that the Joint Commission tried to implement were deemed "human rights," but they did not rest on any express conception of ethics, medical ethics, or morality, nor do they invoke the rights enumerated in the UN's 1948 Universal Declaration of Human Rights; instead, they were formulated as consumer rights to be implemented in response to accreditation standards. In a June 1970 speech, Wiley offers the following advice with respect to enforcing patients' rights. "If I have any message for the poor, it is this—do not expect the American Medical Association to move over and give you a role in health services voluntarily. Do not expect [medical] institutions ... to voluntarily give up their positions of power and comfort so that you can have a role in determining what is to be done, [all] of our experience in welfare rights, all of our experience in the struggle of black Americans for dignity and justice in this country is that the man does not move over; he has got to be pushed."[33] Thus, patients' rights were to be secured by political pressure and enforced by regulatory fiat, not as an assertion of moral respect or of medical ethics (e.g., physicians' ancient Hippocratic commitment to act to benefit their patients), nor on the basis of the human rights enumerated in the UN's Universal Declaration of Human Rights.[34] Even the nondiscrimination clause in the NWRO's formulation of patients' rights demanded is stated in socioeconomic terms. Thus, Right 11 is formulated as: "**No economic discrimination.** Degree or kind of service provided a patient may not be determined solely because of the patient's source of payment, social, economic, racial, or religious status."[35] To reiterate, these "rights" were comparable to other standards regulating hospital or other medical institutions—ethical or medical ethical ideals were not expressly invoked.

Wiley and the NWRO negotiated another official affirmation of rights, A Patient's Bill of Rights, with the American Hospital Association (AHA). On its website, the AHA describes itself as

> the national organization that represents and serves all types of hospitals, health care networks, and their patients and communities . . . [and that] through [its] representation and advocacy activities, ensures that members' perspectives and needs are heard and addressed in national health policy development, legislative and regulatory debates, and judicial matters. [Its] advocacy efforts include the legislative and executive branches and include the legislative and regulatory arenas. Founded in 1898, the AHA [also] provides education for health care leaders and is a source of information on health care issues and trends.[36]

As self-described, the AHA is a lobbying organization for hospitals that also sets self-regulatory standards for them. One way the AHA strives to improve the quality of hospital care is by providing model mission statements stipulating goals and visions for member hospitals. In 1969, as the AHA began to develop a new mission statement for hospitals, Wiley and the NWRO engaged with the AHA to incorporate a statement of patients' rights into the hospital regulatory environment. After negotiation, the NWRO managed to get some of its 1970 demands stipulated in a 1972 public document that the AHA Board of Trustees "affirmed" as "A Patient's Bill of Rights" (see appendix 5).

"Affirm" is, as the British say, a "squirrely" word suggesting a less than robust commitment. This said, the first right that the AHA affirmed is, word for word, the same right Wiley and the NWRO demanded from the Joint Commission three years earlier: "The patient has a right to considerate and respectful care." In Right 10, the AHA recommended a way of alleviating the "interminable" waits inflicted on charity patients. It states that "the patient has the right to know in advance what appointment times and physicians are available and when."[37] The practice of forcing the underclass to stand and wait as respectably employed (predominately white) middle-class people are afforded advance appointments structurally reinforced social class differences and, in the American South, racial caste differences. The seemingly innocuous innovation in AHA Right 10, if implemented, would alleviate the endless waiting periods that frittered away time from the lives of African Americans, disabled people, the elderly, and single mothers, as if it—or they—were of little or no value. Right 10 implicitly recognizes former charity patients not as "the deserving poor" but rather as having the same status, at least in the context of health care, as employed white people. It could, therefore, implement a foundational ideal expressed in a famous line in the earliest known Western formulation of medical ethics, the Hippocratic Oath: "Into as many houses as [physicians] may enter, [physicians] will go for the benefit of the ill . . . both of the free and of the slaves."[38] Hospitals are the physicians' house, and in its straightforward simplicity, Right 10 reaffirms the ideal that physicians should use their skills to benefit the sick, irrespective of whether the sick are slave or free, black or white, female or male, poor or rich, disabled or abled, old or young, unwed or wed, gay or heterosexual, transgender or cisgender. The AHA's Patient's Bill or Rights also affirms NWRO Rights 2 and 3 with minor changes in wording. In addition to the right to considerate and respectful care, these state a patient's (or a patient's surrogate's) right to receive the information needed to give informed consent.

The remaining twenty-three of the NWRO's twenty-six demands relate to issues of institutional governance and are not replicated in the AHA's rights statement. They deal with such matters as accreditation, community representation, provision of a patient ombudsman/patient advocates, catchment areas, medical staff privileges, medical records and data, community-based needs assessment, nondiscrimination, utilization review, antidumping standards, laboratory services, patient privacy, external review, triage, communication with patients, surgical education, diagnostic tests, admission standards, free care for people lacking health insurance, emergency department acceptance of ambulance patients, inspection reports, and nursing responses. A few of these are indirectly echoed in the AHA's rights statement, including a patient's right to privacy and confidentiality (AHA Rights 5 and 6).

In 1651, the shrewdly cynical English philosopher Thomas Hobbes (1588–1679) observed about "covenants"—a word Hobbes used to designate commitments involving future actions in contrast to relatively concurrent transactions—"Covenants without the sword, are but words and of no strength to secure a man at all."[39] Hobbes's dictum about covenants without the sword applies to the AHA's affirmation of patients' rights. Although the AHA *affirmed* these rights, "in the expectation that they will be supported by . . . hospital[s]," in the document's last paragraph, the AHA offers hospitals an excuse for not implementing them. "No catalogue of rights can guarantee a patient the kind of treatment he has a right to expect," the AHA proclaims. Why not? Because, as the AHA explains condescendingly, "A hospital has many functions to perform, including the prevention and treatment of disease, the education of both health professionals and patients, and the conduct of clinical research."[40] In other words, hospitals' medical, educational, and research commitments are more important than honoring patients' rights: in the proverbial nutshell, scientistic paternalism prevails over patients' rights. Actions not taken speak as loudly as words. But words do matter, especially if, to paraphrase Martin Luther (1483–1546), the AHA's affirmation of a "Statement on A Patient's Bill of Rights" affirms "simple aims . . . but their giving is of such a character, that the right hand gives but the left hand takes."[41] Except for providing patients with brochures listing their rights, with some notable exceptions, hospitals did not respect the rights affirmed by the AHA.

A Feminist Endorsement of A Patient's Bill of Rights

Despite the AHA's equivocal affirmation of patients' rights, a resurgent feminist movement thought that A Patient's Bill of Rights was worth disseminating, and it was included in their 1971 feminist self-help manifesto and home health guide. Some sense of the frustration motivating women to disseminate A Patient's Bill of Rights is captured by this speech delivered at the 1972 national meeting of the American Psychological Association (APA, founded 1892) by Carol Downer (1933–), a founder of the women's self-help health care movement.

> In what has been described as "rape of the pelvis," our uteri and ovaries are removed often needlessly. Our breasts and all supporting muscular tissue are carved out brutally in radical mastectomy. Abortion and preventive birth control methods are denied us unless we are a certain age, or married or perhaps they are denied completely. Hospital committees decide whether or not we can have our tubes tied. Unless our uterus has "done its duty," we are often denied. We give birth in hospitals run for convenience of the staff. We're drugged, strapped, cut, ignored, enemaed, probed, shaved—all in the name of superior care. How can we rescue ourselves from this dilemma that male supremacy has landed us in? The solution is simple. We women must take women's medicine back into our own capable hands.[42]

Similar sentiments course through the first edition of the Boston Women's Health Book Collective's bestselling book, *Our Bodies, Our Selves: A Book by and for Women* (1971). Prefacing its first edition is a chapter on "Women, Medicine and Capitalism" that opens with a quotation from the refugee German-Jewish Frankfurt school philosopher, Herbert Marcuse (1898–1979), who taught at Brandeis, a suburban Boston university (founded in 1948), from 1954 to 1965.

> Marcuse says that "health is a state defined by an elite."[43] A year ago [i.e., before engaging in this project] few of us understood that statement. . . . We believed that all people want to be healthy . . . But [we] have found that health is one more example of the many problems we as people, especially as women, face in this society. We have not had power to determine medical priorities; they are

determined by the corporate medical industry (including drug companies, Blue Cross, the AMA, and other profit-making groups) and academic research.

... We have decided that health can no longer be defined by an elite group of white, upper middleclass men. It must be defined by us, the women who need the most health care, in a way that meets the needs of all our sisters and brothers—poor, black, brown, red, yellow, and pink.

Perhaps the most obvious indication of this ideology is the way that doctors treat us as women patients. We are considered stupid, mindless creatures, unable to follow instructions (known as "orders"). While men patients may also be treated this way, we fare worse because women are thought to be incapable of understanding or dealing with our own situation. ...

The doctor preserves his expertise and powers for himself. He controls the knowledge and thereby controls the patient. He maintains his status in a number of ways. ... including dressing in whites. ... Another much more important way doctors set themselves off from other people is through their language. Pseudo-scientific jargon is the immense wall which doctors have built around their feudal (private) property, i.e., around that body of information, experience, etc. which they consider as medical knowledge (e.g., [saying] "epistaxis" for nosebleed, "thrombosis," for blood clot, "scleral icterus" for yellow eyeballs, etc.). ... Doctors insulate themselves from the rest of society by making the education process (indoctrination) so long, tedious, and grueling that the public has come to believe that one must be super human to survive it. (Actually, it is like one long fraternity "rush" after which you've made it and can do what you like.). ... Thus, a small medical elite preserves his own position through mystification, buttressed by symbolic dress, language, and education.[44]

Downer and the Boston Women's Health Book Collective offer sharp critiques of the American medicine of the 1970s. Yet, as Downer came to recognize, although critiques are a necessary step toward moral change, in themselves, they are insufficient to overthrow male-dominated scientistic paternalism. As Kuhn discovered in analyzing scientific revolutions, a community will rationalize and defend problematic practices and incoherent paradigms—such as an astronomy that yields incorrect calendars or a health care system that fails to benefit most of the people in the communities it serves—unless the system is reformed to effectively rectify its inadequacies, or unless someone invents an alternative paradigm that seems, in the eyes of those holding the reins of power, to resolve vexatious problems apparently unresolvable in terms of the established paradigm.

By the mid-1970s, some of the basic conditions for moral reform or moral revolution in biomedical practices were in place. (i) A dissident intelligentsia of civil rights/welfare rights activists, feminists, reform-minded health care professionals, medical educators, moral theologians, and philosophers were coalescing around a critique of the established physician/researcher scientistic paternalist paradigm. Some, like Beecher and Clegg, both morally sighted by Pappworth's exposés, came to recognize (ii) anomalous aspects of practices justified in the name of scientistic paternalism. More generally, dissidents came to see that some practitioners were inverting their professions' publicly proclaimed ideals, putting their career interests ahead of their patients' health and safety. Thus, Beecher and Clegg championed the reforms specified in the Declaration of Helsinki and encouraged stronger patient/subject protections. Clegg also sought to ban experiments in total institutions, such as orphanages. Beecher advocated prior peer review, monitoring by journal editors, and fully informed consent by subjects, yet he still objected to proposals for external regulatory oversight believing that reform should be undertaken and enforced by medical professionals themselves.

The Patients' Representatives/Ombudsman Reform Movement

If we characterize reformers as people who accept an established system but who nonetheless seek to rectify its problematic aspects and deleterious effects, then unlike Downer and other feminist health care moral revolutionaries, Beecher, Clegg, and Wiley were moral reformers. Admittedly, Wiley believed that "in the struggle of black Americans for dignity and justice in this country ... the man does not move over; he has got to be pushed."[45] Yet, unlike the feminists in the Boston Women's Health Book Collective, Wiley was not challenging physicians' authority; instead, he was pushing "the [medical] man" to become a better Hippocratic doctor. Wiley never challenged physicians' authority or their expertise, or even their paternalistic scientism. His objective was to modify an established paradigm to address class and racial inequities: to have physicians offer people receiving Medicaid and Medicare the same level of respect and care accorded other patients. In contrast, the feminist self-help critics were attempting to deconstruct the medical mystique, to democratize medicine, to offer patients an effective voice not only in decisions about their care but to redefine what counts as "health care."

Establishments prefer paradigm-preserving reformers to paradigm-displacing revolutionaries. Thus, Beecher was ultimately elevated to the status of a hero of moral reform, and Wiley could work with the medical establishment because he never challenged the authority of that establishment or the dominance of the scientistic paternalist paradigm. Revolutionaries, in contrast, attempt to displace dominant paradigms, but they usually succeed only after reforms prove insufficient to meet pressing needs of the moment. At that point, some of those implementing or justifying the dominant paradigm abandon it to embrace a revolutionary alternative. Since the biomedical community controls the discourses of medical knowledge and practices, a revolutionary alternative paradigm of medical morality could only displace the ethics of scientistic paternalism if the biomedical community itself came to accept this alternative. So, although Downer proclaimed that "we women must take women's medicine back into our own capable hands,"[46] and although the Boston Women's Health Book Collective declared "that health can no longer be defined by an elite group of white, upper middleclass men. It must be defined by us, the women who need the most health care, in a way that meets the needs of all our sisters and brothers." Yet as they eventually realized they "have not had power to determine medical priorities; these are determined by the corporate medical industry," including the AMA and other health care professional organizations, Blue Cross and other insurance companies, the pharmaceutical and other profit-making company groups, and academic researchers.[47]

In response to the crisis of distrust roiling American medicine in the 1970s, some in the medical community attempted to implement one of the NWRO's reforms by creating a system of patient advocate/representatives/ombudsmen to voice and defend patients' rights. The modern idea of an ombudsman seems to have originated in Norway, where communities turned to respected neutral men to mediate conflicts. As defined in the NWRO's demands, ombudsmen were to champion patients' rights by "resolv[ing] patient complaints about billing, quality of care, or conditions under which care is rendered." One of the oldest surviving implementations of the ombudsman/patient representative movement was initiated at The Mount Sinai Hospital in New York City in 1966.[48] The hospital's website states that it was founded 1855 as "The Jews Hospital," but it became nondenominational in 1866 to care for Civil War casualties of all religions. Since its original name no longer described its patients, in 1872, the hospital was renamed "The Mount Sinai Hospital."[49] About a decade shy of a century later, in 1963, the hospital founded a medical school becoming an educational and research center located at the border of Hispanic East Harlem, Central Park, and the Upper East Side—at that time, it was located between one of the poorest and the wealthiest ZIP codes in the United States.[50]

According to Ruth Ravich (1921–2001), founder of the hospital's patient representative program, "The concept of patient advocacy began to take shape during the 1960s as an outgrowth of the civil rights and consumer movements. At that time, it was believed that ethnic minorities and the 'poor' required special help in accessing and negotiating the healthcare system." As implemented at The Mount Sinai Hospital, a patient representative service "was (1) established to bring people and services into closer contact" and (2) to "reduce alienation" of African American and Hispanic patients and their families who found themselves receiving health care from primarily white middle-class health care professionals. The program later became a model for other programs and "was a major force in the organization of the [National] Society of Patient Representation and Consumer Affairs of the American Hospital Association" in the 1970s.[51]

By 1993, the program had evolved into a "quality improvement initiative."[52] Ravich describes the role of ombudsmen/patient representatives in this initiative as

> essentially problem solvers—for individuals, departments, and systems. The issues they address range from diet to dying, from conflict arising from interpersonal communication to major organizational flaws. They interpret policies and procedures to patients, families, and visitors and help guide them through the elaborate maze of the hospital world. They act as liaison among patients, families, and staff in cases of dissatisfaction with services and potentially litigious situations. They support staff by intervening in response to difficult and complicated patient-family problems and concerns and provide linkages that bridge the inevitable gaps within the system.
>
> ... The roles of the patient representative as ally and advocate for the patient are performed in a collegial context, using persuasion and negotiation rather than confrontation and demand. By considering the needs of the patient and family, as well as the concerns of the staff and the capacities of the institution, a workable solution can usually be found, even in complex situations.
>
> A critical element of the patient representatives' job is to facilitate communication between patient and staff. . . . Resolutions range from a room change or clarification of dietary orders to calling a physician who, according to an irate patient or family, has not visited for several days. If a patient's health care team—doctors, nurses, social workers—is making decisions with which the patient or family disagrees, that disagreement may stem from a misunderstanding of the team's rationale or because no one has elicited, the patient's perception of his or her needs.[53]

Note that the title of Ravich's article is "Quality Improvement Tool[s]": that is, the program focuses on facilitating better communication and better responses to patients' or families' felt needs. Thus, the "patient representative [serves] as ally and advocate for the patient [in] a collegial context, using persuasion and negotiation rather than confrontation and demand." As such, patient representatives are unlikely to challenge a health care team or a medical facility's discriminatory practices, or to protest drafting patients for unconsented experiments. Recognizing this limitation in 1974, founding bioethicist George Annas (1945–) and Joseph M. Healey Jr. (1947–1993), both of Boston University School of Law (founded 1872), proposed a stronger model of patient advocacy. Their analysis rested on two fundamental propositions: "(1) the American medical consumer possesses certain interests, many of which may properly be described as 'rights,' that he [or she] does not automatically forfeit by entering into a relationship with a doctor or a health care facility; and (2) most doctors and health care facilities fail to recognize the existence of these interests and rights, fail to provide for their protection or assertion, and frequently limit their exercise without recourse for the patient."[54] To enforce patients' rights, Annas and Healy proposed a system of patient rights advocates, "individual[s] whose primary responsibility is to assist the patient in learning about, protecting, and asserting his or her rights within the health care context."[55]

Perhaps the closest anyone came to implementing this strong patient advocacy model was in 1975 when Arthur Fleming (1905–1996), US Commissioner on Aging, provided funding for states to create ombudsman programs under the Older Americans Act of 1965.[56] Many nurses volunteered for this role because the American Nursing Association (ANA, founded 1896) viewed patient advocacy as an aspect of nursing ethics.[57] Funding, however, was insufficient to support salaries beyond the grant period, so the ombudsman role was soon filled by state employees, by volunteers from the community, or by nonprofit organizations. Thirty years later, in 1995, these patient advocacy programs were reviewed by the Institute of Medicine (IOM, founded 1970, now part of the National Academies of Science, Engineering, and Medicine, which dates from 1863). The reviewers found that the program was riddled with unacknowledged conflicts of interest and loyalty "that arise from the structural location of . . . ombudsman programs in [or funded by] state offices. . . . [These] conflicts of interest work to the disadvantage of the vulnerable client."[58] More to the point, the IOM found that an ombudsman's obligation to "speak out against government laws, regulations, policies and actions . . . [was] antithetical to the hierarchical rules of government . . . [and that institutions] offer limited guidance with respect to conflicts of loyalty, commitment, and control."[59] Thus, fundamental conflicts of interest and loyalty are built into the role of patient advocate/ombudsman/representative because they are required to represent the interests of the patient rather than those of the interests of health care practitioners, health care institutions, or governmental bodies that fund their position and empower them to act.[60]

This appears to be one reason why the strong model of patient advocacy, of the sort envisioned by Annas and Healey, was unlikely to endure. Any role requiring incompatible commitments to both employers' and to patients' interests and rights is likely to prove problematic. Thus, surviving patient representative organizations tend to focus on activities self-described as "quality improvement," to eschew direct conflicts. This said, the most effective advocacy organizations are those organized by patient or familial or disease-based external advocacy groups because they have no conflicts of interest or loyalty and so can, and do, effectively challenge institutional and governmental practices by, for example, lobbying to increase patients' access to experimental treatments or drugs—in effect, they act on Wiley's tenet that "the [medical] man does not move over; he has got to be pushed."[61]

9
Reforming Modern Medical Ethics

We shall frequently have to notice the manner in which great discoveries thus stamp their impress upon the terms of a [field]; and, like great political revolutions, are recorded by the change of the current coin which has accompanied them. . . . The great changes which thus take place in history [are] the revolutions of the intellectual world.
—William Whewell, 1837[1]

A party of order or stability, and a party of progress or reform, are both necessary elements of a healthy state of political life, until the one or the other shall have enlarged its mental grasp as to be a party equally of order and progress, knowing and distinguishing what is fit to be preserved from what ought to be swept away.
—John Stuart Mill, 1859[2]

Dissident Intelligentsia: A Prelude to Moral Change

The Victorian Age (1837–1931), not unlike our own, was an era of conceptual, moral, and technological change. As Charles Dickens (1812–1870) famously described his era, "It was the best of times, it was the worst of times, it was the age of wisdom, it was the age of foolishness, it was the epoch of belief, it was the epoch of incredulity, it was the season of Light, it was the season of Darkness, it was the spring of hope, it was the winter of despair, we had everything before us, we had nothing before us . . . in short, . . . some of its noisiest authorities insisted on its being received, for good or for evil, in the superlative degree of comparison."[3] The preeminent English philosophers of the Victorian Age, William Whewell (1794–1866), master of Cambridge University's Trinity College, and his archrival, John Stuart Mill, agreed on few things, but, like Dickens, they recognized theirs as an era of superlative change. Moreover, both recognized that conceptual change was a harbinger of substantive change in governance, morality, and the sciences. Mill was particularly concerned with crises of instability and, in exploring them, became the philosopher who introduced the expression "moral revolution"—by which he meant a revolution in morality, not a social or political revolution conducted morally—into the Anglophone philosophical lexicon. He did so in an 1833 review, "Observations on the French Revolution," in which he commented that "all political revolutions not effected by foreign conquest originate in moral revolutions."[4] The implication was that, as in the French political revolution—in which the French people rejected one form of governance, monarchy (the rule of one person), for an incompatible form of governance, republican democracy (governance by the people, without a monarch)—so too, in a prior moral revolution, the French community rejected an established inegalitarian conception of morality in which an aristocracy and a priesthood determined what counts as good or bad, right or wrong, moral or immoral, replacing it with an egalitarian conception of morality declared by populace, or their representatives, and based on an egalitarian notion that all men are born free and have equal rights.[5]

Like Mill, Whewell recognized moral change, but unlike Mill, he sought stability by positing empirically discernable laws of morality, comparable to the laws of physics. Yet Whewell also held it to be an epistemological truth that in morality, as in physics, the conceptual frameworks through which people view the moral laws limit their understanding of them, and so there can be, and have been, epistemological revolutions in morality, just as there have been epistemological revolutions in the biological and physical sciences.[6] His student, Charles Darwin, summed up Whewell's views on scientific laws with a quote from Whewell's 1834 *Bridgewater Treatise*, "*On Astronomy and General Physics*," that Darwin used as prefatory quotation in the six editions of *On the Origin of Species* that he personally edited. The quotation read, "But with regard to the material world, we can at least go so far as this—we can perceive that events are brought about not by insulated interpositions of Divine power, exerted in each particular case, but by the establishment of general laws."[7] Whewell believed that the same holds in morality: that laws governing morality are discoverable, but in morality, as in the sciences, the supposed truths and self-evident axioms of one era, often "recorded by some familiar maxim, or perhaps by some new word or phrase, which forms part of the current language of the [intellectual] world are subject to change ... [creating] the revolutions of the intellectual world."[8] He also held that revolutions in the intellectual world "involve the overthrow of received truths by those that were at first considered 'strange' but that later come to be accepted as 'almost self-evident axioms'"—conversely, truths once considered almost self-evident come to be viewed as not evident at all.[9]

Assembling and Organizing a Dissident Intelligentsia

Whewell's thoughts about scientific revolutions vanished from most later editions of *On the Origin of Species* and lay dormant until Thomas Kuhn rediscovered the concept of scientific revolutions.[10] Using the Copernican Revolution as his paradigm case, Kuhn compiled an inventory of the basic elements structuring a conceptual revolution in the sciences. If we assume, as Whewell did, that conceptual revolutions in morality are analogous to conceptual revolutions in the sciences, then a Kuhnian-type moral revolution could begin only after a dissident intelligentsia[11] noticed anomalies, something that did not make sense in the established paradigm. In the late 1960s and early 1970s, a dissident intelligentsia of medical educators noticed such anomalies, prompting them to form the Society for Health and Human Values (SHHV, 1969–1997). The SHHV's objective was "to identify explicitly the human values that are lacking or inadequately represented in the study and practice of medicine and to begin to remedy the deficit."[12] The words "lacking," "inadequate," and "deficit" implied a critique of the current state of medical education and medical practice. The SHHV also recognized the advent of issues generated by the morally disruptive "advances in medical science" and saw their "task as identifying these problems, [and] forming groups that will develop and assist in solving them, and in developing change in both professional attitudes and public awareness in relation to them."[13] The society met annually from 1972 to 1998, at which time it merged with other bioethics societies to form the American Society for Bioethics and the Humanities (ASBH, founded 1998). During the quarter century prior to the merger, the SHHV was a forum for faculty from academic medical centers involved in teaching medical ethics, medical history, and representations of health care and illness in various artforms (dance, film, literature, etc.).

During the same period, two disaffected Roman Catholic intelligentsia, the philosopher Daniel Callahan (1930–2019), a lapsed Catholic, and André Hellegers (1926–1979), a biologist-physician disillusioned with his faith, each organized bioethics institutes. Both men had been deeply committed Roman Catholics. From 1961 to 1968, Callahan served as editor of the lay Roman Catholic magazine *Commonweal* (founded 1924) and was "committed to the church, to [his] work in the church, to the world which Christ came to redeem"; he "could not imagine being anything but Catholic."[14] Hellegers served as scientific advisor to Pope John XXIII (1881–1963, pope 1958–1963), earning the nickname "The Pope's Biologist." Callahan and Hellegers both supported John XXIII's Second Ecumenical Council of the Vatican (1962–1965), more popularly known as "Vatican II." Citing new scientific information about contraception and fetal development, the majority on the commission, including Hellegers, recommended that the Church abandon its traditional condemnation of birth control. Their recommendation would have made a monumental change in Church doctrine had not John XXIII died during Vatican II.

In 1968, Pope John's conservative successor, Pope Paul VI (1897–1978, pope 1963–1978), published an encyclical letter *Humanae Vitae* (Human Life) reasserting traditional Catholic teachings that condemned birth control as unnatural and prohibited Catholics to use any form of artificial contraception.[15] Hellegers protested that as a "scientist [he was] struck by the absence of biological considerations in the entire encyclical [*Humanae Vitae*]. . . . [It] nowhere acknowledges that there might have been new biological facts of importance discovered since the [earlier] encyclical *Caste Connubi* [Chaste Wedlock, 1930]. . . . [It] is made clear that nothing . . . a scientist could possibly contribute in terms of scientific data could have any pertinence to the subject. . . . Such wording pronounces the scientific method of inquiry irrelevant to Roman Catholic[s]."[16] Hellegers would later remark that the Church "would someday repudiate [Pope Paul VI's decision] just as it had . . . official pronouncements condemning religious liberty and freedom of conscience."[17]

Callahan also protested against *Humanae Vitae*, resigning from his editorial post and joining with Reverend Bernard Haring (1912–1998) and Professor Charles Curran (1934–) in writing a rebuttal to the pope.[18] As Callahan and Hellegers came to appreciate in writing their critiques of *Humanae Vitae*, advances in biology and the life sciences could be morally disruptive in ways that required reconsideration of traditional ethical views. Independently, each began to consider bringing together groups to discuss issues related to values and the life sciences. Callahan's plans became concrete at a Christmas Eve party in 1968, where he mentioned to his next-door neighbor, a Jewish Columbia University psychiatry professor Willard Gaylin (1925–), that he was thinking about founding a "Center for the Study of Value and the Sciences of Man." Gaylin liked the idea, and in March 1969, they cofounded the Institute of Society, Ethics, and the Life Sciences. Initially bilocated in its founders' homes in Hastings-on-Hudson, the institute would later assume the name "The Hastings Center."[19] After working on these issues for a few years, Callahan also recognized the imperative of challenging the dominant scientistic paternalistic paradigm. As he told an interviewer in 1977, "Doctors want to make all the choices. Well, we're saying to them, 'no.' There are some public interests at stake here and some general principles you have to abide by. . . . You are playing in a public ballpark now . . . and you've got to live by certain standards . . . like it or not."[20]

Around the same time, Hellegers, who was then on the faculty of Georgetown University, a Jesuit college in Washington, DC (founded as Georgetown College in 1789), also came to appreciate the need for addressing issues arising at the interface of ethics and the life sciences. So, he applied to a Kennedy family foundation for funding for the Joseph and Rose Kennedy Center for the Study of Human Reproduction and Bioethics. His proposal was eventually funded, and later the new Center's name was shortened to the Kennedy Institute of Ethics (KIE). The KIE opened its doors in 1971 with several moral theologians as founding research scholars.

Influenced by their different geographical locations and institutional affiliations, or lack thereof, each institute developed its own distinct focus. Callahan hired as the Center's first full-time employee a pharmacist, Robert M. Veatch (1939–2020), who had been a Peace Corps volunteer in Nigeria and was a recent Harvard PhD (1970) whose doctoral thesis focused on medical ethics. As Veatch tells the story, he was the first person to enter the new "Center," a small space above a dentist's office. Within a few years, Veatch's public statements were being cited by feminists and other dissenting intelligentsia. In 1973, for example, feminists noted Veatch's condemnation of an experiment in which Mexican American women had been enrolled, some unknowingly receiving dummy "birth control" pills. One feminist wrote that "while ignored by the medical community . . . [the experiment was publicly criticized] by Robert Veatch of the Institute of Society Ethics and the Life Science [i.e., the Hastings Center]. . . . Veatch's criticism . . . was widely discussed in the women's health movement."[21]

As an independent secular institution located in a New York City suburb within commuting distance of the resource- and talent-rich Boston–New Haven–New York City corridor, the Hastings Center developed a topics focus, dividing its staff and fellows into four research groups: death and dying, behavior control (brain washing, psychosurgery), genetic engineering, and population control.[22] As chair of the Center's death and dying focus group, Veatch also served as an informal advisor to Joseph and Julia Quinlan in their efforts to assert a right to refuse mechanical respiratory support on behalf their twenty-two-year-old daughter Karen Ann Quinlan (1954–1985), who was in a permanent vegetative state—setting a precedent for bioethicists acting as advisors in legal cases.[23] The Center also sponsored conferences and workshops, inviting nearby luminaries, like

Henry Beecher, to participate. These events produced articles and reports that soon filled the pages of the Center's premier publication, the *Hastings Center Report* (founded 1971).

Geographical location and institutional affiliation also influenced the Kennedy Institute's objectives. Situated in a Jesuit educational institution, it originally looked to theologians and religiously oriented intellectuals as founding research fellows. One of its primary goals was to promote bioethics education. It also sought to promote in-depth theoretical research, and since it was located near a cornucopia of governmental and government-supported institutions, it sought to inform public policy and to develop a comprehensive information center for the nascent field of bioethics. To implement this last objective, Hellegers worked with moral theologian Leroy Walters (1940–) to start a research library. As Walters told the tale in an interview with sociologist Judith Swazey (1928–2020),

> André had the great gift of building on people's interests and strengths, and so within days of my coming [to the newly formed institute], we had agreed that a good research institute had to have a library to support the research. We bought an unfinished 3-foot-wide pine bookshelf and he and I traipsed over to the Med[ical] Center and bought some textbooks and came back and put them in the shelves.
>
> [Swazey] You didn't need a very big bookshelves then for the bioethics literature.
>
> [Walters] That's right. Now, we had started our library, and I thought, well let's break this list down into topics and have books and articles on each topic. Using the state of the art magnetic-card technology, we spliced in new references as new book and article items were published. By the fall of that first year in '71 we had a list of what we called a Core Ethics Library—30 pages or so of materials we had identified that we were in the process of collecting for our library. The list included the most important works in bioethics at that time. A few months into my work here I said, "You know, we need to have somebody who's responsible for responding to inquiries that come into the Institute and requests for copies of this bibliography." Andre agreed. . . . So sometime in the fall of 1971, I received the title, "Director of the Center for Bioethics" within the Kennedy Institute of Ethics.[24]

In this one-thing-leads-to-another series of events, the institute offhandedly began two seminal projects: its bioethics research library—which today proudly proclaims itself "The world's largest collection of bioethics resources"[25]—and the Bibliography of Bioethics (1975–2009). Also, by sheer happenstance, the title "director" was bestowed on Walters.

A third seminal project, the *Encyclopedia of Bioethics*, began in a similarly informal manner. Warren T. Reich had taught moral theology at

Catholic University in Washington, DC (founded 1887 by Roman Catholic bishops) but was dismissed after publicly dissenting from *Humanae Vitae*. Soon thereafter, Hellegers recruited him as a founding research scholar in the Kennedy Institute of Ethics. Reich recounts thinking as he began his new job that "something very new is happening here. We are using a new word 'bioethics,' which is clearly a new field. . . . I thought: 'What bioethics will need is a reference work, a basic standard work that will pull together what we know from the sciences, ethics, and other fields of learning that can be used for defining the ingredients of the field.'" Hellegers supported Reich's idea, but Reich had been thinking of creating a *Handbuch*, a multivolume compendium of expert knowledge common in German-speaking academic cultures. Yet when Reich discussed his idea with Hellegers, he observed that what Reich was really describing was known in English as an "encyclopedia." Reich responded, "Yes, I think so, but I was hesitant to call it that." Hellegers replied, "You *should* call it that, that's what it is."[26] Reich applied to, and eventually received funding from, the National Endowment for the Humanities, and the first edition of *The Encyclopedia of Bioethics* was published in 1978.

Unbeknownst to Reich or Hellegers, a variant of the word "bioethics" had been used in a 1927 article by German theologian Fritz Jahr (1895–1953), who envisioned a *bio-ethik*: a quasi-Buddhist morality that extended Immanuel Kant's categorical imperative to apply to all living beings, "So that the rule for our actions may be the bio-ethical demand: Respect every living being on principle as a goal in itself and treat it, if possible, as such!"[27] Some four-and-a-half decades later, in 1971, the word "bioethics" also appeared in the title of an article and a book by research oncologist Van Rensselaer Potter (1911–2001). In these publications, Potter defined bioethics as a "science of survival based on biology and wisdom in action," that is, a fusion of biological knowledge and human values to save the future of the biosphere and its inhabitants.[28] Apparently unaware of these earlier uses of some variant of the word "bioethics," in 1973, Callahan wrote an article, "Bioethics as a Discipline," that appeared in the first issue of a short-lived publication, *Hastings Center Studies*. In the article, Callahan describes a new field focused, not on extending the categorical imperative to all living creatures and not on the need for wisdom to ensure survival of the biosphere, but as "an interdisciplinary field in which the purely 'ethical' dimensions neither can nor should be factored out without remainder from the legal, political, psychological and social dimensions."[29] Citing Callahan's article, in 1974, the Library of Congress made "bioethics" a subject heading, thereby canonizing the word as denoting a separate biomedical discipline in which ethics was an ineliminable element. As the influential Austro-English philosopher Ludwig Wittgenstein (1889–1951) once remarked, "The meaning of a word is its use in a language."[30] As we use the word "bioethics" today, it designates the field of study pioneered by Callahan, Gaylin and Hellegers, not the field envisioned by Jahr or Potter.

Another group of bioethics intelligentsia was formed by the Council of Philosophical Studies, which sponsored a 1974 Summer Institute on Morals and Medicine. Philosopher Samuel Gorovitz (1938–) directed the summer institute at Haverford College with the goal of training philosophers to participate in the emerging field of bioethics: among the attendees were Tom Beauchamp, Tris Engelhardt, Bob Veatch, and the author of this book.[31] Thus, by the end of the 1970s, bioethics had emerged as a new field of scholarship and research supported by two research institutes, a bibliography, an encyclopedia, and several journals. By virtue of the Society for Health and Human Values and Gorovitz's Summer Institute, academics began to introduce the subject at law schools, liberal arts colleges, and research universities, as well as in academic medical centers and medical schools. Around the same time, occupational space for the new field as a health care subspecialty had been created by Kelsey's insistence that the FDA enforce the Kefauver–Harris amendment. Further demand for experts who could serve on, or staff IRBs was created about a decade and a half later, after the *Belmont Report* recommended IRBs for all US government-funded research on humans. Serendipitously and inadvertently, a judge's misuse of the expression "hospital ethics committee" (HEC) in the *Quinlan* decision (discussed in the next chapter) had a similar impact on academic medical centers and eventually on hospitals, nursing homes, and other health care facilities throughout the United States.

These new committees, HECs and IRBs, required staffing by people who could be called "ethicists." This, in turn, created a cliental for various forms of credentialing, which led to certificate programs at bioethics institutes and later to fellowship and master's programs at academic institutions. Even as they sought credentialing, bioethicists tried to reform or transform medical practices and to address various ethical problems created by morally disruptive technologies. In practice, however, bioethicists' proposals would have had little if any impact if they could not communicate their ideas to biomedical researchers, health care practitioners, regulators, powerbrokers, and influencers. This required bioethicists to communicate with the rest of the biomedical community. But first, bioethicists had to learn how to communicate with each other.

A Translation Problem Leads to a Lingua Franca and a Shared Secular Ethics

Initially, the coalescing amalgams of bioethicist groups differed about the nature of the issues confronting biomedicine. Wiley, speaking for the NWRO, defined it as the limited access to health care by "people of the ghettos and barrios of our country" who have been required to wait "interminable lengths of time for the most meager and inadequate and most stingily given out medical treatment."[32] For Downer, the problem was rescuing women from the "dilemma that male supremacy has landed us in."[33] The Boston Women's Health Book Collective concurred, defining the problem as wresting health and health care from the hands of an elite group of white, upper-middle-class men. The medical educators of the SHHV defined the issue as a deficit of "human values . . . lacking or inadequately represented in the study and practice of medicine."[34] Callahan, on the other hand, spoke to the need to recognize ethical issues prompted by advances in the life sciences that were part of a "biological revolution."[35]

Complicating matters further, the two most prominent bioethics institutes had been founded by Roman Catholic dissidents, Hellegers and Callahan (in partnership Professor Willard Gaylin of Columbia University, who was Jewish). Like Beecher—and most other Americans of that period—they looked to religious leaders or theologians for moral guidance.[36] Quite naturally, therefore, Hellegers initially populated the Kennedy Institute with moral theologians; even the first philosopher associated with the Institute, Tom Beauchamp, had a theology degree from Yale. Consequently, albeit for different reasons, philosophers like Beauchamp and physicians like Beecher were comfortable with the way moral theologians thought, spoke, and wrote. As Callahan remarks, "When [he] first became interested in bioethics in the mid-1960s, the only resources were theological or those drawn from within the traditions of medicine, themselves heavily shaped by religion. One way [or another], that situation was congenial enough."[37]

Moral theologians were naturally comfortable speaking the language of their discipline. As Methodist theologian Paul Ramsey explained in the preface to his influential 1970 book, *The Patient as Person*, he "shall not be embarrassed to use as an interpretative principle the Biblical norm of *fidelity to covenant*, with the meaning it gives to *righteousness* between man and man."[38] These words have profound associations to fellow Methodists, like Beecher and Veatch; however, those untutored in the language of Methodism, or from different faith traditions, might not comprehend why "covenant" and "righteousness" were relevant to experiments in nature involving "Negro syphilitics" or live cancer cell implants in posthysterectomy patients and dying Jews. So, moral theologians had to translate theological moral concepts into language that health care practitioners, investigators, and government officials could understand. As Roman Catholic moral theologian M. Therese Lysaught (1963–) observes, to communicate to a secular audience, theologians, like Ramsey, had to translate "'the scriptural narrative of God's covenant with . . . all of humanity' . . . into the secular (Kantian) concept of 'respect for persons.'"[39]

Complicating matters further, as newly founded commissions, institutes, and societies tried to encourage cross-disciplinary conversations between health care practitioners, moral theologians, lawyers, philosophers, and policy makers, communication across linguistic siloes proved challenging. The emerging field was confronted by frustrating Tower of Babel issues that sociologist John Evans dubbed a "translation problem." In an interview conducted by LeRoy Walters, moral theologian Karen Lebacqz (1945–), a commissioner on the National Commission for the Protection of Human Subjects, described the commissioners' difficulties in communicating with each other:

> I have several early memories that are very strong. One was how difficult it was for us to talk to each other in those first meetings. We were given four months in which to come to a conclusion on what has to be one of the most difficult topics ever. . . . I remember that it took us almost four months to develop a common language so that when the scientist talked, the rest of us understood what they were saying. And, when the lawyers, talked the ethicist understood what they were saying. And I still remember that Al Jonsen and I as the ethicists on the Commission would say something on the order of, "that's not a morally relevant difference." And, after about three months of this, I forget who it was, but I think maybe one of the scientists, finally thumped his fist down on the table and said, you keep using this phrase, "morally relevant," and I don't know what it means.
>
> So, just finding a common language and being able to understand each other at the beginning was very difficult. We were diverse in terms of our academic fields. While most of us were academicians, and in that sense, maybe not very diverse, we were certainly diverse politically and in terms of our faith, traditions. . . .
>
> So, trying to come to common ground was not easy in those early months. I do remember one of my strongest memories of the Commissions work overall, was how quickly Pat King, who was one of our lawyers on the Commission, Pat King and I would hear a comment and both of our hands would go up at the same time. . . . And, so often we were trying to make the same point in spite of

the fact that her legal language and my language coming out of the field of ethics were slightly different. So, I really felt early on that I had a good colleague in Pat King. That she understood what I was trying to say, and that if I had difficulty getting others to hear it, she was able to step in and say it again in a different language that maybe others could really appreciate it. So, I have memories of both the difficulty of coming to common ground and the difficulty of talking with each other.[40]

Eventually, this communication problem was resolved through development of a bioethical pidgin: a simplified form of language used for communication between people who do not share a common language. As founding bioethicist H. Tristram Engelhardt Jr., MD, PhD (1941–2018), observed, this emerging lingua franca also involved creation of

a secular ethic. Initially, individuals working from within particular religious traditions held the center of bioethical discussions. However, this focus was replaced by analyses that span traditions, including particular secular traditions. As a result, a special tradition that attempts to frame answers in terms of no particular tradition, but rather in ways open to rational individuals, has emerged. . . . Bioethics is developing a lingua franca of a world concerned with health care, but not possessing a common ethical viewpoint.[41]

LeRoy Walters, a moral theologian himself, came to a similar conclusion: that whatever one's religious background, a common bioethics lingua franca was needed to express a common secular ethics. Thus, in response to sociologist Judith Swazey's query, "What role is there for the different faith traditions, for moral theology?" Walters replied,

I'll answer in a couple of ways. I think that we can find a lot of common ground without resort to religious discussion or religious arguments if there's . . . agreement on a very fundamental thesis: that human beings are very worthwhile creatures and deserving of respect. If all discussion partners agree on that fundamental belief, then one can go a long way in finding a substantial measure of agreement among people from widely divergent religious backgrounds and traditions.

. . . At the same time, the religious beliefs of a particular tradition, I think, are generally a poor basis for mandatory laws, and so I think that a humanistic, non-religious approach has to be the basis for public policy rather than the views of any particular religious group. That approach works most of the time. For instance, for research involving all human subjects that are already born, it works just fine, and people from a variety of religious and non-religious traditions can agree. We can all look at Tuskegee, we can all look at the Nazi medical experiments . . . and say that's just outside the pale. We condemn those activities; there was no consent, there was no respect.[42]

Not everyone was pleased with moral theologians, like Walters, who acceded so readily to secular discourse and nontheological ethics as a basis of bioethical reflection. James Gustafson (1925–2021), an influential professor of Christian ethics at Yale Divinity School, decried the fact that Christian theologians were calling themselves "ethicists" when discussing moral issues in biomedicine. Such locutions, he argued, render opaque the relation between their moral discourse and their specific theological or religious outlook. "An ethicist," Gustafson observed disparagingly, "is a former theologian who does not have the professional credentials of a moral philosopher."[43]

As it happens, most of the philosophers entering the nascent field of bioethics in the 1970s also lacked the "credentials of a moral philosopher." Irrespective of whether they studied Continental European philosophers, like Heidegger or Husserl, or were reading works in the analytic philosophical tradition by Ayer, Moore, or Quine, Americans studying philosophy in the 1950s and 1960s seldom encountered serious works on practical or applied ethics. Some may have taken a course on metaethics, but these courses focused on the epistemological, metaphysical, or phenomenological foundations of morality; they seldom addressed practical issues of integrity, exploitation, or oppression. More to the point, except for the occasional anecdote, their professors typically disclaimed expertise on real-life issues. To quote from the bestselling 1954 book, *Ethics*, by Oxford philosopher Patrick Nowell-Smith (1914–2006), "Moral philosophy is a practical science; its aim is to answer questions in the form 'What shall I do?' But no general answer can be given to this type of question. . . . The questions 'What shall I do?' and 'What moral principles shall I adopt?' must be answered by each man for himself; that at least is part of the connotation of the word 'moral.'"[44] Despite its claim to being "practical," this (male chauvinist) approach would have been useless to Boston Women's Healthcare Book Collective, or to Downer, or to Wiley, and it was certainly useless to Callahan (who had studied philosophy at Harvard), Gaylin, Hellegers, or any investigator or physician probing the morality of experiments or asking whether "a patient with an artificially sustained beating heart, but no detectable brain function, is living or dead?" Their questions were precisely of the form, "What should we do?" and "What moral principles should we adopt?" If the authors of major books on moral philosophy would not answer these questions, health care practitioners and scientists—and philosophers working with them—had no use for moral philosophy as it was then understood.[45]

In an interview with sociologist Renée Fox (1928–2020), Ruth Macklin (1938–), one of the first female philosophers to enter the nascent field of bioethics, remarked that when she was an undergraduate at Cornell University, studying ethics was a low-status endeavor:

> Ethics was so far down the line, down the hierarchy, the only one that was lower was aesthetics. . . . What happened in 1968 and 1969 was the civil rights movement, the anti-war movement, the focus on people's rights, the focus on the oppressed.
>
> [Later, when she was] teaching one of the first courses for philosophy . . . that used an anthology [that] had sections in this book on abortion, on capital punishment . . . and euthanasia. . . . This anthology [be]came for me . . . transformation because the appearance of that kind of teaching material, for the first time, brought my personal values in conjunction with my professional life. [And she began] to take moral philosophy seriously.[46]

The convergence of these forces—moral theologians searching for a more secular language; humanist health care professionals seeking to emphasize care, compassion, and communication;[47] philosophers' questing for a practical ethics applicable to current moral issues; and physicians and scientists developing concerns with the moral issues stirred by morally disruptive techniques and technologies (such as dialysis machines, closed chest cardiac resuscitation, and ventilators)—created a need for ethical expertise on HECs and IRBs that culminated in the production of a lingua franca for professionals concerned with health care ethics but not possessing any common training, methodology, or ethical viewpoint. Eventually, it also prompted them to develop new moral precepts foundational to a new version of medical ethics that they called "bioethics."[48]

As Public Trust in Doctors Erodes: The Roads to Reform or Revolution Open

Revolutions, whether political, scientific, or moral, typically occur after a period during which reforms prove inadequate to address current controversies, thereby eroding opinion leaders' and influencers' support for an established paradigm and creating the conceptual space for them to consider alternatives.[49] As it happened, media attention to scandals revealed by whistleblowing moralists had the effect of prying open such a conceptual space. Thus it is significant that discussions of Southam's cancer experiments were no longer confined to professional publications, like *Science*; they were spilling into publications like the *New York Times*, which ran a half-dozen articles on the Southam's experiments between January 21, 1964, and July 8, 1964.[50] The next year, the *Los Angeles Times*, the *New*

York Times, the *Wall Street Journal*, and various local papers and popular magazines covered Beecher's Brook Lodge exposé.[51] The *New York Times* and the *Washington Post* also reported on Beecher's "Ethics and Clinical Research" exposé in the *New England Journal of Medicine*.[52]

Example 16 in Beecher's article involved "artificial induction of hepatitis ... in an institution for mentally defective children in which a mild form of hepatitis was endemic. The parents gave consent for the intramuscular injection or oral administration of the virus, but nothing is said regarding what was told them concerning the appreciable hazards involved." As in Beecher's Southam index case, this meant that in seeking consent, researchers misrepresented a significant factor that, if revealed, would likely have dissuaded many "volunteers." Moreover, as Beecher pointed out, the researchers were violating a Helsinki principle, which stated that "'under no circumstances is a doctor permitted to do anything which would weaken the physical or mental resistance of a human being except from strictly therapeutic or prophylactic indications imposed in the interest of the patient.' There is no right to risk an injury to one person for the benefit of others."[53]

The institution Beecher described was later publicly identified as the New York State School at Willowbrook, Staten Island (a borough of New York City). The lead investigator was a New York University (NYU, founded 1831) medical educator, Dr. Saul Krugman (1911–1995). The case generated controversy within the medical community, and Krugman found support from eminent figures like Dr. Franz J. Ingelfinger (1910–1980), Garland's successor as editor of the *New England Journal of Medicine* (1966–1976), who proclaimed that the children were better off being cared for as subjects of an experiment "under the guidance of Dr. Krugman [than others less able to provide] good medical care."[54] Beecher dismissed this as poppycock, noting a series of discrepancies in Krugman's attempts to justify his experiments. For example, Beecher found that the published data were inconsistent with Krugman's claims about the inevitability of hepatitis infections at Willowbrook. Beecher also pointed out that the consent forms Krugman used failed to mention that hepatitis's long-term effects were damaging to people's liver. Furthermore, he observed that the tactics used to solicit parental consent to enroll their minor children into the Willowbrook hepatitis experiments seemed coercive (i.e., parents were led to believe that if they did not permit their children to serve as experimental subjects, their children would be denied admission to the school[55]). Additionally, Beecher observed that Krugman's claim that the experiment complied with the 1962 draft version of the Declaration of Helsinki was patently false because that version of the Declaration (the version published by Clegg) forbade experiments in total institutions, like Willowbrook. Beecher also dismissed Krugman's claim that the experiments were approved by an NYU ethics committee because, when the experiments began, "no such committee then existed."[56]

Ramsey also criticized the Willowbrook experiments. He focused on the absence of "any attempt to enlist the adult personnel of the institutions . . . in these studies." He observed further that "nothing requires that major research into the natural history of hepatitis be undertaken in children. Experiments have been carried out in the military and with prisoners as subjects. There have been fatalities from the experiments." He goes on to comment that "there would have been no question of the understanding of consent [had it] been given by the adult personnel at Willowbrook."[57] Ramsey also notes that since gamma globulin had been established as a known effective preventative and treatment for hepatitis since the 1950s,[58] the goal of improving the use of gamma globulin "was a desirable goal; but it does not seem to warrant withholding gamma globulin [as a] controlled trial, to show that one procedure is as likely to succeed as another."[59] Finally, Ramsey asks the more fundamental question: "If this had been an orphanage for normal children or a floor of private patients, instead of a school for mentally defective children, one wonders whether the doctors would so readily have accepted the hepatitis as a 'natural occurrence' and even as an opportunity to study."[60] As bioethicist Arthur Caplan later observed, "The Willowbrook studies were a turning point in how we thought about medical experiments on retarded children. . . . Children inoculated with hepatitis virus had no chance to benefit from the procedure—only the chance to be harmed."[61]

Kennedy Hearings: 1974 National Research Act Creates a Bioethics Commission

As media reports about Willowbrook and similar experiments eroded public trust in physicians and hospitals, they also raised questions in the minds of funders. At the same time, civil rights issues were attracting media, political, and public attention. So, once the AP's Jean Heller reported out Buxtun's claim that the USPHS, an agency of the US government, had deprived some 400 black men of treatment for syphilis for four decades, major newspapers, like the *New York Times*, headlined the article on the front page. These and similar headlines prompted Senator Edward (Ted) Kennedy (1932–2009, senator 1962–2009) to convene committee hearings on the conduct of research on humans. Buxtun, Beecher, and newly minted bioethicists like Jonsen and Veatch testified before the committee, whose findings culminated in the National Research Act of 1974. This act, in turn, mandated the establishment of America's first bioethics commission, the National Commission for the Protection of Human Subjects of Biomedical and Behavioral Research (1974–1978). The Commission was mandated to recommend public policy on government-funded research on human subjects. It issued eleven reports: five addressed research on human populations vulnerable to exploitation (as in the Tuskegee syphilis experiment). Other reports addressed fetal research (1975); research on prisoners (1976); research involving children, as in the Willowbrook case (1977); and psychosurgery, as in the *Kaimowitz* case (1977).

These reports were supplemented by a 1978 report, *Ethical Guidelines for the Delivery of Health Services by DHEW*, that focused on dispelling "the traditional aura that since 'charity' patients are receiving their [medical] care, free [of charge], they should not ask too many questions, or they are not sufficiently intelligent or educated to understand much about their own treatment ... recipients of federally assisted or provided health services are entitled to the same information, respect, and concern as are private patients."[62] The Commission also endorsed several of the patients' rights that Wiley had negotiated with DHEW. Among them was the right to be informed, not only of the risks and benefits of any treatment offered to them, but also about possible alternative treatments, the right of access to medical records, and the right of effective access to a grievance committee. Among the works the Commission cited as "Reviewed" (i.e., documents consulted by its members) were Wiley's Joint Commission agreement on patients' rights and the AHA's A Patient's Bill of Rights.[63] Wiley had died five years earlier, but his advocacy still influenced the Commission's report, reasserting equal rights for Medicaid, Medicare, and Indian Health Service patients—and thus serving as a fitting epitaph to Wiley's efforts at creating a bill of rights for all patients.

Institutional Review Boards (IRBs) as Satellite Regulators

The Commission also recommended establishing a system of institutional review boards (IRBs) for all US government-supported research as a way of addressing an inherent conflict of interest permeating the traditional scientistic paternalistic practice of allowing "investigators [to] have sole responsibility for determining whether [their] research involving human subjects fulfills ethical standards." The necessity of satellite regulation became evident to the NIH after it surveyed "its grantee institutions in

1962 [and] discovered that only nine of its fifty-two departments had any policy regarding the rights of research subjects; [and only] sixteen stated that they used written consent forms."[64] Prior review of experiments had been optional, even at the NIH's own Clinical Center, a relatively new facility that admitted its first patient in 1953. As the Commission described the situation at the NIH,

> NIH investigators were not obliged by internal rules or their own sense of propriety to consult with colleagues in order to make certain that their evaluation of risk was not biased by their eagerness to do the research. They were not required to obtain another investigator's opinion, let alone approval. . . . [The] NIH did have a Medical Board Committee [whose mission was to review] "any non-standard potentially hazardous procedure." . . . However, as one deputy director explained, "It is not necessary to present each project to any single central group." Investigators who wanted a consultation . . . had the option of seeking the advice of the Medical Board Committee; but if the investigators believed that their protocols were not hazardous, they were free to proceed. The choice was the investigator's alone, and so, not surprisingly, the board was rarely consulted.[65]

The committee deemed this morally problematic, "because investigators have a potential conflict by virtue of their concern with the pursuit of knowledge as well as the welfare of the human subjects of their research." Thus, "research involving [human subjects] should be reviewed by Institutional Review Boards [IRBs] operating pursuant to federal regulation and located in institutions where the research is to be conducted."[66] This provision outsourced review to the institutions that received NIH funding, allowing a federal agency operating with limited resources to set research standards for any institution receiving funding from it for research involving human subjects. The net effect was to extend the regulatory authority of thinly resourced federal bureaucracies (FDA, NIH, etc.) into the thousands of American facilities conducting government-funded research on humans (and later on laboratory animals).[67] To use political scientist Daniel Carpenter's apt coinage, these "satellite regulators"[68] applied and enforced governmental regulations, policed the practice of informing and obtaining consent from human subjects, ensured that researchers explained how they would respect their subjects' rights and attend to their welfare, and advised investigators on all of these subjects.

Although the concept of satellite regulation was considered a pioneering initiative when first proposed by the NIH and later adopted by the FDA, the idea of prior committee monitoring of proposed research dates to the early nineteenth century, not as a form of governmental regulation but as a form of professional self-regulation to enforce professional ethical standards. Noting this, Beecher, always a champion of professional self-regulation, comments that such committees were first proposed in Thomas Percival's 1803 work, *Medical Ethics*, in which Percival recommend that[69]

> whenever cases occur, attended with circumstance not hitherto observed, or in which ordinary modes of practice have been attempted without success it is for the public good, and in an especial degree advantageous to the poor (who, being the most numerous class of society, are the greatest beneficiaries of the healing art) that *new remedies* and *new methods of chirurgical treatment* be devised. But in the accomplishment of this salutary purpose, the gentlemen of the faculty should be scrupulously and conscientiously governed by sound reason, just analogy, or well authenticated facts. And no such trials should be instituted without a previous consultation of the physicians or surgeons according to the nature of the case.[70]

To rephrase this in twenty-first-century language: Percival was proposing formation of ad hoc committees of physicians or surgeons to review proposed clinical trials of new remedies or surgical (*chirurgical*) methods or techniques. These trials of innovative therapies were to be approved only if the "ordinary modes of practice have been attempted without success." However, no such investigation was to be conducted without prior review and approval of an ethics committee whose purpose was to ensure that the proposed innovation was justified by "sound reason, just analogy, or well authenticated facts" (i.e., only if the review committee found that a protocol rests on a scientifically sound hypothesis and data). Writing in 1970, Beecher cites these lines as precedent setting, remarking that "echoes of all [of Percival's] points are present in the most up-to-date codes."[71] Note, however, that Beecher and Percival were proposing research protocol review committees as a form of professional self-regulation; in contrast, IRBs, as Carpenter astutely observed, are really a form of satellite governmental regulation, even if they are staffed primarily by health care professionals.

10
The Bioethical Turn in Modern Medical Ethics

Renée Fox: [Professor Veatch please] describe . . . the strongest aspects of early bioethics.

Robert Veatch: Well, the very earliest aspects of bioethics took on what seemed to be the first major offense of medicine of the time, which is its incredible paternalism. And that meant we latched on to the principle of autonomy and we affirmed the right of individual patients to make choices in the most individualistic Protestant priesthood-of-all-believers sense . . . in the early period . . . patients weren't informed, and patients had no rights of choice. In my view that lasted five years as the dominant focus. In my own personal history from 1970 to 1975 that was a big issue. . . . By 1975 at least some of us were thinking of medicine as overly individualistic. I wrote a piece for The Hastings Center called, "Autonomy's Temporary Triumph."[1]

If I am right the principle of autonomy is nothing more than a footnote on a full theory of medical ethics. . . . Autonomous are self-legislating but that means legislating for themselves, if this is the case, autonomy's triumph is truly temporary.
—Robert Veatch, 1975[2]

Dissident Intelligentsia: A Prelude to Moral Change

As noted in the quotations published in the last chapter, Whewell believed that revolutions in the intellectual world "involve the overthrow of received truths by those that were at first considered 'strange' but that later come to be accepted as 'almost self-evident axioms.'" These truths, he claimed, are "recorded by some familiar maxim, or perhaps by some new word or phrase, which forms part of the current language of the world; and thus asserts a principle, [that] while it appears merely to indicate a transient notion [it] preserves as well as expresses a truth."[3] As indicated in the quotations prefacing this chapter, founding bioethicist Robert Veatch correctly identifies the concept and word involved in the overthrow of scientistic paternalism as "autonomy." This concept encapsulated the earliest stages of the bioethics revolution against "the major offense of medicine of the time, which is its incredible paternalism," because in that "period . . . patients weren't informed, and patients had no rights of choice."[4] By 1975, however, Veatch was arguing that "autonomy's triumph is truly temporary."[5]

On the Triumph of Autonomy

Autonomy was a Greek term for "self-rule." It was originally used to distinguish self-governing city-states, like Athens and Sparta, from the colonies they ruled in Italy and elsewhere. In the 1780s, a German philosopher of the Enlightenment, Immanuel Kant, repurposed the word to mean "self-ruling persons," that is, people rationally ruling, or controlling, their own conduct. Kant believed that such self-regulation was moral only if a person could universalize, that is, could consistently will that the principles governing their conduct would become universally applicable to all persons. Kant also argued that conforming one's conduct to such principles would lead to a world in which all persons would be respected as ends in themselves (moral agents) and would never be used as merely a means to attaining other persons' ends (or even their own). In the early stages of the bioethics moral revolution, Beauchamp, Veatch, and other bioethicists weaponized a variant of Kant's conception of autonomy to challenge paternalistic presumptions endemic in American medicine from the mid-nineteenth century through the 1970s.

As Veatch relates, in the 1960s and 1970s, revelations about scandalous research and the abuses of patients and research subjects made it evident that medical paternalism had run amuck: that one could not trust doctors' claims of acting in patients' interest or investigators' affirmations that subjects had given oral consent and were informed of risks. So, to address the mounting distrust of American researchers by funders and the erosion of the public's confidence in the entire health care system, influencers (e.g., media), powerbrokers (e.g., foundations), and some government officials (e.g., Kelsey) began to contemplate ways to safeguard the system against abuse. Since professional peer reviews of the USPHS's Tuskegee Study and of Southam's cancer experiments validated both studies as morally acceptable, peer review committees in themselves were patently inadequate. The problem was that when investigators reviewed conduct of fellow investigators, they naturally viewed it from the entrenched perspective of scientistic paternalism as distorted by systemic ableism, ageism, antiblack racism, anti-Semitism, classism, ethnocentrism, homophobia, sexism, vestigial puritanism, and xenophobia. This was evident in the Southam and USPHS syphilis studies since peer review committees saw nothing wrong with consent forms "signed" by functionally illiterate syphilitic black men, or in undocumented oral "consent" from terminally ill Jews, or from semi-conscious women recovering from hysterectomies (another of Southam's cancer implant experiments). Consent forms and peer committee reviews would be hopelessly ineffective safeguards without a new perspective on what counts as ethical conduct. So, founding bioethicists slimmed down and repurposed Kant's concept of autonomy (shedding Kant's universalizability constraint, for example) and, as Veatch testified in his interview with Fox, used it to challenge "the major offense of medicine of the time, which is its incredible paternalism. And that meant we latched on to the principle of autonomy and we affirmed the right of individual patients to make choices in the most individualistic Protestant priesthood-of-all-believers sense."[6]

As Whewell would have put this point, "From being asserted at first as strange discoveries, such truths come at last to be implied as almost self-evident axioms. They are recorded by some familiar maxim, or perhaps by some new word or phrase, which forms part of the current language of the world; and thus asserts a principle."[7] As it happened, this seeming truth was initially preserved in the Belmont trinity of respect for persons, beneficence, and justice, but that "truth," in turn, would later be supplanted by soon-to-become-standard maxims formulated by Beauchamp and Childress in their mnemonic quadrivium of four principles: "autonomy, beneficence, non-maleficence, and justice."

As in any other moral revolution, the bioethical turn in modern medical ethics produced a shift in what counted as moral, transforming informed consent, for example, from "waiver and release" forms designed to protect *investigators* and their funders from lawsuits resulting from any *harms* they might inflict on their subjects into documents attesting to an investigator's moral commitment to protect *subjects' rights* to be given *respect as persons* (i.e., *autonomous agents*). As this revolution unfolded, it stirred the moral imagination of health care practitioners and patients in ways that can be described as a transformation of the world within which medicine and medical research were practiced. Such changes, together with the controversies that almost always surround them, became the defining characteristics of a revolutionary bioethical turn in modern medical ethics.[8]

And since, as Whewell observed, archeologists can detect changes in past political regimes by tracking changes in the portraits on coins, so, too, words being the coinage of the intelligentsia, the "revolutions of the intellectual world" can also be detected by examining changes in terminology. As evident from Veatch's interview, the new term that came to represent the bioethical turn in modern medical was "autonomy," and its virtues came to be viewed as self-evident truths—except in the minds of dissidents, like Veatch himself. Yet, as it turned out, Veatch's prediction that the reign of autonomy would be short-lived was inaccurate.

The Belmont Principles: Respect for Persons, Beneficence, Justice

Creating a new ethics for the NIH, the USPHS, and other government agencies required more than just satellite regulation; it also required a new understanding of ethical standards written in a lingua franca comprehensible to physicians, investigators, and bureaucrats alike. Fortuitously, such a lingua franca was naturally emerging in the nascent field of bioethics, and one early means of disseminating this mode of discourse and the new ideas was the National Commission's *Belmont Report*: which, as Jonsen explained to an interviewer was

> a document which did give some philosophical grounding to the kinds of things that were said in the codes. [Based on] the big cases . . . that were around at the time, like Tuskegee . . . [which] set a certain direction in which [commissioners'] thinking was likely to go . . . there's no question that [Tuskegee] was a primary example of the problems that research ethics must deal with. It dealt with the problem of respect for persons. There was no consent. It had a problem with beneficence: there was no benefit whatsoever to the subjects. And it clearly was a problem of justice, because it was dealing with an already badly deprived minority population who were exploited. So, as we thought through those things, Tuskegee was a constant echo that informed the way in which we viewed research ethics.[9]

Thus, for Jonsen and other commissioners, the USPHS's Syphilis Study was an index case of research immorality. At one point in his hearings, Senator Kennedy asked Peter Buxtun, "What bothered you most about the [syphilis] study?" In his response, Buxtun did not mention the harms suffered by the experimental subjects or the danger of allowing an STI to spread unchecked in the black community. What bothered Buxtun most was "the fact that the participants really did not seem to be consulted. They were being used."[10] Nothing in the standard Anglophone medical or scientific lexicon relating to human subjects research of that era—nothing in the Nuremberg Code, in the Declarations of Geneva and Helsinki, or in the FDA regulations—linked the concept of informed consent with the notion of "being used."[11] It is not to be found in Percival's code or in any AMA code or other code of medical ethics issued by a health care profession organization prior to the bioethics moral revolution of the 1970s, nor can it be found in Pappworth's exposés, or in Beecher's Brook Lodge speech, or in his subsequent *New England Journal of Medicine* exposé. These all focused on the immorality of subjecting subjects to potentially harmful consequences without fully informing subjects of the risks of being harmed.

The idea of "not using people" was novel. As Whewell might have remarked, it signals an important conceptual shift from earlier legal conceptions of the consent process that treated consent as a mechanism for shielding *investigators* from lawsuits—as exemplified in the once-standard US government waiver and release contracts. Waiver and release forms were designed to transfer responsibility for any future harms or deaths to the subjects who had agreed to accept these risks (although some consent forms were coupled with some form of insurance to compensate subjects, or their heirs, for harms suffered or for a subject's death). To reiterate, nothing in the biomedical literature before the bioethics turn prohibited subjects from "being used."

Now the raison d'être of moral philosophy, its fundamental purpose, is to articulate our often-inchoate moral sensibilities and interrogate their rationality: that is, to make clear to ourselves and to others why that which we take to be morally praiseworthy is worthy of our praise, and that which we condemn as morally blameworthy merits our condemnation. The moral philosopher's job is to probe deeper than simply explicating our conceptions of morality and its justificatory framework; philosophers must also attempt to open the blocked doors of the mind so that we can envision alternative concepts and courses of action. Wittgenstein once remarked, "A man will be *imprisoned* in a room with a door that's unlocked and opens inwards; as long as it does not occur to him to *pull* rather than push it."[12] To open new doors in exploring these deeper issues, the Commission recruited several moral philosophers tasking them with analyzing the fundamental principles underlying ethical research on humans.

As Jonsen tells the tale, after reviewing the philosophers' papers, the commissioners met at his house. Those attending, besides Jonsen himself, were lawyer and staff director Michael Yesley, moral philosopher Stephen Toulmin (1922–2009), moral theologian Karen Lebacqz (1945–), and psychologist and neuroscientist Joe Brady (1922–2011). They were

charged with doing a final draft [of the *Belmont Report*] and we spent two
days at my place and worked our way through all the text that we had and
argued out the arguments. . . . We . . . had a general consensus that it ought
to be very brief, and so forth. I typed it. I sat there and typed language as
people were talking. . . . "Respect for persons," the first principle, as I remember it, got its wording largely because of my favoring that language in a book
that I had recently read. A relatively recent book on Kant's philosophy called
Respect for Persons . . . I liked the language and Karen liked that language
too. I think Tris Engelhardt used that in his essay. So, we chose "respect for
persons" rather than "autonomy."[13]

The key feature of that report, in my mind, was that it linked each principle
with a practice, so where there is discussion of "respect for persons" in kind
of a broad, moralist, philosophical way, then it says, "informed consent is the
practice that manifests respect for persons. Beneficence and nonmaleficence
are the principles which lie behind risk-benefit assessment. And justice is the
set of principles that lie behind the allocation of benefits in research."[14]

On this analysis, the fundamental issue raised by the USPHS's Syphilis
Study, Southam's cancer studies, and Krugman's hepatitis experiments at
Willowbrook was that investigators denied their subjects the respect they
deserved as persons: as Buxtun put this point, the investigators "used"
them as if they were not worthy of moral respect.

Writing about victory, Roman historian Tacitus (56–120 CE) observed
that failure is an orphan but success is claimed has many fathers. Philosopher
Tom Beauchamp also claims authorship of the *Belmont Report*, but with
one exception; Beauchamp acknowledges that the section on "Boundaries
between Practice and Research," was written by Yale medical school professor Robert Levine, (1934–2021). In his section on boundaries, Levine
differentiated *therapeutic innovations* from *research* on the grounds that
therapeutic innovations aim at benefiting an individual patient, whereas
research is designed to test hypotheses generalizable to other cases.[15] As
to the rest of the *Report*; as Beauchamp tells the tale, Director Yesley had
informed him that to satisfy a congressional mandate that the Commission "identify the basic ethical principles that should underlie the conduct
of biomedical and behavioral research involving human subjects." So
Beauchamp was to flesh out three principles that the commissioners had
deemed fundamental for ethical research on humans. These principles were
(i) respect for persons, (ii) beneficence, and (iii) justice.[16]

Yesley also pointed out that doing so "is really quite simple for our
purposes": the principle of "respect for persons applies to informed consent, beneficence applies to risk-benefit assessment, and justice applies to
the selection of subjects."[17]

Beauchamp understood Yesley as outlining the following schema:

Principle of	Applies to	Guidelines for
Respect for persons		Informed consent
Beneficence		Risk–benefit assessment
Justice		Selection of subjects[18]

Beauchamp recalls thinking that "this schema may seem trifling, [but] no one at that time had articulated the schema in precisely this way, and Yesley's summary was immensely helpful in peering through countless hours of discussion to see the underlying structure and commitment at work."[19] Beauchamp also realized that "each principle made moral demands . . . the principle of respect for persons demands valid consent, the *purpose* of consent provisions is not protection from risk [of harm or injury] . . . but the protection of autonomy and personal dignity."[20] Beauchamp then set out to write a draft fleshing out the details of each of the three principles, supplementing each with rich details about its philosophical lineage.

After commissioners Brady, Jonsen, Lebacqz, and Toulmin reviewed Beauchamp's draft, they asked him to delete the philosophical references and rewrite the report to make it readily accessible to the typical IRB member. Beauchamp complied. However, in his final revision Beauchamp cashed out the principle of respect for persons as respect for autonomous agency. Thus, whereas Engelhardt and Jonsen had rendered the principle of respect for persons as "one should respect human subjects as free agents out of a duty to such subjects to acknowledge their right to respect as free agents,"[21] Beauchamp shifted the emphasis away from the duty of respect owed to free agents to respecting autonomous beings who are "capable of deliberation about personal goals and of acting under the directing of such deliberation." He then follows this by characterizing not what constitutes respecting *persons* but rather what constitutes *not respecting* an agent's *autonomy*. Thus, he wrote that "to show lack of respect for an autonomous agent is to repudiate that person's considered judgments, to deny an individual the freedom to act on those considered judgments, or to withhold information necessary to make a considered judgment, when there are no compelling reasons to do so."[22]

Not coincidentally, Beauchamp and Childress gave the same characterization in their textbook, *Principles of Biomedical Ethics*, which was published in the next year, 1979: "The autonomous person is one who not only deliberates about and choses [plans] but who is capable of acting on the basis of such deliberations."[23] In this version, they forthrightly replace "Respect for Persons" with "autonomous persons," retaining the concepts of "beneficence" and "justice" and adding a new "do no harm" concept, "nonmaleficence." Thus, with some easily overlooked exceptions, the three Belmont principles for ethical experiments on humans—respect for persons, beneficence, and justice—can be seen as interchangeable with Beauchamp and Childress's four principles of clinical and experimental biomedical ethics—autonomy, beneficence, nonmaleficence, and justice.[24] This gave the appearance of a uniform theory of biomedical ethics applicable to both medical experiments on humans and to the health care practitioner–patient relationship.

In a careful analysis of various ways of construing "respect for persons," from Ramsey to Beauchamp and Childress, moral theologian M. Therese Lysaught observes that Beauchamp hijacked the concept of "respect for persons." As Lysaught characterizes the hijacking—"hijacking" is my terminology, not Lysaught's—Beauchamp first

> established a rhetorical association between respect and autonomy, [and later] reduced the meaning of respect for persons to respect for autonomy. [And next] reduced respect for persons to an "*aspect of* the principle of autonomy . . . often referred to as the principle of respect for persons. . . ."[25] In making this move, the principle of respect for persons has deftly been redefined as a subcategory of the principle of autonomy. . . .
>
> Thus, although for Beauchamp and Childress the principle of autonomy appeared to map the same ground as the principle of respect for persons, they introduced three key changes. First . . . it is not persons as such but autonomy that is to be respected. . . . Second, and somewhat tautologically, the world of persons is delimited to those who are autonomous. . . . [viz.] "It does not apply to persons who are not in a position to act in a sufficiently autonomous manner."[26] Third, respect in this context means "noninterference and correlatively an obligation not to constrain autonomous actions—nothing more but also nothing less."[27] . . . [Thus] "Respect" no longer pertains to the non-autonomous. Instead, their fortunes are determined by the principles of nonmaleficence and beneficence . . . two principles [characterized] in little more than utilitarian terms, [and thus] their protections do not carry the moral security of "respect."[28]

In a 2004 interview with LeRoy Walters, Karen Lebacqz also objected to Beauchamp's substitution of "autonomy" for the Commission's intended wording, "respect for persons." Lebacqz reports that in struggling with issues surrounding research on a nonautonomous group,

> children, we looked carefully at issues around beneficence and non-maleficence. We [also] struggled with research on prisoners. . . . In my way of thinking the principle of respect for persons is absolutely crucial. And it needs to be framed in the language of respect for persons. . . . Part of the reason for that is that we had

a history of some research that was very disrespectful of persons. The Tuskegee Syphilis Study, some of the research that had been done on birth control using women from other countries who were never given an opportunity to consent [they] did not know they were participating in research.

So, we wanted a principle that would require that people be, from the outset, respectful of all the subjects who would be participating in research. In subsequent years the principle of respect for persons became, in the work of Beauchamp and Childress, [in] their very famous book on principles for biomedical ethics, that principle became a principle of respect for autonomy. And, it's my personal view that is unfortunate. That respect for persons is broader than respect for autonomy.

Autonomy, the capacity to make one's own decisions is surely an important part of what it means to be a person. But it is not the only part. Children are persons even though they are not autonomous. So, the principle of respect for persons needs to include dimensions outside of autonomy. And, I would argue for the integrity of the language that the National Commission used, and I think it is the better language. . . .

Al Jonsen and I had team taught a course in ethical theory. And, we used Downey and Telfer's book, *Respect for Persons*, as one of the texts for that class. The book is really an effort to spell out what a continuum perspective on ethics requires. And, it is quite strong on an understanding of what respect for persons means. And, in my view, sees that as going beyond simple respect for autonomy. Though many people do take . . . ethics to be rooted in the concept of autonomy. . . .

If I could go back and do it over again, would I do some things differently? Probably. Though, I think what we did at the time was good for its time. . . . I would try to spell out better than we did the inner meaning of the three principles that we had. For example, I really do believe that respect for persons needs to include a kind of subprincipal of respect for a person's community, respect for the embeddedness of people in their cultures, and so there are ways of getting respect for culture into the principle of respect for persons. That's one that I would like to see either added or at least elaborated.

Similarly, we talked about justice . . . in the language of equal treatment and protection of the vulnerable. A language that we did not use in those days but that that has become very prominent since and very important to me, is the language of oppression. I think there is a difference between populations who are simply vulnerable and populations who are oppressed. And, justice requires rectification of oppression and that might set some structures differently than the way that we did so many years ago. . . . in the Belmont meetings and in our struggles with the Belmont document, we were never able to spell out fully what justice would require. A good bit of my own work since then has been grappling with this difficult principle of justice and trying to determine what it does require.

And, as I've already indicated, I would see a crucial question today as the question of oppression. In part of what was so disturbing about the Tuskegee Syphilis Study is that it was not only a study that was done without the informed consent of people, that alone would be bad enough, but it was a study that was done on an oppressed group of people who were taken advantage of because of their oppression. And, that kind of exploitation is absolutely anathema.

So, justice requires not simply treating people equally, we could treat everybody equally and still be oppressing all of them. Justice requires attention to power issues, how to redress the power imbalances in life. Some of my own work on justice has involved looking at what it means to acknowledge that we have made mistakes, or done things that are wrong, and that we need to rectify those. What is reparative justice, for example. So, from my perspective, the Commission really did a great service to the field of bioethics by lifting up the principle of justice. But we did not give a full enough account of what justice would require. And there is a great deal more work that needs to be done.[29]

As Lebacqz remarks, there is more work that needs to be done on the theoretical foundations of bioethics. Yet, as she also observes, what the commission did "at the time was good for its time." The basic achievement of the 1979 *Belmont Report* was to offer an alternative to what Kelsey called the doctor-as-the-Lord-Almighty scientistic medical paternalist paradigm. The newer bioethics paradigm, its concepts, discourses, and its conception of a therapeutic patient–practitioner partnership predicated on transparency, trust, and mutual respect, ultimately supplanted the scientistic paternalist paradigm. As Veatch once remarked, autonomy, or respect for persons, or respect for human dignity—were the keys: any of these quasi-Kantian conceptions could serve to limit utilitarian ideas of sacrificing-for-other-gains either for medical science, as Krugman claimed in justifying his Willowbrook experiment, or to improve the purity of the future *Volkskörper* gene pool, as Brandt claimed in justifying Aktion T4 at the Nuremberg Trials.

The preordained destiny of reports like the *Belmont Report* is to pass from the keyboards of their authors to the trashcans of congressional assistants, after which they usually lie entombed, unmourned and unread, in the National Archives—save for a few zombie files resurrected by an occasional historian or political scientist.[30] Perhaps not surprisingly, given the typical fate of such documents, neither Beauchamp, nor Jonsen, nor any other commissioner anticipated that anyone would pay serious attention to the *Belmont Report*. As Beauchamp remarked in response to a question about whether he anticipated that "the document that you had completed would still be as widely read and used as it is today?"

No, I think it was unimaginable, absolutely unimaginable, that that would occur. The document had only one purpose: it was really to explain the framework principles that the Commission operated on in delivering what we all thought would be the most important thing we were doing, which is the other 16 volumes, or so . . . those volumes that were on very specific matters that we all knew could be weighty, because they would pass through the Office of

DHEW and then into law, either that or some reason had to be given for them not passing into law.... Basically, we thought of Belmont as a kind of background consideration.

In fact, when Michael [Yesley] first assigned the job to me to write the *Belmont Report*, I thought it was because I was the new kid-on-the-block on the staff. And I was getting the dregs of the assignment. Because it was what nobody else wanted to do.[31]

Beauchamp also noted that at first, "very few scholars paid much attention to it. And those who did, I think, paid just fairly perfunctory attention. There were a few articles—but a very few."[32]

Jonsen concurs. "I don't know that there were news articles about [the *Belmont Report*]. I just don't remember how it was received. It seemed to me it kind of slipped into the world of research, and its acclaim seems to have come later." Jonsen continues,

Certainly, one of the reasons why the *Belmont Report* attained some significant wide acceptance was that Beauchamp and Childress picked up the essential framework—that is the three principles—and they actually made it four, because they broke beneficence down into beneficence and non-maleficence—so they took those principles as the structure of their book on bioethics, generally. So, outside the field of research, the interest was growing in bioethics as a general field within the world of medicine, and by using that framework of *Belmont*, they essentially promoted *Belmont*, which was really written just for the field of research. Also, both of those authors had been part of the Commission's development of those ideas. So, they weren't stealing anything by any means. They were, in fact, contributors to the report itself.[33]

In 1979, shortly after the official publication of the *Belmont Report*, Beauchamp and his colleague, James Childress, published the first edition of *Principles of Biomedical Ethics*. At that time, medical ethics courses and textbooks would devote one week or one chapter to topics such as the history of medical ethics, or the physician–patient relationship, followed the next week by chapter about an entirely unrelated subject, such as abortion or euthanasia. *Principles* offered a radically different approach: a coherent analysis of different topic areas systematically analyzed from the perspective of four overarching ethical principles. In a stroke of pedagogical brilliance, Beauchamp and Childress used a synecdoche to represent a more nuanced description of each principle (i.e., they used one part of a name to represent the whole description). So, just as we say "Yankees" to represent the baseball team named the "New York Yankees," they used "autonomy" to represent the "principle of respect for autonomy." Reducing the core of bioethics to a synecdoche quartet—autonomy, beneficence, nonmaleficence, and justice—was a mnemonic triumph that made the complexities of the emerging field of bioethics accessible and memorable for novices, even as it allowed discussions of the underlying principles sufficiently detailed and nuanced to satisfy experts. As pioneering bioethicist Jonathan Moreno observed, *Principles of Biomedical Ethics* "shaped a field of study for decades, and . . . helped institutionalize [the] field [of bioethics] around the world."[34] *Principles* and the *Belmont Report* soon served as primers, tutoring biomedical practitioners on how to read, understand, and use the emerging bioethical lingua franca and its underlying concepts.

Between them, the *Belmont Report* and *Principles* introduced four neologisms into the Anglophone biomedical ethics lexicon: "autonomy," "beneficence," "nonmaleficence," and "respect for persons." "Justice," in contrast, was a term well established in the Anglophone medical lexicon and Western medical lexicon, tracing back to the Hippocratic ideal of treating the illnesses of the sick equally, irrespective of their social status as slave or free. The *Belmont Report*, however, applied this moral ideal to the treatment of persons who could serve as subjects: thus, a convenience sample of Hispanic women could not justly serve as subjects in an experiment designed to investigate a new drug designed primarily for use by affluent white women in the United States. The concept of "nonmaleficence"—a tongue-twisting neologism—serves as sobriquet (i.e., a short name, or nickname) for a concept so deeply embedded in Western medical ethics that it was Latinized as *primum non nocere*, "first do no harm."[35] More substantively, however, by reducing this multiword English or Latin statement into a one-word synecdoche, "nonmaleficence," Beauchamp and Childress not only fit it into their four-word mnemonic but also removed the notion that it came "first," eliminating the idea that harm prevention is the primary principle of medical research and clinical ethics. Instead, harms must be weighed against any potential benefits that an experimental therapeutic intervention might offer to patients or to society more generally.

Moving from moral principles to the pragmatics of a moral revolution: Beauchamp and Childress's memorable mantra for a new biomedical ethics—autonomy, beneficence, nonmaleficence, justice—could be readily committed to memory, was easily recited, and was ideally suitable for rapid internalization by medical educators, their students, health care practitioners, and biomedical researchers. Just as Thomas Paine's famous 1776 pamphlet *Common Sense* served as the primer of the American revolution against British monarchical governance, *Principles* and the *Belmont Report* jointly became primers for the American bioethics revolution against medical paternalism. They tutored biomedical practitioners on how to read, understand, and use the emerging bioethical lingua franca and its underlying concepts. As the revolution progressed, these usages soon became commonplace at the bedside and benchside and in boardrooms, professional codes of medical ethics, and, most significantly, in HECs and IRBs.

Facilitating this at the clinical level, in 1982, the trio of Albert Jonsen, physician Mark Siegler, and lawyer William Winslade published *Clinical Ethics: A Practical Approach to Ethical Decisions in Clinical Medicine*,[36] which operationalized the new anti-paternalist paradigm for use in clinical decision-making. Eventually, the new ethics was valorized in the AMA's code of ethics and disseminated by the Joint Commission. By these routes it found its way into hospital regulations as well as innumerable forms and chart notes: all reinforcing and assisting the internalization of the new conception of biomedical morality and its justificatory bioethics. Correlatively, once common practices, like soliciting duplicitous or uninformed consent, became so unthinkable that a younger generation of professionals can now exclaim, "How could we have done *that?*"—philosopher Kwame Anthony Appiah's indicator of a successful moral revolution.[37]

The *Belmont Report* and *Principles* as Ascendant Paradigms for a Bioethics Moral Revolution

Moral revolutions, like their political and scientific counterparts, typically follow this pattern described by Kuhn.

> At first a new candidate for paradigm may have supporters . . . [but], if they are competent, they will improve it, explore its possibilities, and show what it would be like to belong to the community guided by it. And as that goes on, if the paradigm is one destined to win its fight, the number and strength of the persuasive arguments in its favor will increase. More . . . will be converted, and the exploration of the new paradigm will go on. Gradually the number of articles and books based on the paradigm will multiply. Still more [adherents], convinced of the new view's fruitfulness, will adopt the new mode of practicing normal [morality or] science, until only a few elderly holdouts remain.[38]

Something like this happened to Beauchamp and Childress's four-synecdoche biomedical ethics as it displaced the established scientist paternalist paradigm. Arguments critiquing and undermining paternalism honed by Beauchamp, Callahan, Childress, Engelhardt, Toulmin, and Veatch, among others, opened a conceptual space for an alternative. Various bioethicists competed for that space by championing alternative paradigms— the three Belmont principles, the four Beauchamp and Childress principles, Veatch's social contract theory, and many others. They also introduced new concepts and terminology—autonomy, clinical equipoise, respect for persons, nonmaleficence—and championed their interpretation of the bio-ethical paradigm by citing new forms of evidence, drawing ethical criteria for moral conduct or decision-making incommensurable with those cited by partisans of the formerly dominant self-policing club morality of Lord Almighty paternalists. In the process, they thereby validated new practices, regulations, and laws (e.g., rules for protecting human subjects established in 1981 as subpart A of 45 CFR part 46 of the HHS regulations, informally known as the Common Rule).

As the new paradigm was disseminated, its principles, theories, prac-tices, and laws gained more adherents because the practices they justified (transparency, therapeutic partnership, and respect for patients and sub-jects' rights) were vindicated by a superior ability at resolving problems that appeared intractable under the tradition paradigm—especially with respect to preventing scandalous experiments on humans and transforming non-resuscitation and other means of discontinuation of life support from covert acts of humanely motivated "civil disobedience" into a legitimate form of documented medical decision-making. This led to general, but not neces-sarily universal, acceptance within the biomedical community and by the media, powerbrokers and influencers, and the public, thereby rendering the older scientist paternalist paradigm and its concepts, associated discourse, and practices obsolete. There followed the phenomenon known as "Kuhn loss," that is, the loss of paternalistic practices and their justifications (e.g., "because the doctor says so") that made sense under the traditional para-digm but were rendered arcane non sequiturs under the new paradigm.

The above Kuhnian portrayal of the progress of the bioethical moral rev-olution is supported by a trail of documentable data. For example, Beecher's 1966 whistleblowing paper in the *New England Journal of Medicine*, "Eth-ics & Clinical Research," received 2,845 citations by the end of 2022 and remains the most cited article in the bioethics literature.[39] Similarly, on its fortieth anniversary in 2019, *Principles of Biomedical Ethics*, in its eighth edition, had been published in four languages, was available in 141 different printings or formats, was held by 6,401 libraries in the WorldCat consor-tium,[40] and, having 32,174 citations, is the most cited book in the bioethics literature.[41]

No moral or scientific revolution has succeeded in displacing a tradi-tional paradigm unless dominant powerbrokers and influencers accept it as advantageous *to them*. To reiterate, a new paradigm will gain influential adherents only if it is vindicated by its superior ability at resolving prob-lems that appear unresolvable under the tradition paradigm *in the eyes of the powerful and the influential*. As will become evident in the sections immediately following, the bioethics paradigm was successful because the biomedical community of investigators and clinicians had good reason to accept the bioethical reconception of their roles, irrespective of whether it was formulated in terms of autonomy, respect for persons, human dignity, or human rights, or formulated as the four-box approach to clinical ethics (i.e., the Jonsen, Siegler, Winslade approach). They had reason to do so because the new bioethical paradigm made DNR orders based on patient or surrogate consent morally and legally permissible and because it also justified prospective placebo-controlled randomized drug trials that had become the gold standard for approving innovative procedures or drugs (assuming volunteers gave informed consent to the possibility of receiving a placebo). More importantly, by giving Medicare and Medicaid patients more respect and by quelling scandals, the new bioethics paradigm made new biomedical technologies and new investigatory and clinical practices acceptable not only to funders and other powerbrokers but also to the community at large, facilitating continued public acceptance and funding of biomedical research and health care delivery. For these practical purposes, the biomedical community had good reason to embrace bioethics as a field of research and scholarship and to welcome clinical ethicists as fellow health care professionals.

A New Paradigm of the Patient–Physician Relationship Emerges

Several court cases also contributed to the success of the bioethics paradigm. Perhaps the most influential of these was a 1975 case involving a comatose twenty-one-year-old woman, Karen Ann Quinlan (1954–1985). According to a neurologist's diagnosis, Quinlan was in a persistent vegetative state: a sometimes "eyes open," seemingly wakeful, but nonetheless irreversibly unconscious state indicative of severe brain damage but not of brain death. The legal case arose because physicians at St. Clare's hospital in Denville, New Jersey, declined a request from Joseph Quinlan (1925–1996) and his wife Julia (1927–), Karen Ann's adoptive parents, to discontinue ventilator support for their daughter. As devout Catholics the Quinlans had consulted their spiritual advisor, Reverend Monsignor Thomas J. Trapasso (1924–2012), about their daughter's condition. He, in turn, reviewed Pope Pius XII's 1957 *Address to Anesthesiologists* in which the pope responded to a question posed by anesthesiologists about physicians' "responsibilities for the welfare of the patient in cases of deep unconsciousness . . . that are considered to be completely hopeless . . . to use modern artificial respiration apparatus, even against the will of the family?"[42] The pope responded, "The rights and duties of the doctor are correlative to those of the patient. The doctor, in fact, has no separate or independent right where the patient is concerned. In general, he can take action only if the patient explicitly or implicitly, directly or indirectly, gives him permission." This means, the pope concluded that "since these forms of treatment go beyond the ordinary means to which one is bound, it cannot be held that there is an obligation to use them nor, consequently, that one is bound to give the doctor permission to use them. . . . Consequently, if it appears that the attempt at resuscitation constitutes in reality such a burden for the family that one cannot in all conscience impose it upon them, they can lawfully insist that the doctor should discontinue these attempts, and the doctor can lawfully comply." Moreover, Pius XII observed, "There is not involved here a case of direct disposal of the life of the patient, nor of euthanasia . . . this would never be licit."[43]

This was a watershed moment in modern medical ethics. Pope Pius's response challenged the scientistic paternalist paradigm by asserting that the patient or the patient's family is responsible for the patient's welfare, and their wishes preempt those of their doctors, who have "no separate or independent right where the patient is concerned."[44] Thus, if a family finds that maintaining someone in a persistent vegetative state is unduly burdensome, it has the right to insist that doctors discontinue ventilation. Moreover, even though the patient may die after discontinuing ventilator support, it is not a form of euthanasia (i.e., mercy killing).

Acting on the advice of their spiritual advisor, who, in turn, acted on that of Pope Pius XII, the Quinlans asked Karen's physicians to discontinue ventilator support for their daughter. Asserting their Lord Almighty "doctors know best" paternalistic prerogatives, Karen's doctors refused. The dispute went to the courts. At first, a New Jersey Court backed the physicians. On appeal, however, the New Jersey Supreme Court judge found that Karen had a right to privacy that permitted her to refuse ventilator support and that her father, as her duly appointed surrogate, could exercise that right on her behalf. The court also conferred on Karen's physicians' legal immunity from prosecution for homicide, provided that they disconnected Karen's ventilator with her father's consent and that an "ethics committee" review Karen's prognosis.[45]

The AMA's lawyer defending the physicians objected that the ethics committee review was unnecessary because "a treating physician is certainly able to determine whether a patient is in a terminal condition . . . [and] most hospitals don't have 'Ethics Committees.'"[46] The lawyer's observation about the scarcity of ethics committees was accurate: hospital ethics committees were virtually nonexistent in 1976.[47] Moreover, the concept of a hospital ethics committee had only been introduced into the legal literature in 1975, and the judge inadvertently mischaracterized the nature of such committees by requiring an ethics committee to assess the accuracy of a neurological diagnosis and prognosis of a persistent vegetative state. Hospital ethics committees had been created as multidisciplinary bodies whose mandate was to discuss controversial ethical issues, to recommend actions in controversial ethical cases, and to recommend future policy; they also served to diffuse the moral and legal responsibility of practitioners carrying out the committee's recommendations. It was not—and still is not—their function to validate medical diagnoses or prognoses.[48]

Yet words can be powerful. Even misnomers may have consequences. A 1983 survey found that a few large hospitals established hospital ethics committees (HECs) the year after the *Quinlan* decision. These were almost all modeled on a committee at Massachusetts General Hospital, described in a seminal article published in 1976.[49] By 1983, HECs had become commonplace, and a survey of hospital staff found that they supported HECs because these committees were "facilitating decision-making by clarifying important issues (73.3%); providing legal protection for hospital and medical staff (60%); shaping consistent hospital policies in regard to life support (56.3%); [and] providing opportunities for professionals to air disagreements (46.7%)."[50]

Yet these newly formed ethics committees had an odd mandate. In contrast to most medical school committees, such as admissions committees or IRBs, "they had no well-defined task to perform; they were ordered to think about ethics, probably the vaguest and most controversial topics," without a "touchstone beyond, perhaps, the skimpy code of the AMA."[51] A second item notably absent from first-generation ethics committees was trained ethicists. Serendipitously, a supply of ethicists was soon at hand from among the health care educators who formed the Society for Health and Human Values in 1969 and from the philosophers trained at the 1974 Council of Philosophical Studies summer institute. Annual courses offered by the Kennedy Institute of Ethics also offered training. And a society founded in 1974, Public Responsibility in Medicine and Research (PRIM&R), provided specialized ethics training for people staffing IRBs. In 1977, PRIM&R began to hold national meetings and soon became the professional society for ethicists and other health care professionals serving on IRBs and IACUCs (institutional animal care and use committees). One way or another, the supply of ethicists increased rapidly and soon filled the occupational space created by the need to find "experts" to staff HECs and IRBs.

Bioethics Also Fills a Void Created by the AMA's Abandonment of Medical Ethics

As Kuhn observed of scientific revolutions, paradigm displacements "are inaugurated by a growing sense . . . often restricted to a narrow subdivision of the . . . community, that an existing paradigm has ceased to function adequately. . . . In a . . . community the sense of malfunction that can lead to crisis is prerequisite to revolution."[52] The malfunctions of medical paternalism were evident in well-publicized controversies surrounding Southam, the USPHS's Syphilis Study, the Willowbrook hepatitis studies, the Quinlan case, and various debates over morally disruptive techniques and technologies, including cardiopulmonary resuscitation (CPR).

Forthright assertions of medical paternalism as stated in the AMA'S 1847 Code of Ethics had become embarrassing by 1903, so the AMA found it politic to leave its presumptions of paternalistic prerogatives implicit but unstated in its codes of ethics. Yet this ethos was so deeply entrenched in American medical practices that the by mid-twentieth century, the AMA lost interest in enforcing its own ethics standards.[53] More surprisingly, the AMA publicized its disinterest in ethics in 1957, a year in which, ironically, it also celebrated the 110th anniversary of its 1847 Code of Ethics. As part of this celebration, the AMA issued a new version of *Principles of Medical Ethics* in which it declared that its principles "are not laws but standards by which a physician *may* determine the propriety of his conduct in his relationships with patients, with colleagues, with members of allied professions, and with the public."[54]

In American English, "may" indicates possibility or permission; in contrast, verbs like "must," "should," or "shall" indicate obligations or duties. In effect, the AMA's proclamation that physicians "may" chose to consult its code to "determine the propriety of [their] conduct," and similar permissive language used throughout the code, demoted the 1957 *Principles* from professional obligations to a mere list of suggested conduct that physicians "may" consult or "consider"—or may choose not to consult or consider. By contrast, about a half-century earlier, the AMA's 1903 *Principles of Medical Ethics*, stated in a section titled "*Duties* of Physicians to Their Patients," that "physicians *should* not only be ever mindful to obey the calls [for attendance] from the sick and injured but *should* be mindful of the high character of their mission and the responsibilities they *must* incur in the discharge of their momentous *duties*" (italics added).[55] This line faithfully echoes the first line in the section "Of the Duties of Physicians to Their Patients," in the original 1847 AMA Code of Ethics. These obligations were reiterated in the AMA's 1912 *Principles*, which states that "in choosing [the medical] profession an individual assumes an *obligation* to conduct himself in accord with its ideals" (italics added).[56]

Neither Moses, nor Percival, nor Hays declared their codes of ethics optional or mere suggestions that an Israelite, a physician, or an AMA member *may* use to determine the propriety of their conduct. They insisted that thou *shalt* honor thine father and mother and that a physician had a professional obligation, *a duty*, to treat patients when called upon, even during epidemics. Yet by the mid-twentieth century, the AMA had abandoned the language of duty and obligation, replacing nouns like "duty" and verbs like "should" and "must" with permissive verbs, like "may," suggesting that it was neither unprofessional nor a violation of physicians' professional obligations for physicians to ignore the AMA's standards of ethical conduct.

Words have consequences. Since the AMA no longer took its ethical principles seriously, neither did American medical schools. Fifteen years after the AMA's 1957 abandonment of professional ethics standards a 1972 survey of 102 American medical schools found that none of the 94 schools responding required medical students to take a course on medical ethics or even to attend lectures on medical ethics. To reiterate, in 1972, a half-decade after Beecher blew the whistle on researchers' abuse of their human subjects, and after Southam's cancer experiments at the Jewish Chronic Disease Hospital and Krugman's hepatitis experiments on children with disabilities at the Willowbrook School had been headline-grabbing scandals, not one American medical school required its students to take a course on medical ethics or even to attend lectures on the subject. Fifteen schools offered no formal medical ethics instruction at all; fifty-six responded that they "touched" on the subject in teaching legal medicine, social medicine, or psychiatry; about one-third (thirty-three) reported that students *may* take an elective course on medical ethics if one was offered.[57]

Just over a decade later, as the bioethics moral revolution was gathering strength, a 1984 survey reported that 84 percent of medical schools required students to take a course or to attend a set of lectures on medical ethics or bioethics during their first two years of instruction. In 1998, the American Association of Medical Colleges (AAMC, founded 1876) adopted as a learning goal "knowledge of the theories and principles that govern ethical decisionmaking and of the major ethical dilemmas in medicine, particularly those that arise at the beginning and end of life and those that arise from the rapid expansion of knowledge of genetics."[58] A survey a decade later, in 2008, reported that "in compliance with the [AAMC learning objectives] all 59 medical schools in the dataset required coursework in bioethics" on average "35.6 hours of instruction."[59] The use of the term "bioethics" rather than "medical ethics" in the 2008 survey is revealing, and so too is AAMC's reference to "principles" and "genethics" in its 1998 statement of learning goals.

One effect of academic medicine's abandonment of medical ethics was that, since there was little demand, publishers supplied few textbooks on the subject. Thus, in the absence of serious competition, Beauchamp and Childress's *Principles of Biomedical Ethics* almost instantly became the dominant ethics textbook for medical schools. Consequently, the founding generation of ethics committee members and successive generations of medical students learned to talk and think in terms of their mnemonic quartet of synecdoches—autonomy, justice, nonmaleficence, and beneficence—and this version of a bioethics paradigm displaced the previously dominant paradigm of scientistic paternalism. As noted earlier, according to Google Scholar, by 2022, *Principles of Biomedical Ethics*, then in its eighth edition, had 33,075 citations and appears to be the most frequently cited book in bioethics.[60]

Rescuing DNR Orders from Staged Disobedience and the Appearance of Homicide

Perhaps the most commonplace of the morally disruptive medical innovations of the 1960s and 1970s was a newly effective technique of non-surgical (closed chest) cardiopulmonary resuscitation (CPR). As this new technique became widespread, it raised questions about whether physicians, acting on their own initiative, could refrain from initiating CPR on terminal patients undergoing cardiopulmonary arrest.[61] As physician Sherwin Nuland (1930–2014) observed in his semi-autobiographical National Book Award–winning 1995 book, *How We Die*—an account of dying in American hospitals—medical teams sometimes initially used CPR

> to indulge the sick person [or her/his family] and [them]selves in a form of medical "doing something to deny the hovering presence of death." This is one of the ways in which [the medical] profession manifests the entire society's current refusal to admit the existence of death's power and even of death itself. In such situations the doctor resorts to a usually ineffective delaying action that uses . . . the busy paraphernalia of scientific medicine, keeping a vague shadow of life flickering when all hope is gone . . . keeping extant certain representatives of life while final and complete death is temporarily frustrated or thwarted. . . . We [doctors do so] to turn our eyes away from the fact that nature always wins.[62]

In the pre–electronic medical record decades of the late twentieth century, patients' medical records were written on paper and kept at a central nursing station or near a patient's bedside. One portion of a patient's medical record would be labeled "Doctor's Orders" or "Medical Orders." As the imperative "orders" indicates, these were doctors' directives to others on the staff indicating how to monitor symptoms and when to start, stop, or modify a medication or other treatment. CPR was different. Any member of the hospital staff who observed a patient "coding"—to use a hospital corridor expression for a patient undergoing cardiopulmonary arrest—could "call a code" to alert colleagues to quickly assemble to perform CPR. Yet as physicians and nurses gained more experience with CPR, they realized that, as Nuland observed, "nature always wins." Consequently, instead of saving terminal patients' lives CPR could become, as Nuland puts it, just "'doing something' to deny the hovering presence of death." Worse yet, for frail elderly patients, CPR was not a benign process: using CPR to momentarily deny the terminality of such patients could inflict unnecessary harm and indignity on them. Torn between what they presumed was their duty—and ever mindful of hospital protocols and the prospect of malpractice litigation—clinicians invented some peculiarly medical forms of staged noncompliance once they realized that, for certain patients, CPR did more harm than good: they acted beneficently by making a show of coding (i.e., seeming to administer CPR, while avoiding anything that could brutalize a patient's death). Alternatively, they reacted slowly in response to a coding alert; they "stopped for coffee and donuts," so to speak, allowing death to follow its natural course—and then to satisfy hospital and legal formalities, they wrote sometimes cryptic notes, in which they would record that CPR had been attempted but failed.

In 1974, the American Heart Association (founded 1924, also, confusingly, abbreviated "AHA," like the hospital association) and the prestigious National Academy of Sciences (NAS, founded 1863) jointly issued a statement asserting that CPR "is not indicated . . . in cases of terminal irreversible illness where death is not unexpected."[63] However, their joint statement did not provide specific guidance on how to implement this advice in hospital protocols or assuage clinicians' fear of violating malpractice and negligent homicide laws. To bring these issues out of the closet, in 1976 two Harvard teaching hospitals, Beth Israel[64] and Massachusetts General,[65] courageously published their non-resuscitation policies in the *New England Journal of Medicine*. Both policies integrate non-resuscitation orders into regular hospital procedures by requiring that formal orders not to resuscitate (ONTR) or do not resuscitate (DNR) orders be entered into the medical record,[66] so that physicians and nurses could discuss, implement, reverse, or object to the DNR/ONTR order.[67,68] The Beth Israel protocol required informed consent from either patients or surrogates before implementing an ONTR. The

Massachusetts General Hospital protocol, on the other hand, treated DNR orders as no different from any other change of treatment order (i.e., as a physician's prerogative that did not require patient or surrogate consent).

As the medical debate over CPR was unfolding, Congress authorized a successor to the National Commission for the Protection of Human Subjects. The new bioethics commission was burdened with the unmemorable name, "The President's Commission for the Study of Ethical Problems in Medicine and Biomedical and Behavioral Research" (1980–1983). President James Carter (1924–, president 1977–1981) appointed the eminent lawyer and civil rights activist, Morris Abram (1918–2000), former president of Brandeis University and a prosecutor at the 1945–1946 Nuremberg Doctors Trial, to chair the commission. Other commissioners included Albert Jonsen, Renée Fox, and various eminent figures in law, medicine, and religion. Alexander Capron, a lawer and founding Hastings Center Fellow who had been a protégé of Jay Katz, was appointed director. He led a staff that included analytic philosophers Dan Brock (1937–2020), Allen Buchanan (1948–), and Daniel Wikler (1946–).

The commission published twelve reports. A 1983 report, *Deciding to Forego Life-Sustaining Treatment*, took on questions about DNR orders and other forms of withholding or withdrawing life-sustaining medical treatments. It forthrightly addressed various issues that seemed morally and legally problematic if viewed through the traditional scientistic paternalist paradigm and sought to normalize the new bioethical paradigm of shared physician–patient decision-making. As Kuhn observed, "The transition from a paradigm in crisis to a new one from which a new tradition of normal science [or morality] can emerge . . . is a reconstruction of a field from new fundamentals, a reconstruction that changes some of a field's most elementary theoretical generalizations as well as many of its paradigm methods and applications. . . . When the transition is complete the profession will have changed its view of the field, its methods, and its goals. . . . [It is] 'picking up the other end of the stick,' a process that involves 'handling the same bundle of data as before but placing them in a new system of relations with one another by giving them a different framework.'"[69]

The Massachusetts General Hospital's DNR policy had construed the DNR order from the scientistic paternalist perspective, but by making physicians solely responsible for these orders, it embroiled doctors in questions about whether, by withholding CPR, physicians were allowing, causing, or permitting a patient's death and thus could be indicted for criminally negligent homicide. Moreover, if one looked beyond medicine and turned to papal authority for guidance, one was faced with the question of whether discontinuing CPR could be excused as an "extraordinary" treatment, even though it was a routine practice. "Extraordinary" was not a helpful concept in the context of critical care medicine because in a critical care unit (CCU, ICU, MICU, NICU, PICU, etc.) vital functions ordinarily performed by patients' hearts, lungs, and kidneys—breathing, circulating, and filtering blood—are routinely exported onto machines. Thus, should a court deem such treatments "ordinary," would doctors be required to brutalize frail elderly patients in a futile effort to prolong their life?

In *Deciding to Forego Life-Sustaining Treatment*, the Commission "pick[ed] up the other end of the stick": "handling the same bundle of data as before but placing them in a new system of relations with one another by giving them a different framework." It favored the Beth Israel model and rejected "such distinction[s] [as ordinary versus extraordinary treatment which are] often difficult to draw in actual practice [and that] fail to provide an adequate foundation for the moral and legal foundations of events leading to death. Rather, the acceptability or unacceptability of conduct turns upon other morally significant factors, such as the duties owed to patients, [and] the patients' prospects and wishes."[70] Moreover, since the new bioethical paradigm required doctors to respect patients' autonomy, the morality—and, eventually, the legality—of withholding CPR properly depended on "patients' prospects and wishes," or those of their surrogates. As the Commission put this point, "Good decision-making about life-sustaining treatments depends on the same processes of shared decision-making that should be part of health care in general. The hallmark of an ethically sound process is always that it enables competent and informed patients to reach voluntary decisions about care." Thus, "a decision to forego treatment was to be considered ethically acceptable when it has been made by suitably qualified decision-makers who have found the risk of death to be justified in the light of all the circumstances. Neither criminal nor civil law ... forces patients to undergo procedures that will increase their suffering when they wish to avoid this by foregoing life-sustaining treatment."[71]

In sum, the prerogative of forgoing life-sustaining treatment, including CPR, morally and legally, lies with patients and their surrogates. As the Commission reiterates elsewhere in their report, "The primacy of a patient's interests in self-determination and in honoring the patients' own view of well-being warrant leaving with the patient the final authority to decide."[72] Furthermore, "the Commission has also found no particular treatments—including such 'ordinary' hospital interventions as parenteral nutrition or hydration [i.e., intravenous hydration and nutrition by a drip into blood vessels], antibiotics and transfusions—to be universally warranted and thus obligatory for a patient to accept."[73] They also observe that end-of-life care should include "care . . . needed to ensure the patient's comfort, dignity and self-determination."[74] To encourage implementation of their recommendations, the Commission appended to their report model laws and hospital DNR policies and encouraged proxy decision-making by emphasizing the value of a durable power of attorney that would allow a family member, or a valued partner or friend, to exercise a patient's wishes for end-of-life care, should the patient lack the capacity to make such decisions.

In an authoritative study of bioethics commissions, sociologist Bradford Gray (1942–), a fellow of both the Hastings Center and the Urban Institute, assessed *Deciding to Forego* as "an unambiguous success in terms of both 'critical impact on public policy' and 'significant impact on bioethics'. . . . [It] filled a tremendous need in a sensible and thoughtful way . . . and is [still in 1995] an authoritative text."[75] What made the report so authoritative and impactful was that it showed doctors that if they accepted the bioethical conception of patient-as-partner in health care decision-making, they circumvented a verbal morass—allowing to die versus killing, extraordinary versus ordinary, withholding versus withdrawing—and avoided issues of malpractice and criminally negligent homicide. By picking up the bioethical end of the stick, physicians could and did turn the ugly duckling of the doctor-alone-decides paternalism, plagued as it was by uncertain moral and legal liabilities, for the increasingly attractive bioethical ideal of shared decision-making—which durable powers of attorney and other advance directives extended to cover patients who are incapable of capacitated decision-making.

As one would expect of a successful paradigm shift, the President's Commission's recommendations were implemented in a 1988 statement by the Joint Commission and thereafter became policy for all US health care organizations receiving Medicare and Medicaid funding from the US government. In 1990, the AMA supplemented its principles of medical ethics, with a statement on the "Fundamental Elements of the Patient-Physician Relationship." As Kuhn (and Whewell) would no doubt have appreciated, this document opens with a statement that obliterated all traces of the older scientistic paternalist paradigm, even as it normalized the newly dominant bioethical paradigm of shared patient–physician decision-making. "From ancient times," the AMA declared, "physicians have recognized that the health and wellbeing of patients depends upon a collaborative effort between physician and patient."[76] With just three well-chosen words, "from ancient times," the AMA clothed the new bioethical paradigm in the mantle of traditional authority, erasing all traces of a moral revolution. It then endorsed patients' rights to be informed, to refuse treatment, to be treated with "courtesy, dignity, responsiveness," to confidentiality and to essential health care.[77]

It is noteworthy that the AMA's commentary on the patient's right "to refuse health care that is recommended by their physicians" echoes the position of the President's Commission's view that

> the freedom to refuse health care includes the freedom to decline all life-prolonging medical treatment. Life prolonging treatment includes medication and artificially or technologically supplied respiration, nutrition, and hydration. When the patient is unable to make medical decisions, the patient's right to decide is exercisable by an appropriate surrogate.[78]

The AMA also adopted bioethical terminology in discussing end-of-life care (italics added to emphasize this point).

2.20 Withholding or Withdrawing Life-sustaining Medical Treatment

> The *principle of patient autonomy* requires that physicians should respect the decision to *forego life-sustaining treatment* by a patient who possesses decision-making capacity. . . . There is no ethical distinction between withdrawing and withholding treatment.[79]

Here the AMA accepts as its own the President's Commission's bioethical construal of the issue as one of patient self-determination rather than of a physician's prerogative, even using the Commission's archaic spelling of "forgo" as "forego" in discussing the right to "*forego* life-sustaining treatments." More practically, the AMA follows the Commission in dismissing the moral relevance of the withholding versus withdrawing distinction. With this statement the AMA accepted the concepts, discourses, and precepts of bioethics as normal. The concepts and discourses of the bioethics paradigm were soon accepted into American medical parlance, signaling the end of a moral revolution and a new normal for American medicine.

Summary Statement: Origins of Modern Medical Ethics and Its Bioethical Turn

Modern medical ethics was first formulated in a series of documents issued by the Nuremberg court and by the World Medical Association from the mid-1940s through the mid-1960s: specifically, (Alexander and Ivy's) Nuremberg Principles/Code and the Declarations of Geneva and Helsinki. These codes and oaths had been created in response to medical scientists' abuse of fellow humans as guinea pigs and physicians' participation in ableist race-based mass sterilizations and exterminations of children and adults with mental and physical disabilities and, most notoriously, in response to genocidal efforts to exterminate gays, Jews, and Roma. Modern medical ethics originated in efforts designed to reject *Rassenhygiene* and prevent its recurrence. Thus, the early versions of modern medical ethics emphasized physicians' moral responsibilities to individual patients and to human subjects, required the informed voluntary consent of human subjects of biomedical research, and prohibited the discriminatory treatment of patients based on their religion, nationality, race, party politics, or social class. Yet, even as they proclaimed these ideals, the Declaration of Geneva and the Declaration of Helsinki perpetuated a scientistic paternalistic conception of medicine as a brotherhood in which "the doctor knows best."

In 1962, Frances Kelsey, a former medical school professor and an FDA bureaucrat, was emboldened by a question raised by New York Senator Jacob Javits to document the absence of informed consent in investigations of new drugs and to offer Senator Javits informed consent language taken from a draft version of the Declaration of Helsinki, which he then introduced into the Kefauver–Harris amendment to the Food, Drug, and Cosmetic Act—thereby transplanting international conceptions of modern medical ethics into modern American law. On a practical level, passage of the 1962 amendment required documented informed consent in any investigation supporting new drugs submitted to the FDA for approval. Kelsey's reward: a President's *Medal* for Distinguished Federal Civilian Service that protected her from being fired outright, as the FDA coped with blowback from media flacks for the pharmaceutical industry and researchers (including the still unawakened Beecher). In a classic bureaucratic response, the FDA sidelined Kelsey as a "bare desk bureaucrat," who was still involved with various efforts at enforcing research ethics standards.

During the same period, the egalitarian antidiscrimination values of the mid-1960s Great Society laws penetrated the American health care system through the Medicaid and Medicare programs that transformed former charity cases into patients paid for by the federal, state, and county governments. In combination with the new civil rights laws, this challenged the American health care system's historically ingrained structural ageism, classism, racism, and sexism. The new funding stream also provided funds for hospitals outside of academic medical centers to purchase new, albeit morally disruptive technologies, like the ventilator keeping Karen Quinlan alive in a permanently comatose human body at St. Clare's Hospital in Denville, New Jersey. As these morally disruptive technologies and techniques became widespread, they further destabilized the prevailing ethos of scientistic medical paternalism. The long-term effect was to undermine scientistic paternalism first in academic medical centers and then in other health care facilities, changing both language and, more important, the day-to-day conduct of health care professionals throughout the United States.

Catalyzing this transformation was a *lumpen intelligentsia* of feminists—like Downer and the women in the Boston Women's Health Book Collective—African American civil rights leaders (Jensen, Wiley), whistle-blowing health care practitioners (Beecher, Buxtun, Pappworth), disappointed and lapsed Catholics (Callahan, Hellegers, Jonsen, Reich), moral theologians (Childress, Ramsey), disaffected analytic philosophers (Baker, Beauchamp, Macklin), lawyers (Capron, Hyman, Kaimowitz), humanist nurses (Gadow) and physicians (Kelsey)—especially psychiatrists (Gaylin, Katz, Kübler-Ross, Lowinger)—scientists (Hellegers), and medical educators (Pellegrino) who recognized the anomalies and inconsistencies in the trust-the doctors-and-researchers-to-know-and-do-what-is-best paradigm. After honing their critiques of established paradigms of biomedicine, many addressing ethical issues raised by the advent of morally disruptive innovations and technologies, newly minted bioethicists developed alternative societies, institutes, and journals, thereby disseminating an interdisciplinary *lingua franca* and a common secular ethics based on an egalitarian paradigm of transparency and shared decision-making that was incompatible with the established doctors-as-lords-almighty paradigm. Their new conceptions required new forms of ethical discourse and new moral concepts ("autonomy," "clinical equipoise," "nonmaleficence," "respect for persons") and new practices (do not resuscitate orders) and legal innovations (during powers of attorney, living wills, and other advance directives). These new bioethical standards of good decision-making emphasized transparency, informed consent, nondiscrimination, and patients' and subjects' rights, often justified by invoking concepts valorized by Beauchamp and Childress's mnemonic synecdoche quartet, "autonomy, beneficence, nonmaleficence, justice." Ultimately, the biomedical professions accepted a paradigm shift because, when leaders of these communities viewed the issues facing them through lens of bioethics, they found that they could resolve a range of problems that appeared unsolvable if viewed through the paternalistic lens of the traditional doctor-knows-best paternalism. As various pre-bioethical revolutionary practices were purged from living memory or dismissed as idiosyncratic, those who recalled, heard, or read about them asked philosopher Anthony Appiah's question, "Whatever could we/they have been thinking?"

These overly longish sentences concisely state the author's view that, to use some of the wording in this book's subtitle, the history of modern medical ethics and its bioethics turn is "A History of How African Americans, Anti-Nazism, Bureaucrats, Commissions, Feminists, Thinktanks, Veterans, and Whistleblowing Moralists Turned Modern Medical Ethics into Bioethics."

11

Making the Unseen Visible: A Metahistorical Analysis of Why the Role of Anti-Nazism and the Names of Significant Blacks, Bureaucrats, Feminists, Veterans, and Most Whistleblowing Moralists Are Absent from, or Minimally Noted in, Standard Histories of Bioethics

In order to determine the moment at which the mutation in discourse took place, we must look . . . to the region where "things" and "words" have not yet been separated, and where—at the most fundamental level of language—seeing and saying are still one. We must reexamine the original distribution of the visible and the invisible insofar as it is linked with the division between what is stated and what remains unsaid.
—Michel Foucault, 1963[1]

The historian's task is to transform presumptively factual data and information into words that form a coherent narrative about the past. Others may question their reconstruction of the past, as did pioneering Canadian bioethicist Benjamin Freedman (1951–1997) when he posed the question, "Where Are the Heroes of Bioethics?"[2] Historian David Rothman placed a wreath of heroism firmly on the head of one whistleblowing moralist, Henry K. Beecher, but there is little else that is heroic in the history of bioethics as standardly constructed. In contrast, the narrative constructed in this book answers Freedman's question by emphasizing the actions of the many heroes and heroines indicated in its subtitle: African Americans (like Jenkins and Wiley), the World War I and II veterans who founded modern medical ethics (Alexander, Ivy, and Pridham), a few bureaucrats (Kelsey), some commissioners (Jonsen and Lebacqz), feminists (Boston Women's Health Book Collective, and Downer), thinktank founders (Callahan,

Gaylin, and Hellegers), and, in particular, veteran health care professionals who served in the American or British military (Alexander, Beecher, Buxtun, Gibson, Ivy, Pappworth, Pridham, and Wiley) and whistleblowing moralists (Buxtun, Gibson, Hyman, Jenkins, Pappworth, and Schatz). This raises the interesting question: why do these people receive little or no attention in Rothman's and Jonsen's histories of bioethics?

How Rothman and Jonsen Constructed
the Story of Bioethics' Origins

Constructing a history involves recovering documents and other sources of information about the past, deciding which aspects of the past to represent in one's narrative and then composing a reconstruction of the past as a narrative. Which still raises the question: why are most of the names that are the focus of this book absent from, or minimally mentioned in, the Rothman and Jonsen narratives? David Rothman was an eminent champion of human rights and a social historian of medicine who constructed his narrative by drawing on resources from the National Archives, the National Institutes of Health, and the medical journals he researched at the New York Academy of Medicine. Not surprisingly, therefore, if one counts the number of pages devoted to various personages, institutions, and events, the National Institutes of Health (NIH) is mentioned on fifty-two pages; Henry Beecher, the hero of Rothman's narrative, is mentioned on forty-five pages; and the committees of three senators—Walter Mondale (twenty-one), Edward Kennedy (eleven), and Estes Kefauver (five)—are the focus of a total of thirty-seven pages. Government commissions are discussed on twelve pages, the Hastings Center on six, and Callahan on four. Curiously, Hellegers and the Kennedy Institute of Ethics get surprisingly short shrift at one page each. The National Welfare Rights organization (NWRO) is also mentioned on two pages.[3] As one would expect from a human rights advocate, Rothman praises A Patient's Bill of Rights as a welcome advance in patients' rights. Yet he never describes the NWRO as an *African American* civil rights organization, nor does he mention that it was led by an African American professor, George A. Wiley, PhD, who receives no attention at all. Insofar as blacks are visible in Rothman's narrative, it is only as victims (one page, the sole reference to this noun in the index), never as heroes of the patients' rights movement. Buxtun, the white whistleblower who made the USPHS's exploitation of black men in its Syphilis Study visible to the media and, through it, to Senator Kennedy, is also relegated to just one page. One might suspect that since Rothman's 1991 narrative was based on research done in the 1980s, he may have lacked information about Buxtun's whistleblowing. Yet, Rothman cites James Jones's 1981 book, *Bad Blood: The Tuskegee Syphilis Experiment*, in which Buxtun is mentioned on eleven pages—which means that information about Buxtun was readily available to Rothman; he just chose not to focus on it in his narrative.[4]

Rothman mentions the Nuremberg Trial on just five pages and the Nuremberg Principles/Code on only four pages. But only mentions them to dismiss their relevance because, Rothman explains, "well into the 1960s the American research community considered the Nuremberg findings and the Nuremberg Code, irrelevant to its own work."[5] Rothman elaborates on this point, eloquently observing that

> neither the horrors described at the Nuremberg Trial, nor the ethical principles that emerged from it had a significant impact on the American research establishment. . . . Over the next fifteen years only a handful of articles in either medical or popular journals took up Nuremberg. . . . The events described at Nuremberg were not perceived by researchers or commentators to be directly relevant to the American scene. The violations had been the work of Nazis, not doctors. . . . Francis Moore [1913–2001] a professor of surgery at Harvard Medical School [said that] "one of the ironic tragedies of the human experimentation by German 'scientists' was that no good science of any sort came from this work." Madness not medicine was implicated at Nuremberg.[6]

It is relevant to Rothman's narrative that empowered Harvard professors, like Moore, subscribed to the comforting untruth that the German medical scientists were not really scientists.[7] Moore's embrace of this falsehood facilitates Rothman's dismissal of the Nuremberg Trials and the Nuremberg Code/Principles as immaterial to the American scientific and medical investigator establishments. By bracketing out the Nuremberg Doctors Trial and the code/principles as irrelevant, Rothman created a narrative focusing on a change in medical decision-making that occurred in American medicine in the 1960s and 1970s, when, as indicated in his book's title, philosophers and others who were once strangers to bedside medicine earned a place at the patient's bedside, even as physicians became increasingly estranged from their patients.

Rothman's narrative focuses on a radical change in the decision-making process that I call "the bioethical turn" in modern medical ethics. We differ because Rothman attributes this change to Henry K. Beecher's whistleblowing article that, quite inadvertently, created conceptual and occupational space for the philosophers, theologians, and other pioneering bioethicists to serve on IRBs and HECS as research ethics and clinical ethics umpires. Rothman thus answers Freedman's question by firmly putting

the crown of herodom on Beecher's head, even though, in point of historical fact, Beecher sought investigators' *self-regulation*, resisting all forms of
externally imposed standards or review by the non-investigators, like the
bureaucrats, philosophers, and theologians, who are the "strangers at the
bedside" celebrated in the title of Rothman's book.

Moreover, as the chronology of events prefacing this book makes clear,
inspired by the inutility of the Nuremberg Code, the World Medical Association published the first practical code of research ethics, the Declaration
of Helsinki in 1964. This, in turn, spurred the FDA's efforts to regulate
experiments on human subjects, which gained further momentum in
the context of the nearly tragic thalidomide episode that facilitated the
Kefauver–Harris amendment and the AMA's endorsement of the Declaration of Helsinki in 1966. Beecher's 1966 article clearly accelerated the pace
of regulatory reform (especially in the NIH, which had initially declined
to regulate the investigators it funded), but it was the Europeans, not the
Americans, who initiated it. Turning from research ethics to clinical ethics,
it was not until a decade later that the Quinlan case initiated widespread
use of hospitals ethics committees (HECs) in clinical contexts. Beecher's
1966 publication was a public bombshell and possibly a private act of contrition, but it did not initiate the bioethical turn in modern medical ethics.

Absence of African Americans, Females, Veterans, and Other Change Agents

Absences speak to an author's conceptions as loudly as their words. Jonsen, for example, omits the NWRO from his index and mentions it just
twice in his narrative. In these mentions, the NWRO is given the anodyne
characterization of "a grass roots advocacy group for persons on welfare"[8] that responded to the Joint Commission's invitation for suggestions
on accreditation standards, "perhaps to the dismay" of the Commission:[9]
an odd description of a nonviolent direct-action African American civil
rights organization and its interactions with the Joint Commission. Blackness, nonviolent militancy, and civil rights are notably absent from this
description, as is Wiley's name. With the notable exception of Beecher (nine
pages), Jonsen glosses over, elides, or neglects to mention moralist whistleblowers: Buxtun and Pappworth are mentioned on two pages apiece, but
no mention is made of Gibson, Hyman, or Schatz. As to women: Kelsey is
mentioned on three pages and the Boston Women's Health Book Collective
on one. Other than on pages mentioning Alexander (four) and Ivy (three),
scant attention is paid to the health care professionals who served in the

American and British armies during the two world wars. As noted above, Pappworth merits two pages, and Gibson is not mentioned at all, nor are Vietnam War or other veterans, except for Buxtun, who, as noted above, is mentioned on two pages. Yet many of these men put their careers at risk to lead a moral insurrection against the abuses of scientistic paternalism: Buxtun, for example, was forced to change professions, while the young Gibson was browbeaten into silence about the Tuskegee Study.

Lesser-known figures are also ignored. No mention is made of William A. Hyman (1893–1966), the lawyer who outed Southam's cancer experiment by successfully suing Jewish Chronic Disease Hospital and petitioning to have Southam's medical license suspended. Yet Hyman may have been the first person to publicly claim that an American physician, Dr. Chester Southam, had conducted experiments that violated a document that Hyman called, or christened, "The Nuremberg Code."[10] This was not the name that Ivy used for his ten principles for "Permissible Research on Humans" that he fashioned to distinguish ethically permissible research on humans from the Nazi doctors' ethically impermissible research (see Appendix). The document retained this name even after it was reshaped by Alexander and appended to the court's 1947 judgment in the Nuremberg Doctors Trial. Yet about two decades later, in 1963, Hyman was calling it "The Nuremberg Code." It is not clear whether he was using a common description of the ten principles or had renamed them to suit the case he was prosecuting against Southam for the unconsented cancer implants at the Jewish chronic disease hospital. Either way, this soon became the standard name for the ten Nuremberg principles.

The three most effective whistleblowers, Beecher, Buxtun, and Pappworth, cite the Nuremberg Code, using that appellation for the newly rechristened code in their critiques of Krugman, Southam, and Tuskegee. For them, the Nuremberg Code and Nuremberg Trial served as a rhetorical bridge in narratives comparing American researchers' conduct with those of Nazi doctors. Yet Jonsen only mentions Buxtun twice (both times on page 147). Other than these brief mentions, he gave only the slightest nod to the story of a conservative Vietnam War combat medic whose Czech-Jewish father and Christian uncle—a World War II veteran of the German army—raised him on tales of Nazi atrocities. Yet Buxtun's ability to recognize similarities between the USPH's Tuskegee Study and Nazi medical experiments at Auschwitz, Dachau, and elsewhere, coupled with his insistence on calling it to the attention of USPHS authorities and his persistence in relating the story to reporters from the Associated Press, was the starting point for the congressional commissions that are featured as central events in Jonsen's historical narrative—and that served as Jonsen's personal entry point into the newly emerging field of bioethics. Had Buxtun's Tuskegee tale not made headlines, there would have been no Kennedy Committee or National Commission. Consequently, Jonsen would likely have been just another Roman Catholic priest, rather than one of the founders of bioethics, and the field of bioethics may have been stillborn.

Rothman's and Jonsen's Treatment of Frances Kelsey

Frances Kelsey is another insurgent activist that Rothman and Jonsen have difficulty accommodating in their establishment-oriented narratives. Jonsen celebrates Kelsey as the savior of American babies but barely acknowledges her as a heroine who stood up to the male-dominated bureaucracy of "Lord Almighty" physicians, the pharmaceutical industry, and the research community as she sought to protect the rights of research subjects and patients. He also fails to mention that, as her reward, a disinformation campaign by media flacks and Beecher led to her ostracism within the federal bureaucracy as a "bare desk bureaucrat." Such gaps may reflect a lacuna in the data out of which Jonsen constructed his narrative since, other than his own memories, most of his sources were memoires from other white male intelligentsia who pioneered bioethics. They too had little interest in the source of the 1962 Kefauver–Harris amendment to the Food, Drug, and Cosmetics Act, which anticipated the bioethical turn by requiring investigators to inform subjects that they are receiving a drug for "investigational purposes and [they] will obtain [subjects'] consent or [that of] their representatives."[11]

Since Kelsey's act was an inflection point that set the precedent for all future government ethics standards regulating medical research on human subjects, one might expect that it would have been lauded by Jonsen or Rothman. Yet neither author credits her with recommending an informed consent requirement. Instead, both focus on Kelsey's role in preventing a thalidomide-induced tragedy, paying scant attention to the 1962 moment when she seized the opportunity created by Senator Javits's inquiry to document the absence of informed consent and to suggest introducing informed consent language into Senator Kefauver's 1962 law, thereby empowering the federal government to require research subjects' informed consent *for the first time*. Because Kelsey's role is often overlooked, it is worth repeating, for clarity and emphasis, that her empirical documentation of the absence of consent laid the foundation for inserting language from a draft of the Declaration of Helsinki into the first piece of federal legislation *mandating* a federal agency to require research subjects' informed consent. Just as importantly, it also gave these agencies power to enforce that requirement. Yet, instead of celebrating Kelsey's achievement, Jonsen attributes it to Senator Javits's procedural move of introducing this language into the draft legislation.

Perhaps Jonsen never found source material documenting Kelsey's role; on the other hand, he may have had an ulterior motive for not searching for this material and/or for excluding it from his narrative: Kelsey's achievement distracts from the preeminent role that Jonsen accords to the National Commission's *Belmont Report*, a document that he himself typed and that he credits as the source of the federal government's requirement for informed consent of research subjects. Distracting readers from Kelsey's role, Jonsen focuses instead on language that opponents inserted to weaken the amendment by allowing investigators to waive consent where they "deem it not feasible, or, in their professional judgment, contrary to the best interests of such human beings."[12] Focusing on this waiver allows Jonsen to dismiss Kelsey's efforts as an inadequate precursor to the more robust reforms recommended by the National Commission for the Protection of Human Subjects in the *Belmont Report*.

Rothman and Jonsen laud Kelsey for saving babies and properly credit her defense of the informed consent requirement against the eviscerating language inserted into the Kefauver amendment, yet they ignore her role in documenting the need to enforce an informed consent requirement.[13] Why? Possibly because saving babies and cleaning up linguistic messes can be viewed as properly feminine activities, but crediting a female with the initiative to challenge paternalist male doctors' scientistic prerogatives as "Lords Almighty" is "uppity," unfeminine, and so, perhaps, not to be mentioned. Both histories also fail to mention that "no good deed going unpunished," the "Lords Almighty" struck back in a campaign against her by investigators, including Beecher, pharmaceutical companies, the business press, and various media flacks, leading to Kelsey's loss of "formal control over investigational drugs and suffer[ing] what one reporter would later describe as a 'humiliating *bare desk* treatment, she was generally ignored and given little to do of consequence.'"[14]

As author of an autobiographically informed history of the origins of bioethics, Jonsen was likely tempted to write a narrative overvaluing those aspects of the history with which he was most familiar and to devalue aspects that might diminish his own accomplishments. Thus, Jonsen observes that Kesley's actions in handling the thalidomide threat are, "of course, a story of moral courage . . . [nonetheless the situation she confronted did] not, in the strict sense, pose an ethical problem. It is rather a story of commercial greed and political collusion, leading to a tragedy. Still, the thalidomide story is an opening event in the federal government's direct involvement in the ethics of experimentation on human subjects."[15]

Jonsen's claim that Kelsey's resistance to regulatory capture did not "pose an ethical problem" is perplexing. Nowhere in his 415-page text does he explain what he means by "ethics"; he just notes that "the word 'ethics' is used in many ways."[16] Recall that when Kelsey requested more clinical data, thalidomide had already been approved virtually everywhere else in the world. Yet Kelsey persisted in her quest for clinical data, even as the pharmaceutical company harangued her and pressured her supervisors at the FDA to overrule her. Still, Kelsey stood by her demand for better clinical data. Eventually, after the Widukind Lenz established the connection between thalidomide and neonatal deformities, the FDA backed Kelsey's denial, saving untold thousands of American babies from disability or death.

It is important to appreciate that when Kelsey denied the pharmaceutical company's application to market thalidomide the FDA had no formal standards for weighing the benefits of pharmaceutical innovation versus potential adverse effects on patients or experimental subjects. These would first begin to be formalized in the Kefauver–Harris amendment and ensuing FDA regulations (1962, 1963). In the absence of formal standards, Merrell Pharmaceutical argued in favor of innovation, claiming that the FDA was obligated to find evidence of potential harm to delay or deny approval of an innovative drug. Kelsey pushed back. She insisted that the burden lay with companies; it was their responsibility to demonstrate the safety and efficacy of their new drugs. Her position would later become official FDA policy. But at the time, she had to sort out this conflict without regulatory guidance.

Dilemmas about conflicting loyalties, principles, and standards of moral conduct are precisely what ethics addresses. Finding herself in a bioethical dilemma, Kelsey responded by developing a bioethical principle about the burden of proof to justify her decision. Defying the drug companies required moral courage, but the underlying principle she asserted is a prime example of bioethical reflection culminating in a bioethical principle. Ethical reflection is not some esoteric activity requiring an advanced degree in philosophy or moral theology. It is the everyday activity that Moll referred to as *Durchschnittsmoral* (i.e., commonsense morality). It is the sort of reflection that the public reads about and debates in public media such as Dear Abby and Ann Landers' columns. Anyone can engage in reflection on morality and its application, and anyone can raise ethical questions about whether certain actions or character traits should be morally praiseworthy or blameworthy.

To reiterate, ethical reflections are not the prerogative of some elite: one does need an advanced degree to ask what makes right actions "right" or wrong actions "wrong." It's a bit like playing ball: anyone can play catch or sandlot baseball, but training and experience help some to hit balls further, catch them more adroitly, and run bases faster—and to get paid as professionals. Kelsey's ethical reflections were so good, they became official FDA policy, and they remain FDA policy to the present day. In a sense, one could argue that she was a pioneering bioethicist, even though she never aspired to that status, and even though neither bioethicists nor historians, like Rothman and Jonsen, acknowledge her as one of the field's founders.

Jonsen and Rothman's Discussions of the Declaration of Helsinki

Rothman never mentions the WMA's Declaration of Helsinki directly. His sole reference to the Declaration is in a quotation from Beecher's 1966 *New England Journal of Medicine* whistleblowing article, "Ethics and Clinical Research." There Beecher cites one of the Declaration's principles to criticize Krugman's Willowbrook School hepatitis experiments.[17] Jonsen, in contrast, does not ignore WMA's 1964 Declaration of Helsinki; instead, he deprecates it as a morally retrograde document. He claims that it "weaken[ed] the Nuremberg Code's strong requirement for consent. It implies that the investigator who is the treating physician is judge of the best course of treatment and is permitted to omit informed consent under special circumstances."[18] Jonsen is referencing Section II, Part 1 of the 1964 Declaration of Helsinki, which reads as follows:

> **II. CLINICAL RESEARCH COMBINED**
> **WITH PROFESSIONAL CARE**
> 1. In the treatment of the sick person, the doctor must be free to use a new therapeutic measure, if in his judgment it offers hope of saving life, reestablishing health, or alleviating suffering.
>   If at all possible, consistent with patient psychology, the doctor should obtain the patient's freely given consent after the patient has been given a full explanation. In case of legal incapacity, consent should also be procured from the legal guardian; in case of physical incapacity the permission of the legal guardian replaces that of the patient.[19]

The plain language of the text restricts its application to a *physician's* desperate therapeutic innovations to benefit an *individual patient*. It does not address the primary focus of the *Belmont Report*, which developed ethical principles for *investigators* (who may be epidemiologists or other non-physicians) conducting generalizable research on *subjects* (who may not be sick patients, e.g., healthy people typically serve as subjects in vaccine research). The Declaration of Helsinki makes it clear that if a physician is attempting last-hope innovations because they seem to offer a patient a chance for life, or for reestablishing health, or alleviating suffering, the physician *should* "obtain the patient's freely given consent [informed by] a full explanation" of the proposed innovative treatment. However, and this is the part that Jonsen, omits, if the patient is legally or physically incapacitated (e.g., unconscious), such innovative life-saving interventions require the consent of a legal guardian. Thus, the Declaration does not permit last-hope medical interventions without informed consent—it just permits surrogate consent if patients are incapable of giving legally valid consent.

As Jonsen observes, the Declaration's language weakens Nuremberg Code Principle 1, which states unequivocally that "the voluntary consent of the human subject is essential." However, the change is intentional because Principle 1 made no provision for surrogacy. It thereby prohibited investigations into new forms of treatment for patients unable to legally or physically consent thus prohibiting the development of new medical treatments for children, unconscious people, and those with incapacitating mental disability or disease, or incapacitating physical disabilities (e.g., comas).[20] The authors of the Declaration of Helsinki sought to remedy this defect and also to addresses the issue of potentially life-saving innovations for patients who are themselves incapable of consent. They did so with a straightforward commonsense morally acceptable solution: permitting surrogate consent. Section III of the Declaration of Helsinki addresses a related subject, "NON-THERAPEUTIC CLINICAL RESEARCH," in a way that addresses and is designed to repair the overprotective language of the Nuremberg Code.

> 3a. Clinical research on a human being cannot be undertaken without his free consent after he has been informed; *if he is legally incompetent, the consent of the legal guardian should be procured.*[21]

It is too easy to read history backward by critiquing a few decontextualized lines written decades earlier and then scolding the authors as retrograde if their pioneering efforts are compared with later, more sophisticated, statements. Yet had pioneers not written earlier lines, more current usages are unlikely to have been as refined. More to the point, Jonsen failed to appreciate the limitations of the Nuremberg Code and so imputed a flaw in the Declaration of Helsinki where none existed. Not coincidently, by alleging this flaw, Jonsen opened the door to spotlighting the accomplishments of his colleagues and himself on the National Commission for Protecting Human Subjects.

Rothman and Jonsen Ignore Wiley and Disparage
A Patient's Bill of Rights

Wiley is neglected in both Rothman and Jonsen's narratives. Yet this uncelebrated veteran, former college professor, and African American civil rights leader successfully negotiated principles of patients' rights with the Joint Commission to ensure that Medicaid and Medicare patients would be treated in the same manner and given the same level of care and respect as other paid-for patients. Both Rothman and Jonsen mention the NWRO's negotiations with the Joint Commission and the AHA.[22] Yet neither reveals that the NWRO was an African American civil rights organization or that it was led by the well-known civil rights activist Professor George A. Wiley, PhD. Moreover, Jonsen's anodyne description of the NWRO as "a grass roots advocacy group for persons on welfare"[23] is about as apt as describing the March 7, 1965, Selma, Alabama, incident known as "Bloody Sunday" as "police rerouting pedestrian traffic on the Edmund Pettis Bridge." It is not as if Wiley was some obscure nonentity. He was a well-known civil rights leader whose actions made media headlines and whose death by drowning was reported in the *New York Times*.[24] It is difficult to imagine that Jonsen or Rothman were unaware of Wiley or did not realize that the NWRO was an African American civil rights organization. It is more likely that they were reluctant to credit authorship of A Patient's Bill of Rights, a foundational document for the bioethical turn in modern medical ethics, to an insurgent afro-wearing African American civil rights activist speaking up for black women and children on welfare. Perhaps Rothman and Jonsen thought that it would somehow tarnish the stature of bioethics as an emerging health care profession and scholarly field to acknowledge its relationship to the civil rights and welfare rights movements.

More surprisingly, Jonsen expressly derogates Wiley's major contributions to bioethics: the insertion of patient's rights documents into Medicaid and Medicare accreditation requirements and the AHA's patients' bill of rights. Jonsen acknowledges that the NWRO managed to insert patients' rights into the Joint Commission's accreditation standards—thereby forcing any health care institution accepting Medicare or Medicaid to acknowledge patients' rights—but he dismisses this accomplishment, even as he admits that, to qualify for Joint Commission accreditation, and hence for Medicare or Medicaid funds,

> hospitals printed the Bill of Rights in brochures provided to patients on admission and proudly exhibited these brochures to the [Joint Commission's] accreditation teams. Otherwise, the Patients' Bill of Rights was hardly a revolutionary

document. It was, in fact, something of a moral fraud, for the rights contained
therein were not wrest from a tyrant by an aroused and offended people but
were defined and granted by *nobles oblige*, with a content, extent, and duration
at the will of the grantor. Dr. Willard Gaylin [cofounder] of The Hastings Center
harshly criticized the Patient's Bill of Rights, describing it as "the thief lecturing
its victims on self-protection. "The Bill of Rights, he said, was nothing more than
hospitals returning to patients the legal rights that hospitals had previously stolen
from them."[25]

In fact, A Patient's Bill of Rights was everything Gaylin and Jonsen
claim it was not. It was a statement issued after the NWRO's negotiations
on behalf of an aroused and offended people—African Americans, Asian
Americans, Latinx, old age pensioners, women with dependent children,
and other welfare recipients—successfully "wrest from a tyrant," the hospitals represented by the AHA. Wiley wrested these rights by threatening
the Joint Commission and the AHA with publicity-attracting picket lines
of women, children, people with disabilities, and old age pensioners (today,
more politely called, "seniors"). These rights were not "defined and granted
by *nobles oblige*, with a content, extent, and duration at the will of the
grantor."[26] Quite the opposite, Jonsen knows that these rights were negotiated by the NWRO initially on behalf of welfare recipients and ultimately
on behalf of all hospitalized patients. Recall the words of an NWRO member: "Most private M.D.'s refuse to take us. We get whatever doctors are at
the clinic and the clinic is crowded, and we are waiting in line a long time,
and then we are pushed through. There's no time to ask questions. WE . . .
need services. That's what my group expects."[27] Speaking on behalf of such
women, Wiley protested that welfare recipients have "been required to wait
interminable lengths of time for the most meager and inadequate and most
stingily given out medical treatment . . . the poor [have] been on the short
end of the medical facilities . . . [they] had access to sophisticated miracle
treatment only as the guinea pigs. [Professional medical organizations] have
dominated the medical establishment [and] have run a colonial empire on
the poor, on the black, on the minorities in this country."[28]

A Patient's Bill of Rights was negotiated to assert the rights of welfare
recipients and later those of all clinic and hospital patients in the context
of a nonviolent populist insurrection against ableist, ageist, classist, racist,
and sexist practices perpetrated by American hospitals that viewed "charity" through the lens of a white middle-class scientistic medical paternalist
gaze. As sociologist Paul Starr (1949–) writes, this bill of rights "raised
radical questions about the prerogatives of the doctor's role. Implicit was
the belief that the interests of doctors and patients frequently diverged and

hence that patients needed protection."[29] Yet, instead of recognizing this statement as progress, Gaylin and Jonsen proffer disinformation about the source of patients' rights and, in the process, erase the African American and the civil rights leadership's role in founding bioethics as a moral insurrection. Again, not coincidently, by default, they implicitly claim credit for these achievements to predominantly white bioethics organizations, like the Hastings Center and the National Commission.

Jonsen also dismisses Americans' everyday ethical reflections as mere "moralism," which he characterizes as an American practice of making quotidian conduct (e.g., "eating ice cream, playing cards, table manners"), into a moral matter.[30] Moralism, he claims, is an artifact of America's Puritan heritage that makes Americans "strongly tempted to endow various aspects of life with moral meaning in a capricious way."[31] Americans thereby "import . . . an energy and passion . . . characteristic of moral sentiment" into everyday matters.[32] According to Jonsen, American moralism tends to mix with American, meliorism (i.e., Americans' urge to make the future better than the past), and American individualism (i.e., a can-do belief that individuals acting freely can make their own fate). In combination, these explain why bioethics was born in America.[33] "Medicine," Jonsen observes, "was particularly easy to interpret within the ethos of meliorism and moralism."[34] He concludes that "bioethics moved with a certain tranquility through the tempestuous moral turmoil of racism and warfare that was contemporaneous with its origins. . . . Still the work of revising commandments of medical morality in such a way as to permit, even encourage, medical progress was done in a relatively quiet way."[35]

Jonsen contrasts his image of the tranquil creation of bioethics, presumably by institutes and government commissions, with the "tempestuous moral turmoil of racism and warfare," of its moments of creation, that is, the massive antiracism protests of 1967 that culminated in riots in Detroit, Newark, and twenty-three other cities.[36] Not surprisingly, given Jonsen's elitist conception of the birth of bioethics, the word "racism" is absent from Jonsen's narrative, and the words "minorities," "racial," and "race" only surface to obliquely designate African Americans. Jonsen mentions just two African Americans by name: the assassinated civil rights leader, Martin Luther King Jr. (1929–1968), and Patricia King (1942–), a law professor at Georgetown University who served with Jonsen on several federal commissions. Given Jonsen's circumlocutory treatment of race and his claim that bioethics was created in a moment of tranquility that elevated it above the moral turmoil of antiracist protests embroiling America during the period of the movement's creation, it is interesting, in retrospect, to revisit the language of the 1978 *Belmont Report*, which Jonsen himself typed as he wordsmithed it with fellow commissioners. The words "race," "racism," and "racist" are notably absent from the *Report*. The word "racial" appears in three places, and the word "black" is used just once—to describe African Americans as victims. Here are the passages:

> In this country, in the 1940s, the Tuskegee syphilis study used disadvantaged, rural *black* men to study the untreated course of a disease that is by no means confined to that population. . . .
> Against this historical background, it can be seen how conceptions of justice are relevant to research involving human subjects. For example, the selection of research subjects needs to be scrutinized in order to determine whether some classes (e.g., welfare patients, particular *racial* and ethnic minorities, or persons confined to institutions) are being systematically selected simply because of their easy availability, their compromised position, or their manipulability, rather than for reasons directly related to the problem being studied. (*Belmont*, 9, 10, italics added)

> *   *   *   *

> One special instance of injustice results from the involvement of vulnerable subjects. Certain groups, such as *racial* minorities, the economically disadvantaged, the very sick, and the institutionalized may continually be sought as research subjects, owing to their ready availability in settings where research is conducted. (*Belmont Report*, 19–20, italics added)

> *   *   *   *

> Injustice may appear in the selection of subjects, even if individual subjects are selected fairly by investigators and treated fairly in the course of the research. This injustice arises from social, *racial*, sexual and cultural biases institutionalized in society. (*Belmont Report* 40, italics added)

These paragraphs misdiagnose the moral pathology underlying the USPHS's Tuskegee Syphilis Study. As Commissioner Karen Lebacqz later pointed out, the underlying issue was not use of a convenience sample of African American men, and it was not even the unfair selection of a functionally illiterate black population because they were easily manipulated: it was researchers' exploitation of an oppressed minority that could be lied to with impunity, because, as blacks in the segregated Jim Crow South of the 1930s, they lacked the status of full citizens and were not deemed worthy of respect. Thus, white investigators could and did deliberately deceive their black subjects, lying to them that they would receive treatment for "bad blood," a common name for syphilis,[37] when they were really being studied to document the progress of untreated syphilis. Worse yet, in the 1970s, after segregation had been declared unconstitutional, at a time when penicillin was readily available, the USPHS still insisted on continuing the study. Even today, the field of bioethics touches on systemic racism only obliquely by discussing "racial disparities," seldom calling out racism, systemic racism, or structural racism by name. Yet if one refuses to name a moral pathology, one cannot address it, much less remedy it, or rehabilitate a society suffering from this pathology, or compensate the communities upon which it has been inflicted.

Jonsen claims that American exceptionalism, individualism, meliorism, and moralism fused to produce bioethics, allowing it rise above the racial strife, antiwar demonstrations, and the assassinations of the 1960s. No doubt this captures his personal experience of creating seminal bioethical documents, such as when he personally typed out the final draft of the *Belmont Report*. Not surprisingly, therefore, Jonsen elides the dark sides of the birth of bioethics, erasing from his history, for example, Kelsey's retributive demotion to bare desk bureaucrat. He prefers to focus instead on the founding of institutes and commissions led by white men while, correlatively, deemphasizing or ignoring the role of blacks, females, veteran military physicians, and whistleblowers. His narrative results in a narrower, less equitable, and less accurate portrayal of America's bioethics moral revolution. One might wonder whether the authors of these "whitewashed" histories might subconsciously favor a narrative centered on founders who, like themselves, were white, middle-class, male intellectuals—and thus be inclined to overlook the role of an afro-hair-styled black civil rights organizer or a female bureaucrat. Yet by narrating an origin story of a field created almost entirely by white male intellectuals who founded institutes and served on committees, their histories offer no role models to people of color, women, or anyone else seeking to become an active change agent, as opposed to serving as an academic, a consultant, a policy analyst, or a satellite regulator. Yet activism is as much the birthright of bioethicists as commissions and thinktanks, and the field needs to recognize and emulate its heroes, heroines, martyrs, moralists, repentant sinners, and whistleblowers, black as well as white, female as well as male, unheralded as well as heralded.

Epilogue

Everything that can be thought at all can be thought clearly.
   Everything that can be said can be said clearly.
—Ludwig Wittgenstein, *Tractatus Logico-Philosophicus* 4.114

One might think that after writing over 100,000 words, I would have thoughtfully and clearly said everything that I thought should be said about the history of modern medical ethics, its bioethical turn, and the construction of histories of bioethics. However, in fairness to the reader, something should be said about how my own narrative was constructed. As I observed in the previous chapter, Albert Jonsen's *The Birth of Bioethics* reflects his personal experience as an omnipresent participant at inflection points in the field's development, whereas David Rothman's history was written from archival materials woven into a narrative by an eminent human rights activist and a social historian of medicine. In constructing my narrative, I relied on a mix of archival materials, many not available to Jonsen or Rothman, and I drew on my experience as a participant observer in bioethics as it evolved over the past half-century.

I became involved with the field that became bioethics in Detroit, Michigan, when my neighbor, friend, fellow civil rights activist, and Wayne State University colleague, Dr. Paul Lowinger (1923–2013), got me off picket lines to work with him on a state government commission.[1] In 1973 we became involved in an early bioethics legal case, *Kaimowitz v. Michigan Department of Mental Health*.[2] The underlying question in the case was whether an incarcerated criminal-sexual-psychopath could give informed voluntary consent to an amygdalectomy, a form of psychosurgery. In assenting, the patient, who was incarcerated in a state hospital for the criminally insane, reportedly told the researchers that they could "cut any part of my brain, just leave my balls alone."

In conducting research in opposition to the proposed amygdalectomy, I discovered Jay Katz and Alex Capron's 1972 book, *Experimentation with*

Human Beings. Their book detailed the Southam cancer implant experiments, the Nuremberg Doctors Trial, and the "Nuremberg Code." I soon taught a seminar on these issues, and Paul and I organized a conference on research ethics for physicians in the Detroit area. We later presented a panel on the rights of incarcerated patients at an American Psychiatric Association meeting—with the formerly incarcerated serial rapist as a co-panelist.

Senior faculty and trustees of my university deemed these and related activities unprofessional: particularly my penchant for organizing picket lines, but they focused on an essay I wrote on the language of racism and sexism that had been published in a local feminist journal, *Moving Out*. After my actions and writings were deemed untenurable, I found a more tolerant academic home at Union College, a small private liberal arts college in Schenectady, New York (founded in 1795). In 1974, I transitioned from dissident civil rights and patients' rights advocate to "bioethicist" after fellow philosopher, Samuel Gorovitz, recruited me for his Council of Philosophical Studies Summer Institute on Morals and Medicine: a training camp preparing philosophers to participate in the emerging field of bioethics.[3] After that, I contributed to the first edition of Warren Reich's *Encyclopedia of Bioethics*, taught a summer seminar at the Kennedy Institute of Ethics, and organized conferences on the history of medical ethics in London and New York. For several decades, I consulted for upstate New York health care practitioners forming hospital ethics committees and IRBs. About twenty years ago, I pioneered a hybrid onsite/online master's program to make graduate bioethics education accessible to working health care practitioners in upstate New York and, eventually, anywhere in the world. I relate these autobiographical details because, whereas the Jonsen narrative draws on his experience at seminal moments in the formation of bioethics, my narrative draws on my personal experiences at the periphery of the field.

On a more personal level, I grew up as an asthmatic youth living in the Bronx projects (public housing for poor people in an outer borough of New York City). So, Wiley's words about standing for interminably long times to see doctors leapt off the page at me. I knew from my childhood experiences what it meant to stand waiting in hallways and watching my widowed mother attempt to deal with condescending physicians and nurses who pushed us through after short, often unhelpful, explanations. Later in my life, I joined picket lines in protest.

In 2019, I drew on my personal experiences in nonviolent protest movements to write *The Structure of Moral Revolutions*, a theoretical analysis of successful moral insurrections that culminate in moral revolutions. In that book, as in this one, I make the case that the conceptual, social, occupational, and professional spaces for bioethics as a research, scholarly, and clinical enterprise were, in large measure, artifacts of a revolutionary moral insurrection that culminated in a moral revolution. In that book, as in this one, I modeled my analysis on Thomas Kuhn's analysis of scientific revolutions. I contend in both books that moral insurrections and successful moral revolutions, like their scientific counterparts, have a structure integral to their success or failure. A primary feature of those moral insurrections successful enough to be counted as moral revolutions is (i) a community's acceptance of a moral paradigm that is incompatible with and displaces a moral paradigm previously accepted by that same community.[4] In the bioethics revolution of the 1970s, the previously accepted scientistic paternalist paradigm ("Doctors as 'Lords Almighty'") was displaced by a more egalitarian conception of therapeutic partnership that supported patients' and research subjects' rights to understand, participate in, and accept or refuse proposed treatments or experiments. As this paradigm displacement was unfolding, bureaucrats (e.g., Kelsey), policy makers (Javits, Kefauver, Kennedy), and foundations (e.g., the Kennedy and the Rockefeller foundations) came to recognize that researchers' conflicted interests and loyalties sometimes overrode their obligations to their patients and research subjects. Instances of this are evident in Krugman's hepatitis experiments at Willowbrook, Southam's cancer experiments in prisons and at Brooklyn Jewish Chronic Disease Hospital, and the USPHS's study of untreated syphilis in hundreds of African American men.

For decades, such morally problematic research had been published in leading medical journals, and neither the journal editors nor the generations of health care professionals reading these articles were woke to, or cognizant of, the ethical questions they raised. In the 1960s and 1970s, (ii) newcomers unacculturated to the cultures of biomedicine (e.g., Buxtun, Hyman, Senator Javits, Jenkins, Pappworth's students) and various dissident medical intelligentsia (Beecher, Gibson, Kelsey, Pappworth, Schatz, Stollerman) saw, or became woke to, (iii) something morally anomalous— that is, that these experiments were inconsistent with health care professionals' claims to be acting in the interests of patients. This was immediately evident to Buxton when he found out that the CDC, the very agency whose mission was to prevent the spread of STIs, had reprimanded a physician for *treating* a patient's syphilis. Prompted by such anomalies, dissident intelligentsia and unacculturated newcomers (iv) developed arguments critiquing the traditional scientistic paternalist paradigm. Kelsey, for example, recognized as anomalous the contention that "the judgment of the investigator is . . . sufficient as a basis for reaching a conclusion concerning the ethical and moral set of questions in [the investigator–subject] relationship."[5] Similarly, Beecher noticed that research subjects would never have "volunteered" for the various experiments he recognized as unethical had they been aware of what they were actually agreeing to.

What effective moralist whistleblowers had in common was not gender, religion, race, or nationality but their status as empowered outsiders: a World War II veteran who became a disciple of Dorothy Day's Catholic worker movement, a Jew in *Wilhelmine* Germany, a female in a male bureaucracy, a full-immersion Methodist from a hick town in Kansas living as an upscale Boston Brahmin, or a black person in the antiblack racist post–World War II segregationist American South or the structurally apartheid American North. Common to almost all of them was a knowledge of the Nuremberg Doctors Trial and/or the document that they referred to as, "The Nuremberg Code." By virtue of their histories, each of them was able to see immorality where their colleagues and contemporaries saw none. Moreover, each of them had the gumption to call attention to them. However, as the history of moral revolutionaries makes apparent in the examples of Moll in Germany and Gibson, Jenkins, and Schatz in the US: although the arguments and protests of morally sighted dissidents are necessary steps toward moral insurrections and potentially successful revolutions, they are insufficient in themselves to displace an established paradigm. Displacement can occur only if some dissidents (v) propose one or more incompatible alternative paradigms (e.g., Beauchamp and Childress's four principles, Belmont's three principles, the Moll/Veatch contractarian model) that (vi) introduce new concepts (e.g., autonomy, respect for persons) or repurpose traditional concepts (e.g., replacing *primum non nocere* with a utilitarian conception of balancing beneficence against nonmaleficence) that (vii) they support by citing new forms of evidence (e.g., Kelsey's documentation that women given thalidomide were not informed that drug was not FDA approved; Beecher's 1966 publication of twenty-two experiments in which researchers' exploitation of patently uninformed research subjects is evident). The dissidents must also (viii) employ incommensurable ethical criteria for assessing morality (e.g., documented informed consent). Successful displacement also requires (ix) effective dissemination of the new paradigm and its principles, whether implicit in the 1962–1964 amendments to the Food, Drug, and Cosmetics Act and attendant regulations, or as A Patient's Bill of Rights (disseminated by AHA hospitals and *Our Bodies, Our Selves*) and, most effectively, in the 1978 *Belmont Report* and Beauchamp and Childress's 1979 *Principles of Biomedical Ethics*—whose first edition included the AHA's A Patient's Bill of Rights and the Nuremberg Code. Successful dissemination of the new paradigm and arguments favoring it allowed dissidents to (x) recruit new adherents who came to believe that these alternative paradigms (xi) demonstrated superior ability at resolving problems that appeared unresolvable under the currently established paradigm (e.g., they reaffirmed biomedical researchers' trustworthiness, thereby securing continued funding for research and new technologies, just as patient or surrogate consent legitimized DNR orders). Resolving issues vexing the biomedical community and their sponsors and funders, in turn, led to (xii) these communities' general, but not entirely universal, acceptance of a new bioethics paradigm, which was also accepted by most media and other influencers and most powerbrokers (e.g., federal agencies, foundations). The new paradigm thereby (xiii) obsolesced the older paradigm (scientistic paternalism), its concepts, and its associated discourse (e.g., "doctor knows best!") and practices (experimenting without fully informing subjects of attendant risks), leading to a phenomenon known as "Kuhn loss" (i.e., the loss of explanations and practices that made sense under the traditional paradigm but became incomprehensible or obsolete under the new paradigm) and Appiah's questions (How could we once have done that? What were we thinking?)

Books dictate what they want to become. This book was originally titled *Oaths, Codes, and Scandals: Moralists and the Making of Modern Medical Ethics*. It had two themes: (1) the scandal reform/revolution cycle and its role in creating modern medical ethics and (2) developing an alternative to American exceptionalist accounts of the birth of bioethics by linking bioethics to WMA's innovations in modern medical ethics and tracing both back to their post–World War II international roots in the Nuremberg Doctors Trial. The phenomenon of moralists, of people who can see immorality where others are blind, fascinated me, as is evident in the book's first chapters. After COVID-19 hit in 2019, however, I retired from my Union College professorship and used the time to review my notes and explore some additional themes. Since COVID-19 precluded visits to archives, I also reread various histories of bioethics and research ethics. As I reviewed these and other materials, I discovered that the person responsible for A Patient's Bill of Rights, George A. Wiley, was an African American civil rights activist. Other sources revealed Kelsey's role in inserting an informed consent requirement into the Kefauver–Harris amendment and her later demotion for insisting that researchers inform subjects and document their consent. I soon began to test my ideas in short papers that I submitted to journals. Yet reviewers repeatedly rejected any notion of a significant African American, female, or feminist role in the history bioethics, so I made what was initially an interesting detail one of this book's major themes.[6]

A word about the relevance of history to bioethics. As founding chair of the American Society for Bioethics and the Humanities' Affinity group on the History of Medical Ethics, I realize that most bioethicists and bioethics conference organizers focus on the present and the future, discounting the past as mere prologue. But the pragmatics of path dependence (i.e., the presumed efficiency of following paths previously blazed and thus retaining past conceptions and extending past practices into new domains) tend to render bioethical policy makers prisoners of the past, blinding them to alternative approaches. History—one of the world's richest data bases—can and should serve as an antidote to path dependence by allowing the deconstruction of the apparent self-evidence of current ideas and practices.

It is a peculiar feature of my personal development that although I started out to become a historian, I earned my doctorate at an interdisciplinary philosophy of science center, and I ended up writing a dissertation on G. E. Moore, a foundational figure for modern analytic ethics. I ultimately became a philosophically trained bioethicist who writes histories (WorldCat lists me as a historian).[7] As both historian and bioethicist, I constructed this alternative narrative because I believe, to use more memorable words penned independently by both an Anglo-Irish statesman, Edmund Burke (1729–1797), and a Spanish American atheist-Catholic philosopher and poet, Jorge Agustín Nicolás Ruiz de Santayana y Borrás, better known in English as George Santayana (1863–1952): "Those who cannot remember the past are condemned to repeat it."[8]

Here is a case in point: historical understanding can be used to highlight the limitations of autonomy as a bioethical concept. During the COVID-19 pandemic, it became apparent to many bioethicists that vaccine-hesitant and mask-refusing members of the public (and, more surprisingly, some health care professionals) justified mask refusals by appealing to the ideal of autonomy. In point of historical fact, the *Belmont Report* tried to eschew this terminology. As Jonsen and Lebacqz testify, and as Beauchamp admits in his personal reflections, the National Commission asked him to write about the principle of *respect for persons*. Yet, as Lysaught demonstrates, Beauchamp eviscerated that concept by transforming respect for *persons* into respect for *autonomy*, which he later represented by a single-word synecdoche, "autonomy," that fit with his three other one-word synecdoche-denominated principles: beneficence, nonmaleficence, and justice.

Beauchamp and Childress's four-word mnemonic soon dominated the field; however, it could be reconciled with the original Belmont principles only if the neo-Kantian concept "autonomy" was a nomological doppel-ganger for the commissioners' preferred Kantian-derived concept, "respect for persons." As became apparent during the COVID-19 pandemic, the doppelganger was an imposter. "Respect for persons"—note the plural—implies both self-respect and respect for other persons—perhaps even, as Lebacqz suggests, respect for others' cultures insofar as they are integral to other persons' identity. In contrast, autonomy, which, as Veatch points out, literally means "self-rule," is inherently inward looking. Unlike "respect for persons," whose plural ending applies to other persons as well as oneself, "autonomy" is morally solipsistic. It does not include the idea of respect for *other* persons, only for the right of autonomous persons to make and act on their decisions. Not surprisingly, therefore, the vaccine-hesitant and mask-refusing public appropriated the ideal of autonomy as a moral jus-tification for their refusals: "My bodies, My choices."[9] Flummoxed, the American bioethics community was mostly at a loss for words, unable to formulate an effective conceptual response.[10] Yet, anyone knowledgeable about the lexigraphic history of bioethical terminology would naturally have concluded that refusals to wear masks violated the duty of respect for other persons. More to the point, had Beauchamp not substituted *autonomy* as a nomological doppelganger for the Engelhardt–Jonsen–Lebacqz–Ramsey concept of "respect for persons," bioethicists would have had better conceptual tools for confronting the issues of vaccine hesitancy and mask resistance. A knowledge of history opens the potential for alternative paths to explore in the future.

A final factor motivating me to develop an alternative narrative was the narrowness of standard histories that I reread during the COVID-19 pan-demic. They presented bioethics as if it was primarily a field of research, regulations, and related scholarship that supports a health care specializa-tion whose origins are to be found in commission reports, court cases, and thinktanks. The only whistleblowing moralist suitable for robust treat-ment in their narratives was Harvard's Professor Henry K. Beecher. But it was not Beecher who incited Senator Kennedy's hearings; it was Buxtun, a USPHS contact tracer, who outed the USPHS's Tuskegee Study to the Asso-ciated Press. And it was the ensuing scandal that led to the Kennedy hear-ings, which, in turn, led to the National Research Act and, ultimately, to the first national bioethics commission and the *Belmont Report*—thereby validating what I call the bioethical turn in modern medical ethics. Looking more intently at history, it should be apparent that it was not some medi-cal luminary whose words prompted Beecher to write "Ethics and Clinical Research"; it was a cantankerous British Jew, Maurice Pappworth, who served as his role model. Moreover, it was a reporter's account of a lawsuit brought by a relatively unknown Jewish lawyer, William A. Hyman, that led to the article in *Science* that inspired Beecher to contact Pappworth, and it was during this correspondence that Beecher decided to emulate Pappworth by blowing the proverbial whistle on unethical experiments.

Speaking of Beecher: I believe it was a mistake for the Hastings Center to expunge the name "Henry K. Beecher" from its newly renamed "Founders Award." It is undeniably true that during the Cold War, Beecher was involved with the CIA's mescaline and LSD experiments and conducted experiments in which he gave psychoactive drugs to uninformed students and others. It is also true that, as David Rothman observes, that in this respect Beecher was like many other American medical researchers during the Cold War. Recall that Beecher had been a frontline medic rationing penicillin during the brutal North African campaign. In that context, he had to deny it to those suffering from war wounds, allocating it instead to those with sexually transmitted diseases (STIs) because those with STIs could be rapidly returned to the front lines, whereas those with deeper wounds could not. When the Cold War broke out, Beecher's country, represented by the CIA and the US Army, called on him once more. It sought his assistance in exploring the utility of psychedelic drugs as truth serums. As he had in the past, Beecher answered his country's call. Working with a secret government contract and apparently hoping to avoid selection bias (nocebo and placebo—i.e., negative, and positive—effects induced by knowledge that one might receive a psychoactive drug), and mindful that the research was subject to national security considerations, Beecher hid a pivotal fact from his subjects: they would be in an experiment in which some or all participants would receive psychoactive drugs. As it turned out, Daniel Callahan—cofounder of the Hastings Center—was one of his subjects in a related army-funded experiment.

Yet Beecher's conscience nagged at him. So, during this period, he published a series of papers on the ethics of research on human subjects. Then he did something that no other well-established American physician-investigator had done. He risked his career, his standing in his profession, everything he had achieved since leaving Peck, Kansas, to denounce unethical experiments to the media. As Beecher anticipated, there was immediate blowback, from his Harvard colleagues and from many others. Nonetheless, in the end, it worked out well for Beecher, who became a hero of bioethics.

I have sympathy for Beecher because, growing up in Bronx public housing, I imbibed a culture that was ableist, ageist, homophobic (anti-LGBTQ+), racist, sexist, and so forth. It was also classist, in the sense that we project kids not only knew that our neighbors living in private housing believed they were better than us; we knew they were right: they were better than us. I spent much of my adult life struggling to overcome my inherited prejudices. So, I am sympathetic to a fellow repentant sinner, Henry Knowles Beecher, who recognized the error of his ways and who attempted to change the culture that made immoral acts seem permissible. None of us are flawless, but I find those who recognize and attempt to remedy their past flaws particularly admirable.

As to Beecher himself, I never met the man. Those who knew him, like Veatch, "fondly remember [his] time with Henry Beecher. He was a giant in stimulating real advances in the relationship between medical scientists and the public, providing major contributions to the pharmacology of anesthesia, the science and ethics of placebos, the definition of death, and, most critically . . . clinical research ethics. He was the profession's conscience, desperately trying to use his considerable influence to get investigators to more conscientiously reflect on the ethics of their projects." In doing so, to reiterate a point because it deserves emphasis, Beecher did something that neither Krugman, nor Southam, nor the USPHS's Tuskegee physicians did; he risked his career, his reputation, everything he had worked for since changing his name from "Unangst" to "Beecher," to reveal and publicly denounce morally questionable experiments to the media. Callahan seemed to recognize the significance of unblinding the medical field to the quotidian immorality of many of its routine research practices, and he pronounced Beecher a hero of the emerging field of bioethics. So did many others until a few bioethicists, judging Beecher's misdeeds and lack of public contrition more significant than his heroic act of outing the quotidian immorality of medical research, condemned him to postmortem ostracism. To conclude with words from another famous sinner, Lord Byron (1788–1824), those who condemn impure reformers seem to forget that "more than Philosophy can preach and vainly preach'd before . . . [most of us] have feet of clay."[11]

Appendix

A. Rules for *Permissible Research on Humans:* The First Draft, 1946

In 1946 the AMA chose Andrew Ivy as its representative at the Nuremberg Doctors' Trial. As Ivy explained in an April 6, 1966, letter to Maurice Pappworth, "I accepted the invitation to serve at the Nuremberg Trials only because I had in mind the objective of placing in an international judicial decision the conditions under which human beings may serve as subjects in a medical experiment, so that these conditions would become the international common law on the subject. Otherwise, I would have had nothing to do with the nasty and obnoxious business. I believe in prevention, not a 'punitive cure' . . . The Judges and I were determined that something of a preventative nature had to come out of the Nuremberg Trial of the Medical Atrocities."[1] These sentiments reflect Ivy's pacifist inclinations which were also evident in his refusal to accept a military rank. On a more practical level Ivy was also concerned that "the publicity of the medical trial does not stir public opinion against the ethical use of human subjects for experiments."[2] To forestall such an outcome he "felt that some broad principles should be formulated by . . . enumerating the criteria for the use of humans as subjects in experimental work."[3]

   To initiate his plan for creating an official statement of humanitarian international law on principles for "permissible experiments on humans," Ivy presented these three principles at a July-August 1946 International Scientific Commission (ISC) meeting at the Pasteur Institute in Paris, which he attended as Special Consultant to the Secretary of War. These, were, in effect, the first draft of the document later called "The Nuremberg Code." Historian Ulf Schmidt comments that "Here, in August 1946, we find the rationale for the creation of principles for permissible experimentation on humans in embryonic form. . . . Ivy's 'broad ethical rules' which he subsequently produced, adopted, and publicized for the purpose of contrasting ethical and unethical experimentation on human subjects."[4] Ivy presented this draft statement of principles to the ISC in August 1946, about six months *before* the commencement of the Nuremberg Doctors' Trial in December 1946.[5]

Rules for Ethically Permissible Experiments on Human Subjects (First Draft of "The Nuremberg Code")

August 1946
I. Consent of the human subjects is required, i.e., only volunteers should be used.
   a. The volunteers before giving their consent should be told of the hazards, if any.
   b. Insurance against an accident should be provided; if it is possible to secure it.
II. The experiment to be performed should be so designed and based on the results of animal experimentation that the anticipated results will justify the performance of the experiment; that is, the experiment must be such as to yield results for the good of society.
III. The experiment must be conducted
   a. so as to avoid all unnecessary physical and mental suffering and injury, and
   b. by scientifically qualified persons
   c. The experiment should not be conducted if there is any *a priori* reason to believe that death or disabling injury will occur.[6]

B. "Permissible Research on Humans," Ivy's Report to the AMA, December 1946

In December 1946, Ivy sent a revised version of his principles to the AMA in a "Report on War Crimes of a Medical Nature Committed in Germany and Elsewhere on German Nationals and the Nationals of Occupied Countries by the Nazi Régime during World War II." This became, in effect, his second draft of "Permissible Experiments on Humans." The AMA officially adopted only three of the proposed principles: I (Informed Consent), II (Prior Animal Experimentation) and III (Supervision by a medically qualified professional). A version of these would later be reformulated in Principles 1, 3, and 8 in the final version of the document later called "The Nuremberg Code."

Ethically Permissible Experiments on Human Subjects
Human experimentation has been conducted according to certain ethical rules in all the countries of the world which have contributed to the progress of medical science, i.e., to the prevention, cure and control of disease and suffering.

I. Consent of the human subjects must be obtained. All subjects must have been volunteers in the absence of coercion in any form. Before volunteering the subjects have been informed of the hazards, if any. (In the U.S.A. during

B. "Permissible Research on Humans," Ivy's Report to the AMA, December 1946 (continued)

the war, accident insurance against the remote chance of injury, disability, and death was provided.)

II. The experiment to be performed must be so designed and based on the results of animal experimentation and a knowledge of the natural history of the disease under study that the anticipated results will justify the performance of the experiment. That is, the experiment must be such as to yield results for the good of society unprocurable by other means of study and must not be random and unnecessary in nature.

III. The experiment must be conducted

(a) only by scientifically qualified persons, and

(b) so as to avoid all unnecessary physical and mental suffering and injury, and

(c) so that, on the basis of the results of previous adequate animal experimentation, there is no *a priori* reason to believe that death or disabling injury will occur, except in such experiments as those on Yellow Fever where the experimenters serve as subjects along with non-scientific personnel.

The involved Nazi physicians and scientists ignored these ethical principles and rules, which have been well established by custom, social usage, and the ethics of moral conduct, and which are necessary to insure [*sic*] the human rights of the individual and to avoid the debasement of a method of doing good and the loss of faith of the public in the medical profession. In fact, the Nazis, working under and in cooperation with [Heinrich] Himmler, Dr. [Karl] Brandt, Dr. [Leonardo] Conti, Dr. [Ernst Robert] Grawitz and others, used human beings with decidedly less considerations, as a rule, as regards pain and hygienic care, than is given to animals when they are used as experimental subjects according to the rules of the American Medical Association, which rules of practice are common to medical research laboratories throughout the world.[7]

Later in the report Ivy comments that the Hippocratic Oath served to facilitate "The faith of the patient and the public in the honorable conduct of the physician and profession [which] must be maintained. It obviously cannot be maintained if experimentation on human subjects without their consent is condoned. Such conduct, if not condemned, would cause the patient and the public to fear the physician and operate to decrease his service to society."[8]

C. "Permissible Research on Humans," Final Version of August 20, 1947[9]

Ivy's proposed principles were revised with extensive commentaries by Leo Alexander in December 1946, and in January and April 1947. (For details see Paul Weindling's analysis in *Nazi Medicine and the Nuremberg Trials*.[10]) Originally the code was called "Permissible Medical Experiments," appropriately reflecting Ivy's concern to distinguish morally permissible experiments on humans from the immoral and impermissible experiments conducted by the Nazi doctors condemned at Nuremberg. The document was rechristened, "The Nuremberg Code," in the 1950s or 1960s when it became an instrument for challenging experiments on humans deemed morally impermissible because they seemed analogous to the experiments condemned by the Nuremberg Tribunal. As early as 1963, William Hyman, the lawyer challenging Southam's cancer experiments at Jewish Chronic Disease Hospital, characterized the list of "Permissible Medical Experiments" as "The Nuremberg Code." Katz and Capon used Hyman's characterization, "The Nuremberg Code," in their discursively encyclopedic groundbreaking 1972 book, *Experimentation with Human Beings*. My assistant, Sue Pickard, and I, in turn, used their book the next year (1973) in preparing materials that Gabe Kaimowitz (1935–), a lifelong civil rights advocate, used in *Kaimowitz v. Michigan Department of Mental Health*.[11]

There may have been earlier uses of this appellation as a convenient way of referencing the ten principles for permissible medical experiments, but because Hyman's usage, as disseminated in Katz and Capron's seminal work, fit the needs of a fledgling movement during the formative period of the field now called, "bioethics," their rechristening remains the name of the document today called, "The Nuremberg Code."

Permissible Medical Experiments
The great weight of the evidence before us is to the effect that certain types of medical experiments on human beings, when kept within reasonably well-defined bounds, conform to the ethics of the medical profession generally. The protagonists of the practice of human experimentation justify their views on the basis that such experiments yield results for the good of society that are unprocurable by other methods or means of study. All agree, however, that certain basic principles must be observed in order to satisfy moral, ethical and legal concepts:

1. The voluntary consent of the human subject is absolutely essential.
This means that the person involved should have legal capacity to give consent; should be so situated as to be able to exercise free power of choice, without the intervention of any element of force, fraud, deceit, duress, overreaching, or other ulterior form of constraint or coercion; and should have sufficient knowledge and comprehension of the elements of the subject matter involved as to enable him to make an understanding and enlightened decision. This latter element requires that before the acceptance of an affirmative

C. "Permissible Research on Humans," Final Version of August 20, 1947[9] (continued)

decision by the experimental subject there should be made known to him the nature, duration, and purpose of the experiment; the method and means by which it is to be conducted; all inconveniences and hazards reasonably to be expected; and the effects upon his health or person which may possibly come from his participation in the experiment.

The duty and responsibility for ascertaining the quality of the consent rests upon each individual who initiates, directs, or engages in the experiment. It is a personal duty and responsibility which may not be delegated to another with impunity.

2. The experiment should be such as to yield fruitful results for the good of society, unprocurable by other methods or means of study, and not random and unnecessary in nature.

3. The experiment should be so designed and based on the results of animal experimentation and a knowledge of the natural history of the disease or other problem under study that the anticipated results will justify the performance of the experiment.

4. The experiment should be so conducted as to avoid all unnecessary physical and mental suffering and injury.

5. No experiment should be conducted where there is an a priori reason to believe that death or disabling injury will occur; except, perhaps, in those experiments where the experimental physicians also serve as subjects.

6. The degree of risk to be taken should never exceed that determined by the humanitarian importance of the problem to be solved by the experiment.

7. Proper preparations should be made, and adequate facilities provided to protect the experimental subject against even remote possibilities of injury, disability, or death.

8. The experiment should be conducted only by scientifically qualified persons. The highest degree of skill and care should be required through all stages of the experiment of those who conduct or engage in the experiment.

9. During the course of the experiment the human subject should be at liberty to bring the experiment to an end if he has reached the physical or mental state where continuation of the experiment seems to him to be impossible.

10. During the course of the experiment the scientist in charge must be prepared to terminate the experiment at any stage, if he has probable cause to believe, in the exercise of the good faith, superior skill and careful judgment required of him that a continuation of the experiment is likely to result in injury, disability, or death to the experimental subject.

D. A Human Right to Healthcare Access (George A. Wiley's Demands)

The following are a summary of Wiley's June 26, 1970, demands for patients' rights standards put forward by the National Welfare Rights Organization (NWRO)[12]

1) The patient has the right to considerate and respectful care.

2) The patient has the right to obtain from his physician complete current information concerning his diagnosis, treatment, and prognosis in terms the patient can be reasonably expected to understand. When it is not medically advisable to give such information to the patient, the information should be made available to an appropriate person in his behalf. He has the right to know, by name, the physician responsible for his care.

3) The patient has the right to receive from his physician information necessary to give informed consent prior to the start of any procedure and/or treatment. Except in emergencies, such information for informed consent should include but not necessarily be limited to the specific procedure and/or treatment, the medically significant risks involved, and the probable duration of incapacitation. Where medically significant alternatives for care or treatment exist, or when the patient requests information concerning medical alternatives, the patient has the right to such information. The patient also has the right to know the name of the person responsible for the procedures and/or treatment.

4) Publicity of accreditation status. Hospital must prominently display JCAH accreditation status—pass or fail—at the entrance of the hospital.

5) Community representation. In defining its relationship to the community (an existing standard), the institution's governing board of any hospital that is publicly owned, or voluntary receiving tax-exemption, must include 51% of its membership from the community it serves.

6) Patient ombudsman/patient advocate. In order "to resolve patient complaints about billing, quality of care or conditions under which that care is rendered," the hospital must provide for an ombudsman or patient advocate to coordinate an administrative hearing mechanism. This ombudsman or patient advocate must not be an employee of the hospital but may be contracted from community organizations.

7) Defined catchment area responsibility. Accreditation must be conditioned on the hospital accepting the responsibility of the delivery of defined health services to the local community ("catchment area"). The delivery responsibility should include emergency ward, outpatient, ambulatory inpatient, and preventive health services. The "catchment area" will be determined by agencies external to the hospital and should not be interpreted to exclude care for populations from outside the hospitals area.

8) Open medical staff. Medical staff privileges will not be denied physicians because of their relationship to or employment by any specific delivery mechanism that is different from the established fee-for-service system.

D. A Human Right to Healthcare Access (George A. Wiley's Demands) (continued)

9) Open records and data collection. All financial records of all accredited institutions must be regarded as matter of public knowledge. Similarly, minutes of the governing board, medical audit, and evaluation and medical staff meeting should be open. Admission and collection policies should be written and made available. Collection of statistics comparing ward and private accommodations should include data on every aspect of patient care (LOS, Dx, Lab tests, staff ratios, etc.).

10) Community-based needs assessment. No plan for or expansion of service may be developed without a determination that the plan meets with the community views on organization and delivery by consultation with groups representing the community the hospital serves.

11) No economic discrimination. Degree or kind of service provided a patient may not be determined solely because of the patient's source of payment, social, economic, racial, or religious status. Private or semiprivate accommodations must be provided for Medicare or Medicaid patients if these accommodations are available.

12) Community-based utilization review. Hospitals must be required to include doctors from the community who have no affiliation or proprietary or financial interest in the hospital to serve on utilization and quality control reviews.

13) Anti-dumping standard. A patient will not be transferred or referred to another facility unless it is "medically indicated" (without regard to mechanism of payment or to social, economic, or racial criteria) and then only when the patient's records are complete and up to date and the receiving institution has consented to the transfer.

14) Timeliness of laboratory services. Accredited hospitals must have laboratory services—not merely the facilities to provide them—available to respond in a timely manner and in the consideration of the physician judgment for the need of the test and the condition of the patient.

15) Privacy. Proper facilities must be ensured by accredited institution for the privacy of patients, their bodies, and disclosures in conversation made during the course of examination in the emergency department, the clinics, and inpatient settings.

16) Investigating agencies. Officials of investigation agencies (presumably, not the JCAH; see #3 above) must not be permitted to interview patients in hospitals without the written consent of the attending physician, and the investigator may not be permitted to be present during actual examination or treatment of the patient.

17) Triage and triage personnel. Accredited hospitals must have an adequate, effective, and expeditious, triage system and personnel appropriate to the minority and foreign-language-speaking population of the local community in order to ensure that each and every applicant for services is understood and receives the same standard of care.

18) Communication. The hospital must have written policy and instruction for personnel to actively advise patient of what is happening to them, what is expected of them, and the nature of the medical tests being performed upon them.

19) Surgical education. No patient shall be operated upon by a resident or intern unless the patient knows and explicitly consents to having these personnel perform the procedures.

20) Medical students. No patient shall be observed by medical students not directly participating in his treatment without his written consent.

21) Diagnostic tests. Hospital regulation will provide that there will be no greater number of diagnostic tests, X rays, or other laboratory procedures performed on ward patients than are performed on patients in other accommodations for the treatment of a particular condition. Nor shall the status of the ward patient be a basis for the solicitation of consent for the purposes of surgical procedures or experimentation.

22) Equal admission. No hospital shall refuse admission to Medicare or Medicaid patients where beds are available.

23) Free care. All hospitals must provide a reasonable quantum of free or below cost patient care.

24) Open emergency department. Hospitals with emergency facilities may not limit those facilities by informing emergency ambulance services in the community not to bring emergency patient to that hospital.

25) Access to JCAH reports. Inspection reports prepared by the JCAH surveyor should be made available to any patient, community group served by the hospital, or staff of the hospital requesting such a report. In keeping with recommendation #3, the surveyors must make explicit their finding with respect to specific complaints and/or allegation made at a hearing or meeting.

26) Nursing response. Hospitals must ensure adequate mechanisms for patients to contact nursing personal from their beds and the prompt response for the nursing staff.

E. Statement on a Patient's Bill of Rights, Affirmed by the Board of Trustees

November 17, 1972

The American Hospital Association presents a Patient's Bill of Rights with the expectation that observance of these rights will contribute to more effective patient care and greater satisfaction for the patient, his physician, and the hospital organization. Further, the Association presents these rights in the expectation that they will be supported by the hospital on behalf of its patients as an integral part of the healing process. It is recognized that a personal relationship between the physician and the patient is essential for the provision of proper medical care. The traditional physician-patient relationship takes on a new dimension when care is rendered within an organizational structure. Legal precedent has established that the institution itself also has a responsibility to the patient. It is in recognition of these factors that these rights are affirmed.

1. The patient has the right to considerate and respectful care.
2. The patient has the right to obtain from his physician complete current information concerning his diagnosis, treatment, and prognosis in terms the patient can be reasonably expected to understand. When it is not medically advisable to give such information to the patient, the information should be made available to an appropriate person in his behalf. He has the right to know by name, the physician responsible for coordinating his care.
3. The patient has the right to receive from his physician information necessary to give informed consent prior to the start of any procedure and/or treatment. Except in emergencies, such information for informed consent, should include but not necessarily be limited to the specific procedure and/or treatment, the medically significant risks involved, and ·the probable duration of incapacitation. Where medically significant alternatives for care or treatment exist, or when the patient requests information concerning medical alternatives, the patient has the right to such information. The patient also has the right to know the name of the person responsible for the procedures and/or treatment.
4. The patient has the right to refuse treatment to the extent permitted by law, and to be informed of the medical consequences of his action.
5. The patient has the right to every consideration of his privacy concerning his own medical care program. Case discussion, consultation, examination, and treatment are confidential and should be conducted discreetly. Those not directly involved in his care must have the permission of the patient to be present.
6. The patient has the right to expect that all communications and records pertaining to his care should be treated as confidential.
7. The patient has the right to expect that within its capacity a hospital must make reasonable response to the request of a patient for services. The hospital must provide evaluation, service, and/or referral as indicated by the urgency of the case. When medically permissible a patient may be transferred to another facility only after he has received complete information

and explanation concerning the needs for and alternatives to such a transfer. The institution to which the patient is to be transferred must first have accepted the patient for transfer.

8. The patient has the right to obtain information as to any relationship of his hospital to other health care and· educational institutions insofar as his care is concerned. The patient has the right to obtain information as to the existence of any professional relationships among individuals, by name, who are treating him.

9. The patient has the right to be advised if the hospital proposes to engage in or perform human experimentation affecting his care or treatment. The patient has the right to refuse to participate in such research projects.

10. The patient has the right to expect reasonable continuity of care. He has the right to know in advance what appointment times and physicians are available and where. The patient has the right to expect that the hospital will provide a mechanism whereby he is informed by his physician or a delegate of the physician of the patient's continuing health care requirements following discharge.

11. The patient has the right to examine and receive an explanation of his bill regardless of source of payment.

12. The patient has the right to know what hospital rules and regulations apply to his conduct as a patient.

No catalogue of rights can guarantee for the patient the kind of treatment he has a right to expect. A hospital has many functions to perform, including the prevention and treatment of disease, the education of both health professionals and patients, and the conduct of clinical research. All these activities must be conducted with an overriding concern for the patient, and, above all the recognition of his dignity as a human being. Success in achieving this recognition assures success in the defense of the rights of the patient.[13]

Notes

Preface

1. Duncan Wilson, *The Making of British Bioethics* (Manchester: Manchester University Press, 2014), 44.

2. David Rothman, *Strangers at the Bedside: A History of How Law and Bioethics Transformed Medical Decision Making* (New York: Basic Books, 1991), 62.

3. Jonathan Moreno, "The Declaration of Helsinki and the American Stamp," in *Ethical Research: The Declaration of Helsinki and the Past, Present, and Future of Human Experimentation*, ed. Ulf Schmidt, Andreas Frewer, and Dominique Sprumont (New York: Oxford University Press, 2020), 351–365 at 35.

4. Rothman, *Strangers at the Bedside*, 31.

Chapter 1

1. Richard Baxter, "A Christian Directory: Or a Sum of Practical Theology and Cases of Conscience," in *The Practical Works of The Rev. Richard Baxter: With a Life of the Author and a Critical Examination of His Writings*, ed. Reverand William Orme in Twenty-Three Volumes (London: Mills, Jowett, and Mills, [1672–1673] 1830, vol. V, part IV, sec. 12), 803.

2. Baxter, "A Christian Directory," 803.

3. This English court office took depositions and oaths and was involved in legal matters involving estates, trusts, and guardianships. It was dissolved by several reform laws passed in 1854 and 1858.

4. John Stuart Mill, "Bentham," *London and Westminster Review*, August 1838, in *"Utilitarianism" and "On Liberty": Including Mill's 'Essay on Bentham' and Selections from the Writings of Jeremy Bentham and John Austin*, ed. Mary Warnock (Malden, MA: Blackwell Publishing, 2003), 55.

5. Miriam Williford, "Bentham on the Rights of Women," *Journal of the History of Ideas* 36, no. 1 (January–March 1975): 167–176; see also Lea Campos Boralevi, *Bentham and the Oppressed* (Berlin: de Gruyter, 1984).

6. Philip Schoenfield, Catherine Pease-Watkin, and Michael Quinn, eds. *Of Sexual Irregularities and Other Writings on Sexual Morality: The Collected Works of Jeremy Bentham* (Oxford: The Clarendon Press, 2014).

7. Allison Stanger, *Whistleblowers: Honesty in America from Washington to Trump* (New Haven, CT: Yale University Press, 2019), 5.

8. Stanger, *Whistleblowers*, 85.

9. Stanger, *Whistleblowers*, 85.

10. Stanger, *Whistleblowers*, 12.

11. Stanger, *Whistleblowers*, 10.

12. Stanger, *Whistleblowers*, 11.

13. Baxter, "Christian Directory," 803.

14. All biblical quotations are from the 1611 King James translation of the Holy Bible.

15. Albert Moll, *Ärztliche Ethik: Die Pflichten des Arztes in allen Beziehungen seiner Thätigkeit* (Stuttgart: Ferdinand Enke, 1902).

16. This maxim appropriates a line from Cicero's *De Legibus* (circa 52 CE, book iii, part iii, section viii) *Salus populi suprema lex esto* (i.e., "the welfare of the people shall be the supreme law"). This medical version of Cicero's maxim was common in German medical ethics of that era. Moll elevated it to foundational moral insight that justified a contractarian reassessment of the physician–patient relationship of the sort that was later explored by founding American bioethicist, Robert Veatch, in his classic work, *A Theory of Medical Ethics* (New York: Basic Books, 1981).

17. Barbara Elkeles, "The German Debate on Human Experimentation between 1880 and 1914," in *Twentieth Century Ethics of Human Subjects Research: Historical Perspectives on Values, Practices, and Regulations*, ed. Volker Roelcke and Giovanni Maio (Stuttgart: Franz Steiner Verlag, 2004), 19–33, at 26.

18. Elkeles, "German Debate," 26.

19. Elkeles, "German Debate," 19–33, 26.

20. Elkeles, "German Debate," "Debate," 26.

21. Elkeles, "German Debate," 26.

22. Moll, *Ärztliche Ethik*, 261–263. Andreas-Holger Maehle, "'God's Ethicist': Albert Moll and His Medical Ethics in Theory and Practice," *Medical History* 56, no. 2 (April 2012): 217–236.

23. Albert Neisser, "Was wissen wir von einer Serumtherapie bei Syphilis und was haben wir von ihr zu erhoffen?" *Archiv für Dermatologie und Syphilis* 44 (1898): 431–539.

24. Elkeles, "German Debate," 25.

25. Elkeles, "German Debate," 25.

26. Elkeles, "German Debate," 25.

27. Jochen Vollmann and Rolf Winau, "The Prussian Regulation of 1900: Early Ethical Standards for Human Regulation in Germany," *IRB: Ethics and Human Research* 18, no. 4 (July–August 1996): 9–11 at 10. The fine was 300 deutsche marks, but Neisser was also required to pay legal costs of 1,245 deutsche marks. His annual income was 2,400 marks; all told, the fine plus court costs amounted to two-thirds of his annual income.

28. Vollmann and Winau, "The Prussian Regulation," 11. This article contains a side-by-side literal translation of the 1900 Prussian ministry edict. All quotations from the 1900 ministry edict are from the Vollmann and Winau translation. Note: in English, any change in treatment or the context of treatment (e.g., a different way of explaining a medical procedure) could be called an "intervention," whereas the German *Eingriffe* suggests something more physical, such as surgery or, in Neisser's case, an injection.

29. Vollmann and Winau, "The Prussian Regulation," 11.

30. Maehle, "God's Ethicist," note 59.

31. Maehle, "God's Ethicist," 20.

32. Elkeles, "German Debate," 26–27.

33. Elkeles, "German Debate," 26–27.

34. Elkeles, "German Debate," 28.

Chapter 2

1. Rudolf Ramm, *Ärztlizche Rechts-und Standeskunde Der Artz als Gesundheitserzieher* (Berlin: De Gruyter, 1943); English edition, *Medical Jurisprudence and Rules of the Medical Profession*. Melvin Wayne Cooper, translator (New York: Springer International Publishing, 2019), 72, 73.

2. John S. Mill, "A Few Observations on the French Revolution: Review of Alison's *History of Europe*," *Monthly Repository*, August 1833, accessed October 7, 2022, http://www.laits.utexas.edu/poltheory/jsmill/diss-disc/french-rev.html.

3. Mill, "A Few Observations."

4. Martin Heidegger (1889–1976) was a member of the National Socialist Party from 1933 to 1945. On assuming the rectorship of Freiburg University in 1933, he delivered an infamous speech filled with the Nazi tropes about the death of liberal culture. Later, as rector, he wrote a letter in the student newspaper in which he tells students, "The Führer himself [i.e., Adolf Hitler], and he alone, is the German reality of today, and of its future, and of its law. Learn to know always more deeply. Starting now each thing demands decision [decisiveness?] and every action, responsibility. Heil Hitler! Rector Martin Heidegger." Cited in Victor Farias, *Heidegger and Nazism*, English translation by Paul Burrell (Philadelphia: Temple University Press, 1987), 108, 118, 119.

5. Ramm, *Rules of the Medical Profession*, 40–41.

6. Cicero, *De Legibus* (Book iii, Part iii, Section viii).

7. Ramm, *Rules of the Medical Profession*, 117.

8. For a more detailed explanation of my use of these terms, see R. Baker, *The Structure of Moral Revolutions: Studies of Changes in the Morality of Abortion, Death, and the Bioethics Revolution* (Cambridge, MA: The MIT Press, 2019), 13–21.

9. Robert Proctor, "Nazi Doctors, Racial Medicine, and Human Experimentation," in *Nazi Doctors and Nuremberg Code*, ed. George J. Annas and Michael A. Grodin (New York: Oxford University Press, 1992), 19–20.

10. Andreas Frewer, "Debates on Human Experimentation in Weimar and Early Nazi Germany as Reflected in the Journal 'Ethik,'" in *Twentieth Century Ethics of Human Subjects Research: Historical Perspectives on Values, Practices, and Regulations*, ed. Volker Roelcke and Giovanni Maio (Stuttgart: Franz Steiner Verlag, 2004), 137–150.

11. Alfred Ploetz, *Grundlagen einer Rassen-Hygiene* (Berlin: S. Fischer, 1895); title in English translates as *Fundamental Principles of Racial Hygiene*.

12. Proctor, "Nazi Doctors," 19–20.

13. Proctor, "Nazi Doctors," 18.

14. Francis Bacon, 1605, "On the Dignity and Advancement of Learning," cited in "Medical Ethics through the Life Cycle in Europe and the Americas," *Cambridge World History of Medical Ethics*, ed. Robert Baker and Laurence McCullough (New York: Cambridge University Press, 2009), 149–150.

15. Christoph Wilhelm Hufeland, *Manual of the Practice of Medicine. The Results of Fifty Years' Experience*, trans. C. Bruchhausen and R. Nelson (London: Hippolyte Bailliere, [1836] 1844), 8. Cited in Baker and McCullough, *Cambridge World History*, 151.

16. Baker and McCullough, *Cambridge World History*, 152.

17. Samuel D. Williams, 1872, *Euthanasia*, cited in Ezekiel Emanuel, "The History of Euthanasia Debates in the United States and Britain," *Annals of Internal Medicine* 121, no. 10 (November 15, 1994): 796.

18. Adolf Jost, *Das Recht auf den Tod* [*The Right to Death*] (Göttingen, Germany: Dietrich), 1895.

19. Karl Binding and Alfred Hoche, *Die Freigabe der Vernichtung lebensunwerten Lebens* (Leipzig: Felix Meiner, 1920), trans. Robert A. Sassone, *The Release and Destruction of Life Devoid of Value* (Robert A. Sassone: Santa Ana, CA, 1975).

20. The characterization of the sick man as a parasite on society is drawn from Friedrich Nietzsche, *The Complete Works*, ed. Oscar Levy and Robert Guppy (New York: Russell and Russell, [1889] 1964), vol. 16, 45–46.

21. Ramm, "Rules of the Medical Profession," 86.

22. Ramm, "Rules of the Medical Profession," 128.

23. Ramm, "Rules of the Medical Profession," 128.

24. Nietzsche, *The Complete Works*, vol. 16, 45–46.

25. Nietzsche, *The Complete Works*, vol. 16, 45–46.

26. Robert Jay Lifton, *The Nazi Doctors: Medical Killing and the Psychology of Genocide* (New York: Basic Books, 1986), 435.

27. Florian Bruns and Tessa Chelouche, "Lectures on Inhumanity: Teaching Medical Ethics in German Medical Schools under Nazism," *Annals of Internal Medicine* 166, no. 8 (April 2017): 591–595 at 593.

28. Thomas S. Kuhn, *The Structure of Scientific Revolutions*, 4th ed. (Chicago: University of Chicago Press, 2012), 137. The word "readers" has been substituted for "scientists," as indicated by brackets.

29. Ramm uses the word "Aryan," to create a Hippocratic heritage for *Rassen-hygiene*. Ramm's translator, Melvin Wayne Cooper, observes that "Aryan" was introduced in a French pro-monarchical book by Joseph Arthur Comte de Gobineau (1816–1882), *Essai sur l'inégalité des races humaines* (*Essay on the Inequality of Human Races*, 1853–1855). The book was translated into German in 1897 as *Versuch über die Ungleichheit der Menschenrassen* (*The Inequality of the Human Races*). The German translation is significant because, "for the first time, race was cast as the primary moving force of world history," claiming that "'racial vitality' lies at the root of all great transformations in history" (Proctor, *Racial Hygiene*, 12).

30. Ramm, "Rules of the Medical Profession," 6, 7.

31. Ramm, "Rules of the Medical Profession," 32–33.

32. E. Friedlander, "Rudolf Virchow on Pathology Education," accessed October 7, 2022, http://www.pathguy.com/virchow.html.

33. Ramm, "Rules of the Medical Profession," 27–64.

34. Melvin Wayne Cooper's footnotes to parts I and II offer a helpful explanation of how Ramm addresses these complaints.

35. Ramm, "Rules of the Medical Profession," 117.

36. Robert N. Proctor, "Nazi Science and Nazi Medical Ethics: Some Myths and Misconceptions," *Perspectives in Biology and Medicine* 43, no. 3 (Spring 2000): 335–346.

37. Ramm, "Rules of the Medical Profession," 118–120.

38. United States Holocaust Memorial Museum. Hadamar Trial, *Holocaust Encyclopedia*, accessed October 7, 2022, https://encyclopedia.ushmm.org/content/en/article/the-hadamar-trial.

39. Herwig Czech, "Hans Asperger, National Socialism, and 'Race Hygiene' in Nazi-Era Vienna," *Molecular Autism* 2018, 9, 29 (April 19, 2018), accessed April 12, 2023, https://pubmed.ncbi.nlm.nih.gov/29713442/. See also Simon Baron-Cohen, Ami Klin, Steve Silberman, and Joseph D. Buxbaum, "Did Hans Asperger Actively Assist the Nazi Euthanasia Program?" *Molecular Autism* 9, no. 29 (April 19, 2018), accessed April 12, 2023, https://molecularautism.biomedcentral.com/articles/10.1186/s13229-018-0209-5.

40. Dr. Karl Brandt cited in Ulf Schmidt, *Karl Brandt: The Nazi Doctor, Medicine and Power in the Third Reich* (London: Continuum, 2007), 376.

41. Adolf Hitler, *Mein Kampf* (Boston: Houghton Mifflin, 1943), 402, cited in Robert Jay Lifton, *The Nazi Doctors: Medical Killing and the Psychology of Genocide* (New York: Basic Books, 1986), 430–431, italics in the original.

42. Ramm, "Rules of the Medical Profession," 89–92.

43. Ramm, "Rules of the Medical Profession," 92.

44. Alexandre Avdeev, Alain Blum, and Irina Troitskaya, "The History of Abortion Statistics in Russia and the USSR from 1900 to 1991," *Population. An English Selection, Institute National d'Etudes Démographiques* 9, no. 7 (1995): 39–66 at 42, accessed April 12, 2023, https://www.academia.edu/1758149/The_History_of_Abortion_Statistics_in_Russia_and_the_USSR_from_1900_to_1991.

234 Notes to Pages 20–25

45. Ramm, "Rules of the Medical Profession," 95.

46. Ramm, "Rules of the Medical Profession," 118–120.

47. Schmidt, *Karl Brandt*, 42–47.

48. Schmidt, *Karl Brandt*, 36, 37.

49. World Medical Association, "Proceedings," *World Medical Association Bulletin* 1, no. 1 (April 1949): 6–12 at 7.

50. Robert Baker, "The Declaration of Helsinki and the Foundations of Global Bioethics," in *Ethical Research: The Declaration of Helsinki and the Past, Present, and Future of Human Experimentation*, ed. Ulf Schmidt, Andreas Frewer, and Dominique Sprumont (New York: Oxford University Press, 2020), 47–69 at 51.

51. In May 2012, the German medical profession's official organization, the *Bundesärztekamme* officially apologized for the many doctors in the Nazi era who were "guilty, contrary to their mission to heal, of scores of human rights violations and we ask the forgiveness of their victims, living and deceased, and of their descendants." Alliance for Human Research Protection, accessed April 12, 2023, https://ahrp.org/german-medical-society-apologizes-for-nazi-era-atrocities-by-doctors/.

52. Schmidt, *Karl Brandt*, 43.

53. Vahkn N. Dadrian, "The Role of Turkish Physicians in The World War I Genocide of Ottoman Armenians," *Holocaust and Genocide Studies* 1, no. 2 (1986): 169–192 at 176.

54. Dadrian, *Genocide*, 176.

Chapter 3

1. Richard Dimbleby, BBC News, April 15, 1945, cited in Bergen-Belsen Concentration Camp, *Wikipedia,* accessed April 9, 2023, https://en.wikipedia.org/wiki/Bergen-Belsen_concentration_camp#Legal_prosecution.

2. Ivy's Letter to Pappworth. Letter 6, April 1966, London: PP/MHP/C5 Wellcome Library, London.

3. Recollections of a member of a British Army Film and Photographic unit. United States Holocaust Museum, Bergen-Belsen, "April 15," accessed October 7, 2022, https://www.ushmm.org/learn/timeline-of-events/1942-1945/liberation-of-bergen-belsen.

4. Richard Dimbleby, BBC News, April 15, 1945, cited in Bergen-Belsen Concentration Camp, *Wikipedia,* accessed October 7, 2022, https://en.wikipedia.org/wiki/Bergen-Belsen_concentration_camp#Legal_prosecution.

5. Dwight D. Eisenhower, April 12, 1945, "The Horrifying Discovery of Dachau Concentration Camp—And Its Liberation by US Troops," History.com, accessed October 7, 2022, https://www.history.com/news/dachau-concentration-camp-liberation.

6. See Ulf Schimdt, *Justice at Nuremberg: Leo Alexander and the Nazi Doctors' Trial* (Houndmills: Palgrave Macmillan, 2004), 108–111. Schmidt notes that Ivy was one of those who found German data "of great value to our Air Forces at a time when aviation medical research is relatively dormant because of rapid demobilization."

Andrew Ivy Papers (8768) 317, cited in Ulf Schmidt, *Justice at Nuremberg: Leo Alexander and the Nazi Doctors' Trial* (Houndmills: Palgrave Macmillan, 2004), 135.

7. Ivy was fearful of such an outcome. See Schmidt, *Justice at Nuremberg*, 136.

8. Paul Weindling, "Human Guinea Pigs and the Ethics of Experimentation: The *BMJ*'s Correspondent at the Nuremberg Medical Trial," *British Medical Journal* 313 (December 7, 1996): 1467–1470 at 1467.

9. Kenneth Mellanby, "A Moral Problem," *The Lancet*, no. 2 (December 7, 1946): 850. Mellanby also attempts "Golden Rule" legitimation of using data from the experiments, declaring that "if I myself had been a victim, and some of the results of value or of interest had been obtained from my death, I am sure that I should have preferred to know that this knowledge would have been used and that I had not died entirely for nothing."

10. Mellanby, "A Moral Problem," 850.

11. Mellanby acknowledges that some of the prisoners might be innocent but argues that the results of the research should be "published." However, he is concerned about bad publicity and noted that "at all costs sensationalism must be avoided, it might be perhaps better to grade [these reports] as 'confidential' and make them available only to bona-fide investigators." This indicates that Mellanby recognizes that, if made public, some people would consider these experiments unethical. Mellanby, "A Moral Problem," 850.

12. Mellanby, "A Moral Problem," 850.

13. Klaus Karl Schilling, "Dr. Klaus Karl Schilling Testifies at the Trial of Former Camp Personnel and Prisoners from Dachau," United States Holocaust Museum, accessed October 7, 2022, https://collections.ushmm.org/search/catalog/pa1069345.

14. War crimes investigator Leo Alexander coined the term "thanatology," invoking Carl Jung's (1875–1951) conception of *Thanatos* as a "death wish." Alexander meant his neologism, "thanatology," to serve as the technological companion to Raphael Lemkin's neologism, "genocide." Alexander defined "thanatology" as the study of scientific or technological means for implementing "genocide." Paul S. Weindling, *Nazi Medicine and the Nuremberg Trials: From Medical War Crimes to Informed Consent* (Basingstoke, UK: Palgrave Macmillan, 2008), 284. The future, however, has a way of repurposing neologisms. In the 1960s, the writings of Elisabeth Kübler-Ross, MD (1926–2004), and those of the French historian of death, Philippe Ariès (1914–1984), inspired a death with dignity movement that changed the meaning of "thanatology" so that today, according to the *New Oxford Dictionary of American English*, the word means the scientific study of death and the practices associated with it, including the study of the needs of the terminally ill and their families.

15. Kenneth Mellanby, *Human Guinea Pigs, by Kenneth Mellanby: A Reprint with Commentaries*, 3rd ed., ed. Lisa M. Rasmussen (Cham: Springer Nature, 2020), 106.

16. Alf Alving et al., "Procedures Used at Stateville Penitentiary for the Testing of Potential Antimalarial Agents," *Journal of Clinical Investigation* 27, no. 3 (1948): 2–5. For a thoughtful ethical reassessment of the experiment, see Franklin G.

Miller, "The Stateville Penitentiary Malaria Experiments: A Case Study in Retrospective Ethical Assessment," *Perspectives in Biology and Medicine* 56, no. 4 (Autumn 2013): 548–567.

17. One inmate died during the study. Although the researchers claimed that his death was not a result of their experiment, Nathan Leopold believed the death was associated with the experiment. Nathan Leopold, *Life Plus 99 Years* (Garden City, NY: Greenwood Press, 1958), 320. See also Miller, "The Stateville Penitentiary," 548–567.

18. Testimony of Dr. Franz Blaha Document Number 3249-PS, Exhibit USA-66 (January 11, 1946), as officially translated into English, accessed October 7, 2022, http://avalon.law.yale.edu/imt/01-11-46.asp. See also Weindling, *Nuremberg Trials*, 95–96.

19. It is noteworthy that, unlike Schilling's experiments, the University of Chicago's Jolliet-Stateville malaria experiments were publicized in a photojournalist story in *Life Magazine* that included photographs of unprotesting prisoners receiving malarial mosquito bites. It was these pictures in *Life Magazine* that alerted the German defense lawyers to the Jolliet-Stateville study. See Anonymous, "Prison Malaria; Convicts Expose Themselves to Disease so Doctors Can Study It," *Life Magazine*, June 1945, 44, accessed October 7, 2022, https://books.google.com/books?id=h0gEAAAAMBAJ&pg=PA43&lpg=PA43&dq=Prison+malaria;+convicts+expose+themselves+to+disease+so+doctors+can+study+it.&source=bl&ots=RcmbOYMzd1&sig=2WrKyi-xWkS1Astv3IdXHCRoc8&hl=en&sa=X&ved=2ahUKEwj5-eTm5YvdAhUkiOAKHVBjBKUQ6AEwAXoECAkQAQ#v=onepage&q=Prison%20malaria%3B%20convicts%20expose%20themselves%20to%20disease%20so%20doctors%20can%20study%20it.&f=false.

20. The 2022 equivalent of $100, in 1945, is $1,645 (https://www.in2013dollars.com/us/inflation/1945?amount=1, accessed October 7, 2022). D. H. Green, "Ethics Governing the Service of Prisoners as Subjects in Medical Experiments," *JAMA* 136, no. 7 (1948): 457. Cited in Pappworth, *Human Guinea Pigs: Experimentation on Man*, 62. For a slightly different version of the form, see Bernard E. Harcourt, "Making Willing Bodies: The University of Chicago Human Experiments at Stateville Penitentiary," *Journal of Social Research* 78 (June 2011): 443.

21. Sydney A. Halpern, *Dangerous Medicine: The Story behind Human Experiments with Hepatitis* (New Haven, CT: Yale University Press, 2021), 4–5.

22. Green, "Ethics Governing Prisoners as Subjects," 457. The central question the commission addressed was whether a parole board should consider a prisoner's participation in an experiment when deciding whether to parole the prisoner.

23. Ivy's role in formulating the document now known as the Nuremberg Code is detailed in historian Paul Weindling's authoritative study, *Nazi Medicine and the Nuremberg Trials: From Medical War Crimes to Informed Consent* (New York: Palgrave Macmillan, 2004). A helpful table on pages 357 to 358, "The Evolution of the Nuremberg Code," charts the respective contributions to the code's ten rules by each of the tribunal's two consultants, Andrew Ivy and Leo Alexander.

24. Ivy's original three criteria for ethical experiments are: 1. Consent of the human subject must be obtained. All subjects have been volunteers in the absence

of coercion in any form. Before volunteering, the subjects have been informed of the hazards, if any. 2. The experiment to be performed must be based on the results of animal experimentation and on a knowledge of the natural history of the disease under study and must be so designed that the anticipated results will justify the performance of the experiment. The experiment must be such as to yield results unprocurable by other methods of study, which are necessary for the good of society. 3. The experiment must be conducted (a) only by scientifically qualified persons and (b) so as to avoid all unnecessary physical and mental suffering and injury and (c) only after the results of adequate animal experimentation have eliminated any a priori reason to believe that death or disabling injury will occur. If there is any a priori reason that accidental death or disabling injury may occur, as in such experiments as those of Walter Reed in which the mosquito was demonstrated to transmit yellow fever, then medical scientists should serve or should have served as volunteers along with nonscientific personnel. Green, "Ethics Governing Prisoners as Subjects," 457.

Ivy recommended these three rules to the Nuremberg Tribunal and in a report that he sent to the AMA and to the US Army: A. C. Ivy, *Report on War Crimes of a Medical Nature Committed in Germany and Elsewhere on German Nationals and the Nationals of Occupied Countries by the Nazi Regime during World War II* (undated manuscript, source, AMA archives). The AMA adopted a simplified version of these guidelines in 1946 as a *pro forma* gesture distinguishing "ethical" American experiments from their "unethical" Nazi counterparts. The AMA's three requirements were (1) voluntary consent, (2) prior animal experimentation, and (3) proper medical protection and management. They were published under the AMA Judicial Council's "Ethical Guidelines for Clinical Investigation," for 1946. These criteria were never seriously promulgated or enforced.

25. Green, "Ethics Governing Prisoners as Subjects," 457.

26. Miller, "The Stateville Penitentiary," 548–567.

27. Inspired by Friedrich Nietzsche's 1883 book *Also sprach Zarathustra* (*Thus Spoke Zarathustra*), Leopold and his companion Richard Loeb (1905–1936) killed a young boy to prove to themselves that they were *Übermenschen*, literally "superior people," who were above morality and the law.

28. Leopold, *Life Plus 99 Years*, 306.

29. Leopold, *Life Plus 99 Years*, 306.

30. Leopold, *Life Plus 99 Years*, 307.

31. Miller, "The Stateville Penitentiary," 558.

32. Miller, "The Stateville Penitentiary," 556.

33. Letter from Roy Hoskins to Richard Lyman, September 28, 1933, Alexander Papers, Boston, cited in Schmidt, *Justice at Nuremberg*, 49.

34. Elkeles, "German Debate," 26.

35. Elkeles, "German Debate," 26.

36. Schmidt, *Justice at Nuremberg*, 32.

37. Schmidt, *Justice at Nuremberg*, 33.

38. Abraham Myerson, Jame Ayer, Tracy Putnam, Clyde. Keeler, and Leo Alexander, *Eugenic Sterilization: A Reorientation of the Problem* (New York: The Macmillan Company, 1936), 56.

39. Myerson et al., *Eugenic Sterilization*, 177–179. Italics is in the original document.

40. Schmidt, *Justice at Nuremberg*, 294.

41. Sam Kean, "Leo Alexander's Unflinching Pursuit: In the Waning Days of World War II, a Psychiatrist Raced across Germany to Uncover the Harrowing Abuses of Nazi Doctors," *Distillations*, Science History Institute, November 17, 2020, accessed October 7, 2022, https://www.sciencehistory.org/distillations /leo-alexanders-unflinching-pursuit#:~:text=Leo%20Alexander's%20Unflinching%20Pursuit%20In%20the%20waning%20days,doctors.%20By%20 Sam%20Kean%20%7C%20November%2017%2C%202020.

42. Robert Lifton papers, New York Public Library; correspondence with Leo Alexander, August 29—Lifton's talk with Alexander cited in Schmidt, *Justice at Nuremberg*, 292.

43. Ulf Schmidt, interview with Cecily Alexander Grable, [Leo Alexander's daughter], Marion, Cape Cod, May 1999. Schmidt, *Justice at Nuremberg*, 264.

44. David Plotz, "The Greatest Magazine Ever Published: What I Learned Reading All of *Life* Magazine from the Summer of 1945," *Slate*, December 27, 2013, accessed October 7, 2022, https://slate.com/human-interest/2013/12/life-magazine -1945-why-it-was-the-greatest-magazine-ever-published.html.

45. Ivy's Letter to Pappworth. Letter 6 April 1966, London: PP/MHP/C5, Wellcome Library, London.

46. Grodin papers, Appendix C, 4, cite note 151, in Schmidt, *Justice at Nuremberg*, 136; see also Weindling, *Nuremberg Trials*, 263.

47. Schmidt, *Justice at Nuremberg*, 136.

48. Schmidt, *Justice at Nuremberg*, 136–137.

49. Ivy, "Report on War Crimes," 1946.

50. For further discussion of the relationship between research ethics and human rights, see Robert Baker, "Bioethics and Human Rights: A Historical Perspective," *Cambridge Healthcare Quarterly* 10, no. 3 (Summer 2001): 241–252.

51. Jonathan Moreno, "The Declaration of Helsinki and the 'American Stamp,'" 353.

52. Michael Grodin, "Historical Origins of the Nuremberg Code," in *The Nazi Doctors and Nuremberg Code: Human Rights in Human Experimentation*, ed. George J. Annas and Michael A. Grodin (New York: Oxford University Press, 1992), 130–131.

53. Jochen Vollmann and Rolf Winau, "The Prussian Regulation of 1900: Early Ethical Standards for Human Experimentation in Germany," *IRB: Ethics & Human Research* 18, no. 4 (July–August 1996): 9–11; also available in Grodin, "Historical Origins," 127.

54. Grodin, "Historical Origins," 131–132.

55. As Grodin notes ("Historical Origins," 143), several scholars, Breuer, Fischer, and Fluss, contend that the 1931 "Guidelines for Research on Patients" remained in effect until 1945. Their claim insinuates that the Nazi use of inmates as human material was an aberration. However, this does not accord with Ramm's account of the reorganization of German medical system in the wake of the National Socialist revolution because the 1931 guidelines could not have remained operative without being reissued, or otherwise formally reaffirmed, by the new Nazi Physicians' Chamber (Ramm, *Rules of the Profession*, 48–52). Moreover, apologists' claims that the 1931 Guidelines remained in effect even after the National Socialist reorganization of German medicine are not supported by known records of bureaucratic or judicial proceedings against physicians for violations of these guidelines. As Weindling observes, "The claim that the 1931 regulations were legally binding until 1945 was refuted at the Nuremberg trial and there is no evidence that the consent procedures were routinely enforced" prior to the National Socialists' reorganization of German medicine. Furthermore, even had these guidelines for experimental therapeutic interventions on *patients* been enforced, since prisoners are not *patients*, their treatment as experimental subjects would have been outside the scope of the 1931 regulations (Weindling, *Nuremberg Trials*, 260).

56. Plato (circa 375 BCE) translation (1888) reprinted online 2017. *The Republic of Plato*, trans. Benjamin Jowett (Oxford: Clarendon Press), book III, 329B.

57. Winfried Schleiner, *Medical Ethics in the Renaissance* (Washington, DC: Georgetown University Press, 2007), 51–60.

58. Thomas Percival, *Medical Ethics: Or, a Code of Institutes and Precepts, Adapted to the Professional Conduct of Physicians and Surgeons* (London: J. Johnson, 1803), 156–168.

59. Ivy, *Report on War Crime*, 11. Italics added.

60. Elkeles, "German Debate," 26.

61. Ivy, *Report on War Crime*, 11.

62. James Carroll, *Report to Surgeon General Sternberg*, August 18, 1906, cited in Robert Baker, *Before Bioethics: A History of American Medical Ethics from the Colonial Period to the Bioethics Revolution* (New York: Oxford University Press, 2013), 258.

Chapter 4

1. J. A. Pridham, "Founding of the World Medical Association," *World Medical Association Bulletin* 3, no. 3 (July 1951), 207.

2. Pridham, "Founding of the World Medical Association," 207–208.

3. Pridham, "Founding of the World Medical Association," 209.

4. World Medical Association, "Proceedings," *World Medical Association Bulletin* 1, no. 1 (April 1949): 7.

5. World Medical Association, "Proceedings," 8. Italics added.

6. World Medical Association, "Proceedings," 8. Emphasis added.

7. World Medical Association, "Proceedings," 7.

8. Ivy, *Report on War Crime*, 21.

9. World Medical Association, "Proceedings," 9.

10. World Medical Association, "Proceedings," 9.

11. World Medical Association, "Proceedings," 10. Emphasis added.

12. World Medical Association, "Proceedings," 10.

13. World Medical Association, "Proceedings," 10.

14. Percival, *Medical Ethics*, chap. I, art. 1, 9.

15. Andreas-Holger Maehle, *Doctors, Honour, and the Law: Medical Ethics in Imperial Germany* (New York: Palgrave MacMillan, 2009).

16. World Medical Association, "Proceedings," 10.

17. World Medical Association, "Proceedings," 10.

18. "Membership in the World Medical Association for the Japan and Western German Medical Associations," *World Medical Association Bulletin* 3, no. 3 (July 1951): 204, 210.

19. Maehle, "'God's Ethicist,'" 2012, note 32.

20. Franz Büchner, "Der Eid des Hippokrates: Wortlaut des am 18. November 1941 in der Aula der Universität Freiburg gehalten öffentlichen Vortages," in *Der Mensch in der Sicht moderner Medizin*, ed. Franz Büchner, 1985, 131–151; cited in Karl-Heinz Leven, "The Invention of Hippocrates: Oath, Letters and Hippocratic Corpus," in *Ethics Codes in Medicine: Foundations and Achievements of Codification since 1947*, ed. Ulrich Tröhler, Stella Reiter-Theil, and Eckhard Herych (Aldershot: Ashgate, 1998), 14.

21. Franz Büchner, lecture at Freiburg University, November 18, 1941, cited in Carol Poore, *Disability in Twentieth-Century German Culture* (Ann Arbor: University of Michigan Press, 2009), 89.

22. Evelyn Shuster, "Fifty Years Later: The Significance of the Nuremberg Code," *New England Journal of Medicine* 337, no. 20 (November 1997): 1436–1440 at 1437.

23. On German medical schools' abandonment of the Hippocratic Oath, see Weindling, *Nuremberg Trials*, 151, 282.

24. Shuster, "Fifty Years Later," 1437.

25. Schmidt, *Justice at Nuremberg*, 206.

26. Testimony of Werner Leibbrand in Schmidt, *Justice at Nuremberg*, 207; see also Schuster, "Fifty Years," 1438.

27. Schmidt, *Justice at Nuremberg*, 206.

28. Schmidt, *Justice at Nuremberg*, 1439.

29. Schmidt, *Justice at Nuremberg*, 1439.

30. Thomas Rutten, *Hippokrates im Gespräch*, Katalog der Ausstellung des Institutes für Theorie und Geschicte der Medizin und der Universitäts- und Landesbibliothek

Münster, 10 December 1993–8 January 1994, Münster. This is a catalogue associated with an exhibit.

31. Chamber of Physicians of West German, June 14, 1947. On October 18, 1947, the Chamber condemned all German physicians who committed crimes against humanity and war crimes. In attempt to right the Nazis' wrongs against fellow physicians, about a month later, on November 29, 1947, it petitioned the German government to "reinstate full medical rights" to all physicians stripped of the right to practice by the Nazis for "reasons of race, religion, or politics." Cited in World Medical Association, "Proceedings," 9.

32. Paraphrase of Corinthians 5:16–18.

33. World Medical Association, "Proceedings," 12.

34. World Medical Association, "Proceedings," 12.

35. World Medical Association, "Proceedings," 12.

36. World Medical Association, "Declaration of Geneva," passed 1948; preliminary publication in *World Medical Association Bulletin* 1, no. 1 (April 1949): 13; official publication as *Serment de Geneve, Declaration of Geneva, Declaracion en Gemebra, World Medical Association Bulletin* 1, no. 2 (July 1949): 35–37. In the original publication, each linguistic rendition of the Declaration of Geneva was presented in its entirety, in parallel columns. The side-by-side version, although designed as a measure to emphasize solidarity, nonetheless highlights subtle differences between the linguistic communities. The oath was originally composed in English and that version prohibits discrimination according to "social standing," an earned status available to all irrespective of heredity or social class. In contrast, the French version prohibits discrimination based on "classe sociale" (i.e., social class), a socioeconomic concept. The French version also demands "absolute respect for human life," whereas in the English original this was pared down to "utmost" respect for human life." There are also differences in the wording between the first and the final versions of the Spanish versions of the Declaration.

37. Scholars today believe that the version of the Oath that they have pieced together is written in a style that appeared only after Hippocrates and his sons died. Imagine, for example, that centuries from today historians discover a message that starts "how r u." Such a message is unlikely to have been written prior to the invention of the cell phone, which made texting such abbreviations standard. Using similar stylistic evidence, classicists believe that neither Hippocrates nor his sons wrote our current version of the Oath, which seems to be a stylized version of a more prosaic earlier oath.

38. Stephan Pow and Frank W. Stahnisch, "Ludwig Edelstein (1902–1965): A German Historian of Medicine in North American Exile and the Emergence of the Modern Hippocratic Oath," *Journal of Medical Biography* 24, no. 4 (November 2015): 527–537; Thomas Rütten, "Ludwig Edelstein at the Crossroads of 1933. On the Inseparability of Life, Work, and Their Reverberations," *Early Science and Medicine* 11, no. 1 (2006): 50–99.

39. Ludwig Edelstein, *The Hippocratic Oath: Text, Translation, and Interpretation* (Baltimore: The Johns Hopkins Press, 1943). Edelstein's translation reflects his intriguing hypothesis that the Oath was written by a Pythagorean medical cult.

However, after a seven-decade-long search for evidence of this cult unearthed no evidence of its existence, classicists now regard this hypothesis, however intriguing, as defunct.

40. It is unclear which English translation of the Oath the WMA used as the basis of the Declaration of Geneva. Another oft-used nineteenth-century English translation was that of the Scottish physician-classicist Francis Adams (1796–1861). As in Edelstein's version, this translation too offered a prolife reading of the Oath as prohibiting abortion and forbidding giving deadly medicines. Accessed October 7, 2022, http://classics.mit.edu/Hippocrates/hippooath.html.

41. A twentieth-century English translation of the Oath by Heinrich von Staden (von Staden 1996) presents a more literal translation of the Oath and is the version currently used by most scholars (e.g., Miles 2004). This version does not support the view that the oath prohibited abortion or euthanasia.

42. World Medical Association, Declaration of Geneva, passed 1948, preliminary publication in *World Medical Association Bulletin* 1, no. 1, 13; official publication as *"Serment de Geneve,* Declaration of Geneva, *Declaracion en Gemebra,"* *World Medical Association Bulletin* 1, no. 2 (July 1949): 35–37.

43. In the original publication, each translated rendition is presented in its entirety, in parallel columns besides the original English version. In later publications, each translation of the Declaration was published as a separate full-page document. Thus, the WMA's *Bulletin* 1, no. 2 (July 1949): 35 contained the English version of the Declaration; page 36, the French *Serment*; and page 37, the Spanish *Declaracion*. In 1951, Abbott Laboratories of North Chicago, Illinois, offered "handsome reproductions" of the Declaration and the International Code of Ethics, "printed in four colors on fine Vellum stock . . . 18×24 inches and containing no advertising . . . to grace the wall of [physicians] office or home . . . free" in all three languages. This insert was published in *World Medical Association Bulletin*, vol. 3, 1951.

44. WMA General Assembly, Chicago, United States, Declaration of Geneva, October 2017, accessed October 7, 2022, https://www.wma.net/policies-post/wma-declaration-of-geneva/.

45. Jacques Jouanna, *Hippocrates,* trans. M. B. DeBevoise (Baltimore: The Johns Hopkins University Press, 1999), 46.

46. Plato, *Protagoras* 311b-c, cited in Jouanna, *Hippocrates,* 46.

47. Jouanna, *Hippocrates,* 46–48.

48. Edelstein, *The Hippocratic Oath.* Italics added.

49. Edelstein, *The Hippocratic Oath,* 7.

50. Italics added. Ramm concurs on this point—at least with respect to treating members of the *Volk*. See "Rules of the Medical Profession," 89–92.

51. "Rules of the Medical Profession," 91–92.

52. "Rules of the Medical Profession," 92.

53. Moll, 1902, *Ärztliche Ethik*, 261–263, Andreas-Holger Maehle translation, in Maehle, "God's Ethicist," 2012.

54. Roberto Lo Presti, "History of Science: The First Scientist," accessed October 7, 2022, https://www.nature.com/articles/512250a#:~:text=Aristotle%20 is%20considered%20by%20many,%2C%20observation%2C%20inquiry%20 and%20demonstration.

55. Aristotle, *Politics 1.1252b*, ed. Gregory R. Crane, *Persus Digital Library*, accessed October 7, 2022, http://www.perseus.tufts.edu/hopper/text?doc=Perseu s:abo:tlg,0086,035:1:1252b.

56. Carl Linnaeus. 1766–1768. Systema naturæ per regna tria naturæ, secundum classes, ordines, genera, species, cum characteribus & differentiis. Accessed April 13, 2023, https://archive.org/stream/systemanaturaepe01linn/systemanaturaepe01linn _djvu.txt.

The sometimes overlooked connection between geography (i.e., a specific land or country) and genetics (i.e., "blood") in conceptions of race is crudely reflected in the Nazi slogan "blood and soil" and more subtly reflected in the words of Heidegger and other intellectual anti-Semites who charged Jews with "Semitic Nomadism." To Heidegger's original readership this meant that Jews, having no land of their own, are "other" in the lands in which they take up residence: thus Jews can be viewed as a dangerous cultural parasitic tribe trespassing on the culture, language, and domain of a proper race [Aryan, German] unified by land, by history, by language and culture—and by genetics.

57. For a brief introduction and historical review of the impact of Darwin's theory of evolution on eugenicist thought, see Diane B. Paul and James Moore, "The Darwinian Context: Evolution and Inheritance," in *The Oxford Handbook of the History of Eugenics*, ed. Philippa Levine and Alison Bashford (Oxford: Oxford University Press, 2010), 27–42.

58. "White coat ceremonies" are so named because medical students first put on their medical uniform (i.e., a white coat) at this ceremony and in doing so swear a Hippocratic-derived oath. These ceremonies originated at the University of Chicago in 1989 and took a standard form after Dr. Arnold Gold introduced them at Columbia University in 1993.

59. Shernaz S. Dossabhoy, Jessica Feng, and Manisha S. Desai, "The Use and Relevance of the Hippocratic Oath in 2015—A Survey of US Medical Schools," *Journal of Anesthesia History* 4, no. 2 (April 2018): 139–146; see also Audie C. Kao and Kayhan P. Parsi, "Content Analyses of Oaths Administered at U.S. Medical Schools in 2000." *Academic Medicine* 79, no. 9 (September 2004): 882–887.

60. Fritz Baumgartner, "Hippocrates and the Dignity of Human Life," *American Journal of Obstetrics and Gynecology* 186, no. 6 (June 2002): 1378–1379; see also Thomas Cavanaugh, *Hippocrates' Oath and Asclepius Snake: The Birth of the Medical Profession* (New York: Oxford University Press, 2018); Wesley J. Smith, *Culture of Death: The Age of "Do Harm" Medicine* (New York: Encounter Books, 2016); and Wesley J. Smith, "The War on the Hippocratic Oath," *NRL News Today*, February, 2, 2018, https://www.nationalrighttolifenews.org/2018/02 /war-hippocratic-oath/October 7, 2022. Finally, see Stephen Miles, *Oath Betrayed: Torture, Medical Complicity, and the War on Terror* (New York: Random House, 2006).

61. Mary L. Davenport, Jennifer Lahl, and Even C. Rosa, "Right of Conscience for Health-Care Providers." *Linacre Quarterly* 79, no. 2 (May 2012): 169–191.

62. Satyaseelan Packianathan, Srinivasanm Vijayakumar, Paul Russell Roberts, and Maurice King III, "Reflections on the Hippocratic Oath and Declaration of Geneva in Light of the COVID-19 Pandemic," *Southern Medical Journal* 113, no. 7 (July 2020): 326–329 at 329.

Chapter 5

1. Henry K. Beecher, *Research and the Individual: Human Studies* (Boston: Little Brown and Company, 1970), 231.

2. Beecher, *Research*, 234.

3. Beecher, *Research*, 279.

4. Maurice Pappworth, *Human Guinea Pigs: Experimentation on Man* (Boston: Beacon Press, 1968), 188.

5. Beecher, *Research*, 231.

6. Beecher, *Research*, 228.

7. Alan Yoshioka, "Use of Randomisation in the Medical Research Council's Clinical Trial of Streptomycin in Pulmonary Tuberculosis in the 1940s," *British Medical Journal* 317, no. 7167 (October 31, 1998): 1220–1223.

8. Henry K. Beecher, "The Powerful Placebo." *JAMA: The Journal of the American Medical Association* 159, no. 17 (December 24, 1955): 1602–1606.

9. Beecher, *Research*, 231.

10. Beecher, *Research*, 234.

11. Rothman, *Strangers at the Bedside*, 62.

12. World Medical Association, "Principles for Those in Research and Experimentation" (approved by The General Assembly of the World Medical Association), *World Medical Journal* 2 (1955): 14–15.

13. From the 1960 Report of the Medical Ethics Committee, cited in Susan E. Lederer, "Research without Borders: The Origins of the Declaration of Helsinki," in *Twentieth Century Ethics: Historical Perspectives on Values, Practices, and Regulations*, ed. Roelcke and Giovanni Maio (Stuttgart: Fritz Steiner Verlag, 2004), 205–206.

14. Hugh Clegg, "Report of the Medical Ethics Committee at the 35th Council Session, Mar. 25 to April 3, 1959." Manuscript 17.1/59 (WMA Archives Geneva), cited in Susan E. Lederer, "Research without Borders," in *Twentieth Century Ethics*, ed. Roelcke and Maio, 204–205.

15. Erving Goffman, "The Characteristics of Total Institutions," in *Symposium on Preventive and Social Psychiatry* (Washington, DC: Walter Reed Army Institute of Research, 1958), 43–84.

16. Anda Baicus, "History of Polio Vaccination," *World Journal of Virology* 1, no. 4 (August 2012): 108–114.

17. Paul S. Appelbaum and Charles W. Lidz, "Twenty-Five Years of Therapeutic Misconception," *The Hastings Center Report* 38, no. 2 (March–April 2008): 5–7.

18. A. Spinelli, "Report of the Committee on Medical Ethics. 44th Council Session, Chicago, Illinois May 6–12, 1962," World Medical Association. Copy is courtesy of Susan Lederer, who, in turn, received it from Jay Katz.

19. By contrast, research is defined in the *Belmont Report* as "an activity designed to test a hypothesis, permit conclusions to be drawn, and thereby to develop or contribute to generalizable knowledge (expressed, for example, in theories, principles, and statements of relationships). Research is usually described in a formal protocol that sets forth an objective and a set of procedures designed to reach that objective. When a clinician departs in a significant way from standard or accepted practice, the innovation does not, in and of itself, constitute research. The fact that a procedure is 'experimental,' in the sense of new, untested or different, does not automatically place it in the category of research" (National Commission for the Protection of Human Subjects of Research, April 18, 1979, *The Belmont Report*, Part A, *Boundaries Between Practice and Research*, accessed October 7, 2022, https://www.hhs.gov/ohrp/regulations-and-policy/belmont-report/read-the-belmont-report/index.html.

20. A. Spinelli, May 6–12, 1962, "Report of the Committee on Medical Ethics."

21. Ladimer, Irving, cited at Laura Stark, *Behind Closed Doors: IRBs and the Making of Ethical Research* (Chicago: University of Chicago Press, 2012), 113.

22. Stark, *Behind Closed Doors*, 113.

23. Susan E. Lederer, "Research without Borders: The Origins of the Declaration of Helsinki," in *Twentieth Century Ethics*, ed. Roelcke and Maio (Stuttgart: Franz Steiner Verlag, 2004), 199–217.

24. Bradford A. Hill, cited in Beecher, *Research*, 280.

25. Beecher, *Research*, 281.

26. Beecher, *Research*, 282.

27. The International Conference on Harmonisation of Technical Requirements of Pharmaceuticals for Human Use (ICH) Glossary, 1.12, "Clinical trial/study: Any investigation in human subjects intended to discover or verify the clinical, pharmacological, and/or other pharmacodynamic effects of an investigational product(s), and/or to identify any adverse reactions to an investigational product(s), and/or to study absorption, distribution, metabolism, and excretion of an investigational product(s) with the object of ascertaining its safety and/or efficacy. The terms clinical trial and clinical study are synonymous." Glossary, 1.12, accessed October 7, 2022, https://ichgcp.net/1-glossary.

28. World Medical Association, 1964. *Declaration of Helsinki*, accessed October 7, 2022, https://www.wma.net/wp-content/uploads/2018/07/DoH-Jun1964.pdf.

Chapter 6

1. The paradigm case of conversion from blindness to sightedness is Saul's transformation while taking the road to Damascus from physically and metaphorically blind anti-Christian inquisitor, until he converts to Christianity and becomes

physically sighted as the apostle newly baptized as Paul (Acts of the Apostles, 9:7–9, 17). More metaphorical conversions have been experienced by Christians ever since (see, e.g., Cizik 2007). These metaphorical accounts of metaphorical "blindness" have been recounted in ableist cultures that often stigmatize people with disabilities. I recognize this and I apologize for using ableist-infused terminology to capture the phenomenon of people's inability to perceive what their minds are not prepared to conceive.

2. The biblical line from Jeremiah was reformulated by John Heywood in *A dialogue conteinyng the nomber . . . of all the prouerbes in the englishe tongue* (1546). It was later used by Jonathan Swift in his 1738 *Polite Conversation. Dialogue III.* The actual line is spoken by *Lady Smart*: "but, you know, there's none so blind as they that won't see." Dialogue III, 174, accessed October 7, 2022, https://www.gutenberg.org/files/60186/60186-h/60186-h.htm. Immanuel Kant used the blindness metaphor in the *Critique of Pure Reason* (1787) to explain the relationship between thought and perception. Pasteur's remark was part of a December 7, 1854, ceremonial speech delivered at the opening of a brand-new Facility for Sciences at Lille, France. Twentieth-century American philosophers Norwood Hanson and, independently, the philosopher Wilfred Sellars summarized Kant's notion of the relation between perception and conception as "percepts without concepts are blind." A similar line was also used by Ray Stevens in his 1970 hit song, "Everything Is Beautiful," accessed October 7, 2022, https://www.google.com/search?client=safari&rls=en&q=Ray+stevens+everything+is+beautiful+lyrics&ie=UTF-8&oe=UTF-8.

3. Robert Andrews, ed., *"Boston Toast" Famous Lines: A Columbia Dictionary of Familiar Quotations* (New York: Columbia University Press, 1996), 53, accessed April 13, 2023, https://books.google.com/books?id=MtciwlIG3sMC&pg=PA53#v=onepage&q&f=false.

4. The book referenced is Marcus Garvey, *The Philosophy and Opinion of Marcus Garvrey*, ed. Marcus Garvey, Amy Jacques Garvey, William Loren Katz (New York: Universal Publishing House, [1923], 3rd ed. 2020).

5. Aja Romano, October 9, 2020, "A History of 'Wokeness': Stay Woke: How a Black Activist Watchword Got Co-opted in the Culture War," *Vox*, accessed October 7, 2022, https://www.vox.com/culture/21437879/stay-woke-wokeness-history-origin-evolution-controversy.

6. *Oxford English Dictionary*, June 2017, "New Word Notes June 2017," accessed October 7, 2022, https://public.oed.com/blog/june-2017-update-new-words-notes/. The song cited was "Master Teacher" by Erykah Badu, 2009, which has a repeated refrain "I stay woke," accessed October 7, 2022, https://songmeanings.com/songs/view/3530822107858756121/.

7. Garvey, *The Philosophy and Opinion of Marcus Garvrey*, 29.

8. Romano, "A History of 'Wokeness.'"

9. "libtard," *Cambridge Dictionary*, accessed October 7, 2022, https://dictionary.cambridge.org/dictionary/english/libtard.

10. Guido Palazzo, Franciska Krings, and Ulrich Hoffrage, "Ethical Blindness," *Journal of Business Ethics* 109, no. 3 (2012): 323–338. See also Barbara Kump

and Markus Scholz, "Organizational Routines as a Source of Ethical Blindness," *Organizational Theory* 3 (February 26, 2022): 1–24.

11. "Unwoke," *Urban Dictionary, Wiktionary,* January 21, 2021, characterization, accessed October 16, 2022, https://www.urbandictionary.com/define.php ?term=Unwoke, https://en.wiktionary.org/wiki/unwoke.

12. Appiah, *The Honor Code,* xii.

13. The convention at that time was that "Francis" spelled with an "i" was a male name, whereas "Frances" spelled with an "e" was a female name.

14. Linda Bren, "Frances Oldham Kelsey: FDA Medical Reviewer Leaves Her Mark on History," accessed October 7, 2022, https://web.archive.org/web/2006 1020043712/https://www.fda.gov/fdac/features/2001/201_kelsey.html.

15. Morton Mintz, "'Heroine' of FDA Keeps Bad Drug Off Market," *Washington Post,* July 15, 1962, accessed October 7, 2022, https://www.washingtonpost .com/wp-srv/washtech/longterm/thalidomide/keystories/071598drug.htm.

16. Stephen Phillips, *Medicine on the Midway,* 2011, reissued March 9, 2020, "How a Courageous Physician-Scientist Saved the U.S. from a Birth-Defects Catastrophe," accessed October 7, 2022, https://www.uchicagomedicine.org /forefront/biological-sciences-articles/courageous-physician-scientist-saved-the -us-from-a-birth-defects-catastrophe2.

17. Phillips, *Medicine on the Midway.*

18. Phillips, *Medicine on the Midway.*

19. Phillips, *Medicine on the Midway.*

20. R. L. Brent, "The Contributions of Widukind Lenz to Teratology and Science: Comments on "Thalidomide Retrospective: What Did the Clinical Teratologist Learn?" *Teratology* 46, no. 5 (November 1992): 415–416. See also W. G. McBride, "Thalidomide and Congenital Abnormalities," Letter to the Editor, *The Lancet* 2 (December 16, 1961): 1358.

21. *Menschliche Erblichkeitslehre und Rassenhygiene (Human Heredity Theory and Racial Hygiene),* published in five editions from 1921 to 1940.

22. Phillips, *Medicine on the Midway.*

23. Mintz, "Heroine of the FDA," July 15, 1962.

24. Frances Oldham Kelsey, *Autobiographical Reflections,* U.S. Food and Drug Administration, accessed March 13, 2022, https://www.fda.gov/media/89162 /download.

25. Federal Food and Cosmetic Act of 1962. PL 87–781, 76 Stat, 780–788, accessed October 7, 2022, https://www.govinfo.gov/content/pkg/STATUTE-76/pdf/STATUTE -76-Pg780.pdf.

26. Mark S. Frankel, *The Public Health Service Guidelines Governing Research Involving Human Subjects* (Washington, DC: George Washington University Program of Policy Studies in Science and Technology, Monograph 10, 1972), 20–21, cited in Rothman, *Strangers at the Bedside,* 87.

27. Rothman, *Strangers at the Bedside,* 87.

28. Headlines from the *New York Times* and the *Saturday Review* in 1962, cited in Daniel Carpenter, *Reputation and Power: Organizational Image and Pharmaceutical Regulation at the FDA* (Princeton, NJ: Princeton University Press 2010), 247; for a brief but thorough account of Kelsey and the thalidomide story, see also 238–256.

29. Frances O. Kelsey, "Patient Consent Provisions of the Federal Food, Drug, and Cosmetic Act," in *Clinical Investigation in Medicine: Legal, Ethical, and Moral Aspects: An Anthology and Bibliography*, ed. Irving Ladimer and Roger W. Newman (Boston: Law-Medicine Research Institute, Boston University, 1963), 336–338, as cited in Austin Connor Kassels and Jon F. Merz. "The History and Policy Evolution of Waivers of Informed Consent in Research," *Journal of Legal Medicine* 41 (2021): 1–2, 1–28 at 7.

30. Carpenter, *Reputation and Power*, 319.

31. Carpenter, *Reputation and Power*, 481, 481, note 22.

32. Herbert L. Ley Jr., "Federal Law and Patient Consent," *Food, Drug, Cosmetic Law Journal* 24, no. 11 (November 1969): 520–526.

33. Frankel, *Public Health Service Guidelines*, cited in Rothman *Strangers at the Bedside*, 86.

34. Karen Geraghty, "Protecting the Public: Profile of Dr. Frances Oldham Kelsey," *Virtual Mentor: American Medical Association Journal of Ethics* 3, no. 7 (July 1, 2001): 252–254.

35. Maurice H. Pappworth, "'Human Guinea Pigs'—A History," *British Medical Journal* 301, no. 6766 (December 22–29, 1990, *Clinical research* ed.), 1456–1460 at 1456.

36. Pappworth "'Human Guinea Pigs'—A History," 1456.

37. Pappworth "'Human Guinea Pigs'—A History," 1456.

38. Joanna Seldon, *The Whistle-Blower* (Buckingham: The University of Buckingham Press, 2017), 111, 112. The biographer, Lady Joanna Seldon, is Maurice Pappworth's daughter. She wrote this biography during the last two years of her life while being treated for what turned out to be a terminal case of cancer of the liver.

39. Seldon, *The Whistle-Blower*, 111, 112.

40. Seldon, *The Whistle-Blower*, 3.

41. Pappworth, "'Human Guinea Pigs'—A History," 1456.

42. Maurice Henry Pappworth, "Human Guinea Pigs: A Warning," *Twentieth Century Magazine* 50, no. 4 (1962): 66–75. For more about Pappworth, see Christopher Booth, "Obituary: M H Pappworth," *British Medical Journal* 309 (December 10, 1994): 577; see also, Paul J. Edelson, "Henry Beecher and Maurice Pappworth: Honor in the Development of the Ethics of Human Experimentation," in *Twentieth Century Ethics*, ed. Roelcke and Maio, 219–233; Jenny Hargrove, "British Research Ethics after the Second World War," in *Twentieth Century Ethics*, ed. Roelcke and Maio, 181–197.

43. Hugh Clegg, [Unsigned editorial with prefatory comment] "Draft Code of Ethics on Human Experimentation," *British Medical Journal* 2 (October 27, 1962): 119, accessed October 7, 2022, https://www.bmj.com/content/2/5312/1119.

44. Clegg, unsigned editorial. Records shared with the author by Susan Lederer indicate that Clegg's 1962 version of the Declaration accurately reproduces the WMA Committee on Ethics' working draft with the exception that the preface was deleted.

45. Lederer, "Research without Borders," 199–217.

46. Seldon, *The Whistle-Blower*, 2, 3.

47. Gaw, "Exposing Unethical Human Research," 150–155; Lancaster and Rathwell, "The Moralist."

48. Jonathan Beecher, personal communication, March 10, 2006. Cited in Michael Gionfriddo's contribution to E. Lowenstein, MD, and Bucknam McPeek, MD, eds., "Enduring Contributions of Henry K. Beecher to Medicine, Science, and Society: Part I," *International Anesthesiology Clinics* 45, no. 4 (Fall 2007): 18–19.

49. Gionfriddo, "Enduring Contributions," 19–20.

50. Gionfriddo, "Enduring Contributions," 35.

51. Gionfriddo, "Enduring Contributions," 18–19.

52. Lancaster and Rathwell, "The Moralist."

53. Beecher, "The Powerful Placebo."

54. Alfred E. McCoy, "Science in Dachau's Shadow: Hebb, Beecher, and The Development of CIA Psychological Torture and Modern Medical Ethics," *Journal of the History of the Behavioral Sciences* 43, no. 4 (Fall 2007): 401–417 at 410.

55. Beecher, *Research and the Individual*, 209, 210.

56. Michael L. Gross, *Bioethics and Armed Conflict: Moral Dilemmas of Medicine and War* (Cambridge, MA: The MIT Press, 2006), 15.

57. Henry K. Beecher, "Letter to Dean George P. Berry," February 1959, cited in Vincent J. Kopp, "Henry Knowles Beecher and the Development of Informed Consent in Anesthesia Research," *Anesthesiology* 90, no. 6 (June 1999): 1756–1765 at 1757.

58. See Robert M. Veatch, "Henry Beecher's Contributions to the Ethics of Clinical Research," *Perspectives in Biology and Medicine* 59, no. 1 (Winter 2016): 3–17.

59. George A. Mashour, "Altered States: LSD and the Anesthesia Laboratory of Henry Knowles Beecher," *Bulletin of Anesthesia History* 23, no. 3 (July 2005): 11–14 at 12.

60. Beecher, *Research*, 210. For an intriguing analysis of the impact of G. E. Moore's *Principia Ethica*, see Daniela Donnini Macciò, "On Economics and Philosophy; G.E. Moore's Imprint on the Cambridge Apostle's: From *Principia Ethica* to *My Early Beliefs*," PhD Thesis, Università Degli Studi Di Macerata, 2013. The author wrote a more pedestrian doctoral dissertation on G. E. Moore in 1967.

61. Beecher, *Research*, 209–210. It is not clear whether this is a firsthand report. In his article "Scarce Resources and Medical Advancement" for the spring 1969 issue of *Daedalus: Journal of the American Academy of Arts and Science*, 248–274 at 280, Beecher describes this as an order by Colonel Edward Churchill in

1943 applicable to the North African Theater of Operation. Around this time, Beecher himself was serving as an army physician in North Africa and, in all likelihood, had to comply with this order.

62. Quote from Fletcher, Beecher, *Research*, 210–211.

63. Beecher papers, undated, Francis A. Countway Library of Medicine, Harvard University. Cited in Rothman *Strangers at the Bedside*, chap. 4, note 3, 273.

64. Rothman, *Strangers at the Bedside*, chap. 4, note 3, 273–274.

65. Advisory Committee on Human Radiation, *Final Report of the Advisory Committee on Human Radiation Experiments* (New York: Oxford University Press), 79–80.

66. Daniel Callahan, personal communication to Jonathan Moreno, November 25, 2014, cited in Jonathan Moreno, "'Acid Brothers' Henry Beecher, Timothy Leary, and the Psychedelic of the Century," *Perspectives in Biology and Medicine* 59, no. 1 (Winter 2016): 107–121 at 117.

67. Henry K. Beecher, "Letter to Dean George P. Berry," February 1958, cited in Vincent J. Kopp, "Henry Knowles Beecher," 1757.

68. Beecher, *Research*, 211.

69. Beecher, *Research*, 211–212.

70. Beecher, *Research*, 212.

71. Ralph W. Emerson, "Self-Reliance," 1847, *Essays: First Series*, accessed October 7, 2022, https://en.wikisource.org/wiki/Essays:_First_Series/Self-Reliance.

72. Henry K. Beecher, "Consent in Clinical Experimentation: Myth and Reality" *JAMA* 195 (1966): 34–35, cited in Robert M. Veatch, "Henry Beecher's Contributions to the Ethics of Clinical Research," *Perspectives in Biology and Medicine* 95, no. 1 (2016): 3–17 at 9.

73. Veatch, "Beecher's Contributions," 9.

74. Francis H. Bradley, *A Metaphysical Essay* (London: Schwan Sonnenschein, [1793] 1908, 2nd ed.), xiv.

75. According to Christopher Booth, Pappworth kept this letter on, or over, his desk for the rest of his life ("Obituary: M. H. Pappworth," *British Medical Journal* 309 (December 10, 1994): 1577–1578). Pappworth's daughter tells a variant of this story claiming that in an interview earlier in his career, in Liverpool, the interviewer said to Pappworth, "No Jew can ever be a gentleman and I do not want one working for me" (Seldon, *The Whistle-Blower*, 56). It is not clear whether this is two versions of one incident or accounts of two separate incidents. What is clear is that Pappworth, an observant orthodox Jew, confronted anti-Semitism throughout his life. One documentable fact is that although Pappworth passed the exam for admission to the Royal College of Physicians in 1936, he was not elected a fellow of the college—an indicator of professional achievement normally awarded about decade or so after passing the exam—until 1994, an unprecedented fifty-seven-year delay.

76. Henry K. Beecher, "Letter to Maurice Pappworth," January 7, 1965, The Wellcome Library, London, accessed October 7, 2022, https://wellcomecollection .org/works/sn8mqmuy.

77. Elinor Langer, "Human Experimentation: Cancer Studies at Sloan-Kettering Stir Public Debate on Medical Ethics," *Science* 143, no. 3606 (February 7, 1964): 551–553.

78. Beecher, "Letter to Pappworth," January 25, 1965, Wellcome Library, London, 1.

79. Beecher, "Letter to Pappworth," January 25, 1965, 2.

80. Langer, "Human Experimentation," 551.

81. Lawrence Altman, *Who Goes First? The Story of Self-Experimentation* (Berkeley: University of California Press, 1986), 128. Salk injected his own children with his vaccine, but Altman is suspicious of his claim to have taken it himself.

82. Langer, "Cancer Studies at Sloan-Kettering," 551–553.

83. Langer, "Cancer Studies at Sloan-Kettering," 551.

84. Hyman's papers are in the archives of the State University on New York, accessed October 7, 2022, https://archives.albany.edu/description/catalog/apap258#summary.

85. Langer, "Cancer Studies at Sloan-Kettering," 552, 553.

86. Typescript A Beecher papers, Henry Knowles Beecher Archive, Francis A. Countway Library of Medicine, Boston, MA, cited in Susan E. Lederer, "'Ethics and Clinical Research' in Biographical Perspective," *Perspectives in Biology and Medicine* 59, no. 1 (2019): 18–36 at 28. Note: Beecher's CIA and US Army experiments were officially secret so he could not have confessed them publicly without risking federal prosecution.

87. Lederer, "Ethics and Clinical Research," 28.

88. Lasagna's 1994 interview, cited in Lancaster and Rathwell, "The Moralist." The reference is to John 8:7, when Jesus spoke out to people about to stone an accused adulteress, "He that is without sin among you, let him cast the first stone at her." The punishment of being stoned to death for adultery was prescribed in Deuteronomy 22:22–24 and in Leviticus 20:10–12.

89. Franklin G. Miller, "Henry Beecher and Consent to Research," *Perspectives in Biology and Medicine* 59, no. 1 (Winter 2016): 78–94 at 80. Miller draws on H. K. Beecher, "Experimentation in Man," *JAMA* 169, no. 5 (1959): 461–478.

90. Langer, *Science*, 552, 553.

91. Beecher, "Letter to Pappworth," January 25, 1965, 1.

92. The *New York Times* published articles on Southam's experiments prior to the scandal. The dates and titles are: June 15, 1956, "14 Convicts Injected with Live Cancer Cells" and in an article almost a year later, April 15, 1957, "Cancer Defenses Found to Differ; Tests Indicate Victims Lack Some Mechanisms that Well Human Being Has, Cancer Recurred [when] Deficiency is Noted, Warning by Southam."

93. Beecher, "Letter to Pappworth," January 25, 1965, 1.

94. The question "Why me?" may echo a famous line from Rabbi Hillel (circa 110 BCE–10 CE), "If I am not for myself, who will be for me? But if I am only for myself, what am I? If not now, when?" (Pirke Avot 1:14).

95. Susan E. Lederer, "'Ethics and Clinical Research' in Biographical Perspective," *Perspectives in Biology and Medicine* 59, no. 1 (Winter 2016): 18–36 at 33–34.

96. Beecher, *Research*, 119.

97. Henry K. Beecher, "Consent in Clinical Experimentation: Myth and Reality," *JAMA* 195, no. 1 (January 3, 1966): 34–35 at 35.

98. Miller, "Henry Beecher," 79; see also Franklin G. Miller, "Clinical Research Before Informed Consent," *Kennedy Institute of Ethics Journal* 24, no. 2 (June 2014): 141–157.

99. Pope Pius XII, "The Moral Limits of Medical Research and Treatment," Proposition 12, September 14, 1952, *Papal Encyclicals Online*, accessed October 7, 2022, https://www.papalencyclicals.net/pius12/p12psych.htm. It is unclear why Secretary Wilson wrote this memo. Some speculate that it was a response to the pope's speech, or to the WMA's attempts to develop a code of research ethics, or as an affirmation of human rights stipulated in the UN's Universal Declaration of Human Rights, or because the Nuremberg Code was issued by a US military court—or some combination of these factors.

100. Minutes from a meeting of the Harvard Medical School Administrative Board on October 6, 1961, and in a formal response to the Army some months later. Lancaster and Rathwell, "The Moralist."

101. Lancaster and Rathwell, "The Moralist."

102. Beecher, "Ethics and Clinical Research," 1354. Italics added.

103. Langer, "Human Experimentation: Cancer Studies at Sloan-Kettering Stir Public Debate on Medical Ethics," *Science*, 143, no. 3606 (February 7, 1964): 551–553 at 551.

104. Lederer, "'Ethics and Clinical Research,'" 29.

105. Henry K. Beecher, "Letter to George Burch," June 27, 1966, cited in Vincent J. Kopp, "Henry Knowles Beecher and the Development of Informed Consent in Anesthesia Research," *Anesthesiology* 90, no. 6 (June 1999): at 1758.

106. Anonymous, Letter to Henry Beecher, April 5, 1965, Beecher Archives Countway Library, cited in Kopp, "Henry Knowles Beecher," 1758–1759.

107. Henry K. Beecher, "Letters to John Talbott," August 20, 1965; August 30, 1965, Beecher Papers, Countway Library, cited in Rothman, *Strangers at the Bedside*, 72.

108. Anonymous reviewer. Cited in Rothman, *Strangers at the Bedside*, 73.

109. Beecher Papers, Countway Library, cited in David S. Jones, Christine Grady, and Susan E. Lederer, "Ethics and Clinical Research—The 50th Anniversary of Beecher's Bombshell," *New England Journal of Medicine* 374, no. 24 (June 16, 2016): 2393–2398 at 2395.

110. Beecher Papers, Countway Library, cited in Jones et al., "Beecher's Bombshell," 2395.

111. Kopp, "Henry Knowles Beecher."

112. Letter from Garland to Beecher, March 3, 1966, Beecher Papers, Countway Library, cited in Kopp, "Henry Knowles Beecher," at 1760.

113. "Blurring the line between editor and coauthor, [Garland] helped Beecher revise the manuscript." Jones et al., "Beecher's Bombshell," 2395.

114. Jones et al., "Beecher's Bombshell," 2395.

115. Henry K. Beecher, "Ethics and Clinical Research," *New England Journal of Medicine* 274, no. 24 (June 16, 1966): 1354–1360 at 1354.

116. Henry K. Beecher, "Ethics and Clinical Research," 1354. It is noteworthy that papal authority is also the only moral authority cited in Beecher's other bombshell article, published anonymously as "A Definition of Irreversible Coma: Report of the Ad Hoc Committee of the Harvard Medical School to Examine the Definition of Brain Death," *JAMA* 205, no. 6 (1968): 337–340. This is especially noteworthy because Beecher was a devout Methodist, not a Roman Catholic.

117. Prior to the civil rights and Medicare and Medicaid legislation of the 1960s, medical education, and hospital healthcare generally, tended to be sectarian in nature. Thus, there were Catholic, Jewish, and Protestant hospitals, and similarly, sectarian medical schools were associated with them in various ways.

118. Beecher, "Ethics and Clinical Research," 1354.

119. Beecher, "Ethics and Clinical Research," 1355. Italics added.

120. Beecher, "Consent in Clinical Experimentation," 160.

121. Beecher, "Ethics and Clinical Research," 1354–1360.

122. Beecher, "Ethics and Clinical Research," 1354–1360.

123. Beecher, "Ethics and Clinical Research," 1354–1360.

124. Beecher, "Ethics and Clinical Research," 1354–1360.

125. Beecher, "Ethics and Clinical Research," 1354–1360. As of April 4, 2023, this article was cited 2,963 times in the medical literature. Accessed April 4, 2023, https://www.google.com/search?q=ethics+and+clinical+research+beecher&oq =ethics+and+&aqs=chrome.2.69i59j69i57j0i67i650j46i199i465i512j0i512l2j69 i60j69i61.5296j0j4&sourceid=chrome&ie=UTF-8.

126. Seldon, *The Whistle-Blower*, 124–125.

127. Maurice H. Pappworth, "Letter to HK Beecher," January 19, 1965, Henry K. Beecher Papers (H MS c64), Harvard Medical Library in the Francis A. Countway Library of Medicine, cited in Gaw, "Exposing Unethical Human Research," 151.

128. Maurice H. Pappworth, "Letter to HK Beecher," February 20, 1965, Henry K. Beecher Papers (H MS c64), Harvard Medical Library in the Francis A. Countway Library of Medicine cited in Gaw, "Exposing Unethical Human Research," 152. Italics added.

129. Henry K. Beecher, "Letter to MH Pappworth," July 11, 1966, PP/MHP/C5. Wellcome Library, London, cited in Gaw, "Exposing Unethical Human Research," 152.

130. Henry K. Beecher, "Letter to MH Pappworth," July 29, 1966, PP/MHP/C5. Wellcome Library, London, cited in Gaw, "Exposing Unethical Human Research," 152–153.

131. Gaw, "Exposing Unethical Human Research," 152.

132. Seldon, *The Whistle-Blower*, 145–146.

133. Seldon, *The Whistle-Blower*, 146–147.

134. Michael Halberstam, *The New York Times*, July 14, 1967, cited in Seldon, *The Whistle-Blower*, 147.

135. Christopher Booth, *International Medical Tribune of Great Britain*, cited in Seldon *Whistle-Blower*, 148.

136. Paul J. Edelson, "Henry Beecher and Maurice Pappworth: Honor in the Development of the Ethics of Human Experimentation," in *Twentieth Century Ethics of Human Subjects Research*, ed. Roelcke and Maio (Stuttgart: Franz Steiner Verlag, 2004), 219–233 at 224.

137. Editorial, "Responsibilities of Research," *Lancet* 1:1020 and 1:1144; cited in Edelson, "Henry Beecher and Maurice Pappworth," 225, 226.

138. M. H. Pappworth, "'Human Guinea Pigs'—A History," 1456.

139. Edelson, "Beecher and Pappworth," at 228.

140. "*Human Guinea Pigs: Experimentation on Man*," Google Scholar 539 citations, accessed October 16, 2022, https://scholar.google.com/scholar?hl=en&as _sdt=0%2C33&q=Pappworth+%E2%80%9CHuman+Guinea+Pigs%3A+Expe rimentation+in+Man+citations&btnG=.

141. Martin Gore, "Foreword," in Seldon, *Whistle-Blower*, i, ii.

142. Edelson, "Beecher and Pappworth," 228.

143. In 2021, the Hastings Center changed the name of its Henry Beecher award to "Bioethics Founders Award" because a committee decided that Beecher was not worthy of being honored on the grounds that he himself had conducted unconsented experiments on human subjects and had never publicly apologized for doing so.

Chapter 7

1. Count Gibson Jr., "Letter to Dr. Sidney Olansky," May 28, 1955, cited in Reverby, *Examining Tuskegee*, 75.

2. Reverby, *Examining Tuskegee*, 75, 76.

3. Peter Buxtun, "Letter to William Brown," November 24, 1965, cited in Reverby, *Examining Tuskegee*, 79.

4. Jenn Stanley, "CHOICE/LESS: The Backstory, Episode 4: Tuskegee Was the 'Tip of the Iceberg,'" June 21, 2017, accessed October 7, 2022, https://rewirenewsgroup .com/wp-content/uploads/2017/06/CHOICELESS-The-Backstory-Ep4-Tuskegee .pdf.

5. Alexander Hamilton, *A Full Vindication of the Measures of the Congress, &c*, December 17, 1774 [public letter], accessed October 7, 2022, https://founders.archives .gov/documents/Hamilton/01-01-02-0054#ARHN-01-01-02-0054-fn-0001-ptr.

6. Marc A. Edwards, Carol Yang, and Siddhartha Roy, "Who Dares to Speak Up?" *American Scientist* 109, no. 4 (July–August 2021): 238–242 at 239.

7. Edwards et al., "Who Dares to Speak Up?" 239.

8. Reverby, *Examining Tuskegee*, 77. (The CDC, or Communicable Disease Center, was a branch of the USPHS that, after 1980, became known as Centers for Disease Control and Prevention. It is still called the "CDC.")

9. Peter Buxtun, interviewed by Carl Elliott, "Tuskegee Truth Teller," *The American Scholar* 1 (Winter 2018, December 4, 2017): 44–56 at 49.

10. Buxtun quote in Edwards et al., "Who Dares to Speak Up?" 240.

11. Edwards et al., "Who Dares to Speak Up?" 240.

12. Edwards et al., "Who Dares to Speak Up?" 240.

13. Peter Buxtun, "Letter to William Brown," November 24, 1968, cited in Reverby, *Examining Tuskegee*, 79.

14. James H. Jones, *Bad Blood: The Tuskegee Syphilis Experiment* (New York: The Free Press, Macmillan Publishing Co. 1981), 193.

15. Advisory Committee on Human Radiation Experiments, *Final Report of the Advisory Committee on Human Radiation Experiments* (New York: Oxford University Press, 1996), 100.

16. "Minutes, April 5, 1965," unpublished typescript, Tuskegee Syphilis Study–National Library of Medicine, cited in Allan M. Brandt, "Racism and Research: The Case of the Tuskegee Syphilis Study," *The Hastings Center Report* 8, no. 6 (December 1978): 21–29 at 26.

17. Reverby, *Examining Tuskegee*, 82.

18. Reverby, *Examining Tuskegee*, 80.

19. James B. Dale and Alan L. Bisno, "In Memoriam: Gene H. Stollerman," *Clinical Infectious Diseases* 59, no. 12 (September 2014): 1805–1806.

20. Jones, *Bad Blood*, 1981, 199.

21. Vanessa Northington Gamble, "Under the Shadow of Tuskegee: African Americans and Health Care," *American Journal of Public Health* 87, no. 11 (November 1997): 1773–1778. See also Vanessa Northington Gamble, "A Legacy of Distrust: African Americans and Medical Research," *American Journal of Preventive Medicine* 9, no. 6, Supplement (November 1, 1993): 35–38. In these papers, Gamble argues that African Americans distrust of American medicine was exacerbated by the Tuskegee Study but has deeper historical roots.

22. Hamilton, *A Full Vindication of the Measures of the Congress, &c.*

23. Donald H. Rockwell, Anne R. Yobs, and M. B. Moore Jr., "The Tuskegee Study of Untreated Syphilis; The 30th Year of Observation," *Archives of Internal Medicine* 114 (December 1964): 792–798 at 792.

24. Reverby, *Examining Tuskegee*, 75, 76.

25. Reverby, *Examining Tuskegee*, 75, 76.

26. Jones, *Bad Blood*, 190.

27. Count Gibson Jr., "Letter to Dr. Sidney Olansky," May 28, 1955, cited in Reverby, *Examining Tuskegee*, 75.

28. Reverby, *Examining Tuskegee*, 72.

29. Reverby, *Examining Tuskegee*, 75, based on reports from Gibson's former colleagues, see footnotes 89 and 90, p. 282.

30. Reverby, *Examining Tuskegee*, 83.

31. Jenn Stanley, "CHOICE/LESS: The Backstory, Episode 4: Tuskegee Was the 'Tip of the Iceberg,'" June 21, 2017, accessed October 7, 2022, https://rewirenewsgroup .com/2017/06/21/choiceless-backstory-episode-4-tuskegee-tip-iceberg/accessed.

32. J. E. Moore, "Letter to Taliaferro Clark of the United States Public Health Service, Clark, NA-WNRC," cited in Allan M. Brandt, "Racism and Research: The Case of the Tuskegee Syphilis Study," *The Hastings Center Report* 8, no. 6 (December 1978): 21–29 at 22.

33. Brandt, "Racism and Research," 23.

34. H. S. Cumming, "Letter to R. R. Moton," September 20, 1932, NAWNRC, cited in Brandt, "Racism and Research," 22.

35. Howard Cole et al., "Cooperative Clinical Studies in the Treatment of Syphilis: Latent Syphilis," *Venereal Disease Information* 13 (September 20, 1932): 351. The authors also concluded that the latent syphilitics were potential carriers of the disease meriting treatment. Cited in Brandt, "Racism and Research," 23.

36. Brandt, "Racism and Research," 23.

37. The colloquial use of the expression "bad blood" was not restricted to black communities or to the American South. Historian Susan Reverby observes, about the same time that the Syphilis Study began, US government posters were distributed with messages like "Your Blood is Bad Means You Have Syphilis." An Arkansas poster read "Syphilis is . . . Bad Blood . . . GET A BLOOD TEST." Reverby, *Examining Tuskegee*, 89.

38. US Department of the Interior, National Park Service, 2018. National Register of Historic Places: U.S. Public Health Service Syphilis Study, Macon County, AL, 1932–1973, Section E, p. 4. Capitalization is in the original.

39. O. C. Wenger. "Letter to R. A. Vonderlehr," July 21, 1933, cited in Susan Reverby, ed., *Tuskegee's Truths: Rethinking the Tuskegee Syphilis Study* (Chapel Hill: University of North Carolina Press, 2000), 85.

40. Reverby, *Tuskegee's Truths*, 85. Emphasis added.

41. Susan Lederer, "The Tuskegee Syphilis Study in the Context of American Medical Research," in Reverby, *Tuskegee's Truths*, 266–275 at 266.

42. Lederer, "Tuskegee Syphilis Study," 268–269.

43. Donald H. Rockwell, Anne R. Yobs, and M. B. Moore, "The Tuskegee Study of Untreated Syphilis: The 30th Year of Observation," *Archives of Internal Medicine* 114, no. 6 (December 1964): 792–798.

44. Jean Heller, "Syphilis Victims in U.S. Study Went Untreated for 40 Years," *New York Times*, July 26, 1972, 1 (AP Release, July 25, 1972).

45. Jones, *Bad Blood*, 203.

46. Department of Health, Education, and Welfare Public Health Service, *Final Report of the Tuskegee Syphilis Study Ad Hoc Advisory Panel* (Washington, DC: Government Printing Office, April 28, 1973), 1.

47. Department of Health, Education, and Welfare Public Health Service, *Final Report*, 7, 8.

48. Department of Health, Education, and Welfare Public Health Service, *Final Report*, 11.

49. Department of Health, Education, and Welfare Public Health Service, *Final Report*, 12.

50. Katz misquotes Bernard. The actual quotation from Bernard is "Do we have a right to perform experiments and vivisections on man? . . . It is our duty and our right to experiment on man whenever it can save his life, cure him, or gain him personal benefit. The principle of medical and surgical morality, therefore, consists in never performing an experiment on man an experiment which might be harmful to him to any extent, even though the result may be advantageous to [medical] science, i.e., to the health of others." Bernard invokes the *primum non nocere* (first do no harm) principle. He never mentions informed consent. Claude Bernard, *An Introduction to the Study of Experimental Medicine*, trans. Henry Copley Green (Reprint, New York: Dover Publications, 1957), 101.

51. R. H. Kampmeier, "The Tuskegee Study of Untreated Syphilis," *Southern Medical Journal* 65, no. 10 (October 1972): 1247–1251.

52. Kampmeier, "Tuskegee Study," 14, 15.

53. Jeremy Bentham, *Benthamiana: Select Abstracts from the Works of Jeremy Bentham*: With an Outline of His Opinions on the Principal Subjects Discussed in His Works, ed. John Hill Burton (Philadelphia: Lea & Blanchard, 1844), reprint (London: Forgotten Books, 2018), 115.

54. Beecher, "Ethics and Clinical Research," 1354.

55. Count Dillon Gibson Jr., "Obituary," *Hartford Courant*, July 25, 2002, accessed October 8, 2022, https://www.courant.com/news/connecticut/hc-xpm -2002-07-25-0207242365-story.html.

56. Reverby, *Examining Tuskegee*, 82.

57. Elliott, "Tuskegee Truth Teller," 56.

58. Elliott, "Tuskegee Truth Teller," 56.

59. *Report of the Committee on the Supervision of the Ethics of Clinical Investigation in Institutions*, cited in Duncan Wilson, *The Making of British Bioethics* (Manchester: Manchester University Press, 2014), 48–49.

60. Wilson, *British Bioethics*, 50.

61. Edelson, "Beecher and Pappworth," 228.

Chapter 8

1. Talcott Parsons, *The Social System* (London: Collier-Macmillan Company, 1951), 463.

2. World Medical Association, "Declaration of Geneva," as amended by the 68th WMA General Assembly, Chicago, October 2017, accessed October 8, 2022, https://www.wma.net/policies-post/wma-declaration-of-geneva/.

3. "Deserving poor" is an expression that wears its meaning on its face: it designates poor people who deserve assistance in contrast to its antonym, "the undeserving poor," who are deemed unworthy of assistance. The scope of these expression changed over time. Sometimes the undeserving poor were the sinful: non-churchgoers, deserters, malcontents, unwed mothers, or prostitutes. At other times, worthiness was defined by work status: those unable to work because of illness or disability were deserving; able-bodied malingerers and vagabonds who "chose" not to work were "undeserving." E. Bruenig, "The Undeserving Poor: A Very Tiny History," June 6, 2019, accessed October 9, 2022, https://medium.com/@ebruenig/the-undeserving-poor-a-very-tiny-history-96c3b9141e13.

4. Baker, *Before Bioethics*, 36–61.

5. Percival, *Medical Ethics*, chap. I, sec. I, 9.

6. Percival, *Medical Ethics*, chap. I, sec. III, IV, 10–11.

7. Percival, *Medical Ethics*, chap. I, sec. VIII, 13.

8. Percival, *Medical Ethics*, chap. II, sec. I, 30.

9. For a detailed account of the AMA's adoption of Percival's Code of Ethics, see Robert Baker, *Before Bioethics: A History of American Medical Ethics from the Colonial Period to the Bioethics Revolution* (New York: Oxford University Press, 2013), 131–167.

10. AMA Code of Ethics (1847), chap. I, sec. 1–2; chap. II, sec. VI, in *The American Medical Ethics Revolution: How the AMA's Code of Ethics Has Transformed Physicians' Relationships to Patients, Professionals and Society*, ed. Robert Baker, Arthur Caplan, Linda Emanuel, Stephen Latham (Baltimore: The Johns Hopkins University Press, 1999), 324, 333.

11. Robert Baker, "The American Medical Ethics Revolution," in *The American Medical Ethics Revolution*, ed. Robert Baker, Arthur Caplan, Linda Emanuel, Stephen Latham, 17–51. (See also Baker, *Before Bioethics*, 131–167.)

12. *The American Medical Ethics Revolution*, 326, 327.

13. Parsons, *The Social System*, 464–465.

14. Parsons, *The Social System*, 437.

15. Parsons, *The Social System*, 464–465.

16. Parsons, *The Social System*, 441.

17. Parsons, *The Social System*, 446–447.

18. Parsons, *The Social System*, 446–447.

19. Parsons, *The Social System*, 463.

20. AMA Code of Ethics (1847), chap. I, sec. 6, 9, *The American Medical Ethics Revolution*, 326, 327.

21. AMA Code of Ethics, 326, 327.

22. George Arthur Wiley, "Health Care in the Inner City: Like It Is: Point of View of a Consumer," in *Health Care Problems of the Inner City: Report of the 1969 National Health Forum. New York: National Health Forum* (1969, 12). Cited in

Joseph C. D'Oronzio, "A Human Right to Healthcare Access: Returning to the Origins of the Patients' Rights Movement," *Cambridge Quarterly of Healthcare Ethics* 10, no. 3 (Summer 2001): 285–298, note 23 at 294. This speech, at the seventeenth annual meeting of the National Health Council (founded 1920), is the only known speech by George Wiley on this topic.

23. Robert M. Veatch, *Disrupted Dialogue: Medical Ethics and the Collapse of Physician-Humanist Communication (1770–1980)* (New York: Oxford University Press, 2005), 208.

24. Nick Kotz and Mary Lynn Kotz, *A Passion for Equality: George Wiley and the Movement* (New York: W. W. Norton & Company, 1977), 195.

25. Kotz and Kotz, *A Passion for Equality*, 195, 196.

26. Kazuyo Tsuchiya, *National Welfare Rights Organization* (1966–1975), accessed October 8, 2022, https://Www.Blackpast.Org/African-American-History/National-Welfare-Rights-Organization-1966-1975/.

27. d'Oronzio, "A Human Right," 288–289.

28. G. A. Wiley, "Health Care in the Inner City," cited in d'Oronzio, "A Human Right to Healthcare," note 23 at 294.

29. Pollner, "*Off Our Backs*," cited in Ruzek, *The Women's Health Movement*, 191–192.

30. D'Oronzio, "A Human Right," 291.

31. The male chauvinist word "ombudsman" originally meant a *man* representing someone, accessed October 8, 2022, https://en.wikipedia.org/wiki/Ombudsman#Origins_and_etymology.

32. d'Oronzio, "A Human Right," 297.

33. d'Oronzio, "A Human Right," 293.

34. United Nations, Universal Declaration of Human Rights, 1947, Article 25.1, accessed October 8, 2022, https://www.un.org/en/about-us/universal-declaration-of-human-rights.

35. D'Oronzio, "A Human Right," 297. Bolding is in the original.

36. American Hospital Association website, accessed October 8, 2022, https://www.aha.org/about.

37. AHA, *A Patient's Bill of Rights*, 1972, Right #10.

38. Von Staden, "Hippocratic Oath," 7.ii at 407.

39. Thomas Hobbes, *Leviathan* 1651, *The Second Part: Of Commonwealth Chapter 17 Of the Causes, Generation, and Definition of a Commonwealth.* Excerpts from *Sections* 15–20, Eris Project at Virginia Tech, accessed October 8, 2022, https://history.hanover.edu/courses/excerpts/161hob.html.

40. AHA, *A Patient's Bill of Rights*, 1972. Epilogue statement.

41. Martin Luther, "Commentary on the Sermon on the Mount," by Martin Luther, Matthew 5–7, trans. Charles A. Hay (London: Pantianos Classics (reprints), (1530–1532)), January 1, 1892, 236, accessed October 8, 2022, http://www.godrules.net/library/luther/37luther0.htm.

42. Downer, "Covert Discrimination against Women," 1.

43. This may be from one of Marcuse's lectures since he undoubtedly contended that the elite in advanced industrial capitalist societies defined the meaning of "health" for their societies. See Herbert Marcuse, *One-Dimensional Man: Studies in the Ideology of Advanced Industrial Society* (New York: Beacon Press, 1964).

44. Boston Women's Health Book Collective, "Women, Medicine and Capitalism," in *Women and Their Bodies: A Course* (Boston: Self-published, 1970), 6–7, accessed October 8, 2022, https://www.ourbodiesourselves.org/wp-content/uploads/2020/04/Women-and-Their-Bodies-Free-Press.pdf.

45. George A. Wiley (circa 1969, 1970), "Health Care in the Inner City," cited in D'Oronzio, "A Human Right to Healthcare," 293.

46. Downer, "Covert Discrimination against Women," 1.

47. Boston Women's Health Book Collective, "Women, Medicine and Capitalism," 6–7.

48. Ruth Ravich and Lucy Schmolka, "Patient Representation as a Quality Improvement Tool," *The Mount Sinai Journal of Medicine* 60, no. 5 (1993): 374–378.

49. The Mount Sinai Hospital, "About the Hospital," accessed October 8, 2022, https://www.mountsinai.org/locations/mount-sinai/about/history.

50. At the time of writing this book, the author was teaching courses at Mount Sinai as a faculty member in the Clarkson University–Icahn School of Medicine bioethics program.

51. Ravich and Schmolka, "Patient Representation," 374.

52. Ravich and Schmolka, "Patient Representation," 374.

53. Ravich and Schmolka, "Patient Representation," 375.

54. George J. Annas and Joseph M. Healey Jr., "The Patient Rights Advocate: Redefining the Doctor-Patient Relationship in the Hospital Context," *Vanderbilt Law Review* 27, no. 2 (March 1974): 243–269 at 245.

55. Annas and Healey, "The Patient Rights Advocate," 258.

56. Jo Harris-Wehling, Jill C. Feasley, and Carroll L. Estes, eds., *An Evaluation of the Long-Term Care Ombudsman Programs of the Older Americans Act* (Washington, DC: Division of Healthcare Services, Institute of Medicine, 1995), 44.

57. American Nurses Association, "A Suggested Code," *American Journal of Nursing* 26, no. 8 (1926): 599–601 at 600, cited in Beth Epstein and Martha Turner, "The Nursing Code of Ethics: Its Value, Its History," *Online Journal of Issues in Nursing* 20, no. 2 (May 2015), accessed October 8, 2022, https://portal.savonia.fi/amk/sites/default/files/pdf/eng/savoniauas_nursing_article_for_preliminarytask_2020.pdf.

58. American Nurses Association, "A Suggested Code," 8.

59. American Nurses Association, "A Suggested Code," 8.

60. Two types of conflicts of interest are common in health care: financial and dual loyalty. Financial conflicts arise from "incentives that bias . . . [by] reward[ing] physicians for increasing or decreasing services, or providing one kind of service rather

than others. . . . Divided loyalty conflicts occur when [practitioners] perform [more than one role] . . . divided loyalty or dual roles often overlap." Marc A. Rodman, *Conflicts of Interest and the Future of Medicine* (New York: Oxford University Press, 2012), 15, 16.

61. George A. Wiley (circa 1969), "Health Care in the Inner City," cited in D'Oronzio, "A Human Right to Healthcare," 293.

Chapter 9

1. William Whewell, *History of the Inductive Sciences from the Earliest to the Present Day* (London: John W. Parker, 1837), vol. I, 11. For further discussion see I. Bernard Cohen, *Revolution in Science* (Cambridge, MA: The Belknap Press of Harvard University Press, 1985), 528–532.

2. John Stuart Mill, "On Liberty," in *Utilitarianism and On Liberty*, ed. Mary Warnock (Oxford: Blackwell Publishers [1859] 2002), 88–180 at 124.

3. Charles Dickens, *A Tale of Two Cities* (London: Chapman and Hall, 1859), accessed April 18, 2022, https://etc.usf.edu/lit2go/22/a-tale-of-two-cities/108/book-the-first-recalled-to-lifechapter-1-the-period/.

4. John S. Mill, "A Few Observations on the French Revolution: Review of Alison's *History of Europe*," *Monthly Repository*, August 1833, accessed March 13, 2022, http://www.laits.utexas.edu/poltheory/jsmill/diss-disc/french-rev.html.

5. National Constituent Assembly (France), July 9, 1789, *Déclaration des droits de l'homme et du citoyen*: "The National Assembly doth recognize and declare, in the presence of the Supreme Being, and with the hope of his blessing and favour, the following sacred rights of men and of citizens":

Article 1 "Men are born and remain free and equal in rights." https://avalon.law.yale.edu/18th_century/rightsof.asp (accessed January 22, 2022). As the words in the English translation plainly state, this declaration applies to males and citizens (also males); it does not apply to females.

6. Notable among more recent philosophers writing about moral revolutions are Kathryn Pyne Parsons (Addelson), "Nietzsche and Moral Change," *Feminist Studies* 2, no. 1 (1974): 57–76, and Kwame Anthony Appiah, *Honor Code: How Moral Revolutions Happen* (New York: W. W. Norton, 2010). Several other bioethicists with a philosophy of science background, notably Arthur Caplan (in informal conversations with the author) and Baruch Brody, have suggested that bioethics was an artifact of a moral revolution. See Baruch Brody, 2015 Lifetime Achievement Award, American Society for Bioethics and Humanities, referenced by Craig Klugman in November 10, 2015, column, accessed March 13, 2022, http://www.bioethics.net/2015/11/bioethics-the-revolution-is-over/.

7. Charles Darwin, *The Origin of Species* (London: John Murray, 1850). The sixth edition, in 1872, was the last edition edited by Darwin himself, and it is considered the authoritative edition. The quotation from Whewell heads the prefatory quotations, accessed April 18, 2022, https://www.gutenberg.org/files/2009/2009-h/2009-h.htm.

8. William Whewell, *History*, vol. I, 11.

9. Cohen, *Revolution in Science*, offers an excellent history of notions of "revolution" in science. Cohen observes that in 1620, when Copernicus wrote about "revolutions," he used the term in the sense of "revolve" (i.e., a circular motion as in a merry-go-round, a revolving door, or a wheel). Later, as the word came to be used to describe radical changes in governance, it was also used to describe "scientific revolutions."

10. For Kuhn, scientific revolutions are recurring events, whereas, prior to him, with the notable exception of Whewell, historians of science recognized only one "scientific revolution," which they dubbed "the scientific revolution," as if it were the one and only such revolution—a tendency that also infects many philosophical treatments of morality.

11. "Intelligentsia" originated as a Polish term that characterized artists, composers, journalists, musicians, philosophers, singers, and writers *actively* resisting the Russification of Poland, that is, the Russian government's attempt to replace Polish culture, literature, music, history, and the Polish language itself with the Russian language and culture. As the concept internationalized, it came to reference artists and intellectuals *acting* for or against a cause, in contrast to intellectuals, who just analyze or discuss a cause.

12. Daniel M. Fox, "Who Are We: The Political Origins of the Medical Humanities," *Theoretical Medicine* 6, no. 3 (October 1985): 327–341 at 32, 34, cited in Jonsen, *Birth of Bioethics*, 25.

13. Fox, "Who Are We?" 338, cited in Jonsen, *Birth of Bioethics*, 25.

14. Paul Lauritzen, "Daniel Callahan & Bioethics: Where the Best Arguments Take Him," *Commonweal*, June 1, 2007, 3, accessed March 13, 2022, https://www.commonwealmagazine.org/daniel-callahan-bioethics.

15. The prohibition of birth control reads as follows. "*Unlawful Birth Control Methods*":

Therefore We base Our words on the first principles of a human and Christian doctrine of marriage when We are obliged once more to declare that the direct interruption of the generative process already begun and, above all, all direct abortion, even for therapeutic reasons, are to be absolutely excluded as lawful means of regulating the number of children. (14) Equally to be condemned, as the magisterium of the Church has affirmed on many occasions, is direct sterilization, whether of the man or of the woman, whether permanent or temporary. (15) Similarly excluded is any action which either before, at the moment of, or after sexual intercourse, is specifically intended to prevent procreation—whether as an end or as a means. *Encyclical Letter, Humanae Vitae, of the Supreme Pontiff, Paul VI*, accessed March 13, 2022, https://www.vatican.va/content/paul-vi/en/encyclicals/documents/hf_p-vi_enc_25071968_humanae-vitae.html.

16. Andre E. Hellegers, "A Scientists Analysis," in *Contraception, Authority, and Dissent*, ed. Charles E. Curran (New York: Herder and Herder, 1960), 216–217, cited in Jennifer K. Walter and Eran P Klein, eds., *The Story of Bioethics: From Seminal Works to Contemporary Explorations* (Washington, DC: Georgetown University Press, 2003), 226–227.

17. Robert McClory, *Turning Point: The Inside Story of the Papal Birth Control Commission, and How Humanae Vitae Changed the Life of Patty Crowley and the Future of the Church* (New York: Crossroad, 1995), 141.

18. Daniel Callahan, Bernard Haring, and Charles Curran, *The Catholic Case for Contraception: Leading Catholic Authorities Oppose Pope Paul's Position on Birth Control* (London: Macmillan, 1969).

19. Hastings Center, "Daniel Callahan, 1930–2019, " accessed March 13, 2022, https://www.thehastingscenter.org/news/daniel-callahan-1930-2019/.

20. Transcript of an interview with Daniel Callahan, March 17, 1977, conducted by Allan Brandt, 2–6, 49, in Rothman, *Strangers at the Bedside*, 209.

21. Ruzek, *The Women's Health Movement*, 45.

22. Daniel Callahan, "Institute of Society, Ethics, and the Life Sciences: A Survey of Goals, Plans, and Budgetary Needs," January 1970, in a folder marked, ISELS, p. 1. Rockefeller Archives, Rockefeller Brothers Fund, Sleepy Hollow, New York, cited in M. L. Tina Stevens, *Bioethics in America: Origins and Cultural Contexts* (Baltimore: The Johns Hopkins University Press, 2000), 51.

23. *In the Matter of Karen Quinlan, An Alleged Incompetent*, 70 N.J. 10, Supreme Court of New Jersey. Argued January 26, 1976. Decided March 31, 1976.

24. Leroy Walters Transcript, interview with LeRoy Walters, PhD, Director, Kennedy Institute of Ethics, and Joseph P. Kennedy, Sr. Professor of Christian Ethics, Kennedy Institute of Ethics, Washington, DC, March 13, 2000. The interview was conducted by Dr. Judith P. Swazey at the Kennedy Institute of Ethics, Georgetown University. Acadia Institute Project on Bioethics in American Society, Archives of the Bioethics Library, Kennedy Institute of Ethics, 37–38.

25. Kennedy Institute of Ethics website, accessed January 23, 2022, https://bioethics.georgetown.edu/.

26. Warren T. Reich, STD [Doctor of Sacred Theology], Professor Emeritus and Senior Research Scholar, Kennedy Institute of Ethics, Washington, DC, March 29, 2000. The interview was conducted by Dr. Judith P. Swazey at the Kennedy Institute of Ethics, Georgetown University. Acadia Institute Project on Bioethics in American Society, Archives of the Bioethics Library, Kennedy Institute of Ethics, 24–26.

27. Fritz Jahr, "Bio-Ethik," *Kosmos* 24, no. 4 (1927), cited in Hans-Martin Sass, "Fritz Jahr's 1927 Concept of Bioethics," *Kennedy Institute of Ethics Journal*, 17 no. 4 (2008): 279–295 at 279.

28. Van Rensselaer Potter, *Bioethics: Bridge to the Future* (Englewood Cliffs, NJ: Prentice-Hall, 1971), 1, 2.

29. Daniel Callahan, "Bioethics as a Discipline," *The Hastings Center Studies* 1, no. 1 (1973): 66–73 at 72.

30. Ludwig Wittgenstein, *Philosophical Investigations* (Hoboken, NJ: Wiley-Blackwell Publishing Co., [1953] 2001), § 43, 25e.

31. The author was one of the philosophers recruited to attend the Council of Philosophical Studies Summer Institute. See R. Baker, "From Metaethicist to Bioethicist," *Cambridge Quarterly of Healthcare Ethics* 11 (2002): 369–379.

32. Wiley, "Health Care in the Inner City," cited on D'Oronzio, "A Human Right to Healthcare," note 23, 294.

33. Carol, "Covert Discrimination against Women," 1.

34. Fox, "Who Are We?" 3234, cited in Jonsen, *Birth of Bioethics*, 25.

35. Daniel Callahan, "Institute of Society, Ethics, and the Life Sciences: A Survey of Goals, Plans, and Budgetary Needs," January 1970, in a folder marked, ISELS, p. 1. Rockefeller Archives, Rockefeller Brothers Fund, Sleepy Hollow, New York, cited in M. L. Tina Stevens, *Bioethics in America*, 51.

36. The sole moral authority referenced in Beecher's Ad Harvard Committee report on brain death is a statement from the Roman Catholic pope. See "A Definition of Irreversible Coma. Report of the Ad Hoc Committee of the Harvard Medical School to Examine the Definition of Brain Death," *JAMA* 205 no. 6 (August 5, 1968): 337–340.

37. Daniel Callahan, "Religion and the Secularization of Bioethics," *Hastings Center Report* 20 no. 4 (Suppl., July–August 1990): 2–4.

38. Paul Ramsey, *The Patient as Person: The Lyman Beecher Lectures at Yale University* (New Haven, CT: Yale University Press, 1970), xii.

39. Therese M. Lysaught, "Respect: Or, How Respect for Persons Became Respect for Autonomy," *Journal of Medicine and Philosophy* 29, no. 6 (December 29, 2004): 665–680.

40. Karen Lebacqz, interview with Karen Lebacqz, Robert Gordon Sproul Professor of Theological Ethics Emeritus, Pacific School of Religion, Berkeley, California, by Leroy B. Walters, Professor of Christian Ethics and Professor of Philosophy, Kennedy Institute of Ethics, Georgetown University, 2004. *Oral History of the Belmont Report and the National Commission for the Protection of Human Subjects of Biomedical and Behavioral Research,* accessed March 14, 2022, https://www.hhs.gov/ohrp/education-and-outreach/luminaries-lecture-series/belmont-report-25th-anniversary-interview-klebacqz/index.html.

41. Tristram Engelhardt Jr., *The Foundations of Bioethics* (New York: Oxford University Press, 1986), 5, cited in John Evans, *The History and Future of Bioethics: A Sociological View* (New York: Oxford University Press, 2012), 7.

42. LeRoy Walters, interview with LeRoy Walters, by Dr. Judith P. Swazey at the Kennedy Institute of Ethics, Georgetown University, March 13, 2000, 49–51.

43. James Gustafson, "Theology Confronts Technology and the Life Sciences," *Commonweal* 105, no. 12 (June 16, 1978): 386–392, cited in John Evans, *The History and Future of Bioethics*, 29.

44. Patrick H. Nowell-Smith, *Ethics* (Harmondsworth: Penguin Books, 1954), 319–320.

45. For an autobiographical account of the transformation of an analytically trained metaethicist into a bioethicist, see Baker, "Metaethicist to Bioethicist," 369–379.

46. Ruth Macklin, Acadia Institute Project on Bioethics in American Society, interview with Ruth Macklin, May 18, 1999, 24–26. See also Baker, "Metaethicist to Bioethicist," 369–379.

47. Starting out in bioethics, I was influenced by writings of Swiss American psychiatrist Elisabeth Kübler-Ross, especially her description of a seminar in which terminal patients discussed their approach to death. See Elisabeth Kübler-Ross, *On Death and Dying: What the Dying Have to Teach Nurses, Clergy, and Their Own Families* (New York: Macmillan, 1969). See also Robert Baker, "Kübler-Ross and Bioethics: A Cautionary Tale," *The American Journal of Bioethics* 19, no. 12 (2019): 48–49.

48. H. Tristram Engelhardt Jr., *The Foundations*, 5, cited in Evans, *The History and Future of Bioethics*, 7.

49. Kuhn, *Structure of Scientific Revolutions*, 6.

50. Halpern, *Dangerous Medicine*, 164.

51. Halpern, *Dangerous Medicine*, 165.

52. Halpern, *Dangerous Medicine*, 165.

53. Beecher, "Ethics and Clinical Research," 1359.

54. F. J. Ingelfinger, *Yearbook of Medicine* (Chicago: Yearbook Medical Publishers, 1968), 430, cited in Beecher, *Research and the Individual*, 122.

55. Halpern, *Dangerous Medicine*, 9, 10.

56. Beecher, *Research and the Individual*, 127.

57. Ramsey, *Patient as Person*, 48–50.

58. M. A. Berger, "A History of Immune Globulin Therapy, from the Harvard Crash Program to Monoclonal Antibodies," *Current Allergy and Asthma Report* 2, no. 5 (September 2002): 368–378.

59. Ramsey, *Patient as Person*, 49.

60. Ramsey, *Patient as Person*, 49.

61. Paul A. Offit, *Vaccinated: One Man's Quest to Defeat the World's Deadliest Diseases* (Washington, DC: Smithsonian Books, 2007), 27.

62. National Commission for the Protection of Human Subjects of Biomedical and Behavioral Research, *Ethical Guidelines for the Delivery of Health Services by DHEW* (Washington, DC: US Government Printing Office, 1978), 101, accessed October 8, 2022, https://repository.library.georgetown.edu/bitstream/handle/10822/559347/ethical_guidelines_health_services_min.pdf?sequence=1&isAllowed=y.

63. National Commission for the Protection of Human Subjects of Biomedical and Behavioral Research, Appendix to *Ethical Guidelines for the Delivery of Health Services by DHEW*, 1978, accessed October 8, 2022, https://repository.library.georgetown.edu/handle/10822/559340.

64. Jonsen, *Birth of Bioethics*, 142–143.

65. Rothman, *Strangers at the Bedside*, 56. See also William J. Curran, "Governmental Regulation of the Use of Human Subjects in Medical Research. The Approach of Two Federal Agencies. *DÆDALUS* 98, no. 2 (Spring 1969): 542–594.

66. National Commission for the Protection of Human Subjects of Biomedical and Behavioral Research, *Report and Recommendations Institutional Review Boards,*

1978, accessed October 8, 2022, https://videocast.nih.gov/pdf/ohrp_institutional
_review_boards.pdf.

67. Joseph Catania et al., "Survey of U.S. Human Research Protection Organizations: Workload and Membership," *Journal of Empirical Research on Human Research Ethics* 3, no. 4 (December 2008): 57–69.

68. Carpenter, *Reputation and Power*, 549.

69. Thomas Percival, *Medical Jurisprudence or A Code of Ethics and Institutes, Adapted to the Professions of Physic and Surgery* (Manchester, UK, 1794), sec. 1, art. XII, cited in Beecher, *Research and the Individual*, 218.

70. Percival, *Medical Jurisprudence*, sec. 1, art. XII.

71. Beecher, *Research and the Individual*, 218.

Chapter 10

1. Robert Veatch, interview with Robert M. Veatch, PhD, Professor of Medical Ethics, conducted by Dr. Renee C. Fox and Dr. Judith P. Swazey at Professor Veatch's office, March 26, 1999, 37.

2. Robert M. Veatch, "Autonomy's Temporary Triumph," *Hastings Center Report* 14, no. 5 (October 1984): 38–40.

3. William Whewell, *History of the Inductive Sciences from the Earliest to the Present Day* (London: John W. Parker, 1837), vol. I, 11. "Field" substituted for "science." For further discussion, see Cohen, *Revolution in Science*, 528–532.

4. Veatch, interview, March 26, 1999, 37.

5. Veatch, "Autonomy's Temporary Triumph," 38–40.

6. Veatch, interview, March 26, 1999, 37.

7. Whewell, *History of the Inductive Sciences*, vol. I, 11.

8. This description, taken from *Structure of Moral Revolutions*, is adapted from Kuhn's characterization of scientific revolutions in Thomas Kuhn, *Structure*, 6. See also Cecilie Eriksen, "The Dynamics of Moral Revolutions—Prelude to Future Investigations and Interventions," *Ethical Theory and Moral Practice* 22, no. 3 (2019): 779–792.

9. Albert Jonsen, interview with Albert R. Jonsen by Judith Swazey, June 19 and 22, 1998, 46.

10. Peter Buxtun, "Testimony by Peter Buxton from the United States Senate Hearings on Human Experimentation, 1973," in Reverby, *Tuskegee's Truths*, 154.

11. In his influential proto-bioethical 1970 book, *The Patient as Person*, Methodist moral theologian Paul Ramsey (1913–1988) attempted to translate Christian ethics into a secular medical ethics based on faithfulness or fidelity (Ramsey, *Patient as Person*, xii). On this analysis, "consent" was interpreted "as a canon of loyalty" (Ramsey, *Patient as Person*, 5). At one point, however, Ramsey remarks that "a rule governing medical experimentation on human beings is needed to ensure that no person shall be degraded and treated as a thing or as an animal in order that

good may come of it" (Ramsey *Patient as Person*, 9). Ramsey's neo-Kantian remark comes close to capturing the sense of Buxtun's intuition—although, to reiterate, Ramsey himself tried to justify the ethics of experimentation in terms of the concept of fidelity.

12. Ludwig Wittgenstein, *Culture and Value*, Peter Winch translation of *Vermischte Bemerkungen* (Chicago: University of Chicago Press, 1977), 42e.

13. Jonsen and Lebacqz used *Respect for Persons* by British philosophers Robert Downie (1933–) and Elizabeth Telfer (1936–) as a textbook for a course that they had recently taught. Engelhardt also invoked this concept in his essay for the Commission.

14. Jonsen, *Birth of Bioethics*, 47, 48. In a different rendering, Jonsen credits H. Tristram Engelhard Jr. with the first two principles of bioethics rendered as (1) "respect for persons as free moral agents" and (2) "support[ing] the best interests of subjects" and "benefit[ing] society." Noting that Beauchamp submitted a paper on distributive justice, Jonsen comments that "after much discussion the commission took Engelhardt's first two principles and Beauchamp's principle of distributive justice and crafted 'crisp' principles: respect for person, beneficence, and justice." Jonsen, *Birth of Bioethics*, 103.

15. The National Commission for the Protection of Human Subjects of Biomedical and Behavioral Research, *The Belmont Report: Ethical Principles and Guidelines for the Protection of Human Subjects of Research* (Washington, DC: U.S. Department of Health, Education, and Welfare, DHEW Publication No. (OS) 78–0012, US Government Printing Office, 1978), 2–4.

16. US Department of Health and Human Services, "Summary: The Belmont Report," 1978, accessed October 8, 2022, https://www.hhs.gov/ohrp/regulations-and-policy/belmont-report/read-the-belmont-report/index.html. See also interview with Tom Lamar Beauchamp, PhD, Senior Research Scholar, Kennedy Institute of Ethics Professor of Philosophy Georgetown University, Washington, DC, September 22, 2004; Belmont Oral History Project Interviewer: Dr. Bernard A. Schwetz, DVM, PhD, Director, Office for Human Research Protections.

17. Michael Yesley, quoted in Beauchamp, "The Origins, Goals, and Core Commitments of the *Belmont Report* and *Principles of Biomedical Ethics*," in *The Story of Bioethics*, 21.

18. Beauchamp, "The Origins, Goals, and Core Commitments," in *The Story of Bioethics*, 21.

19. Beauchamp, "Origins," 21.

20. Beauchamp, "Origins," 21.

21. In Englehardt's background paper, he rendered the neo-Kantian precept of respect for persons as, "One should respect human subjects as free agents out of a duty to such subjects to acknowledge their right to respect as free agents" (8–5). This principle entailed that "experimentation upon unwilling human subjects should be regarded as immoral, even if the results of such experimentation would be of considerable general utility. [Thus] with regard to such experimentation that the Nazi use of human subjects would be worthy of condemnation, even

if it had been the case that such experimentation had revealed extremely useful information not otherwise attainable. Such basic rights cannot be outweighed by goods" (8–6). H. Tristram Engelhardt Jr. "Basic Ethical Principles," in "The Conduct of Biomedical and Behavioral Research Involving Human Subjects." The National Commission for the Protection of Human Subjects of Biomedical and Behavioral Research, *Appendix: The Belmont Report: Ethical Principles and Guidelines for the Protection of Human Subjects of Research*, vol. I, sec. 8 (U.S. Department of Health, Education, and Welfare, DHEW Publication No. (OS) 78–0013; Washington, DC: US Government Printing Office, 1978), 2–4.

22. National Commission for the Protection of Human Subjects. *The Belmont Report*, 5.

23. Thomas L. Beauchamp and James F. Childress, *Principles of Biomedical Ethics* (New York: Oxford University Press, 1979), 56–57.

24. National Commission for the Protection of Human Subjects. *The Belmont Report*, 5. As conceived by Engelhardt in his supporting essay, "respect for persons" extends to any being capable of free agency or free moral action, including adolescents, people with psychological problems and mental disabilities, and perhaps even to nonhuman primates. An echo of this broad conception of respecting persons is included in the final edition of the *Belmont Report*, where "respect for persons" extends to "persons with diminished autonomy [who] are entitled to protection. The principle of respect for persons thus divides into two separate moral requirements: the requirement to acknowledge autonomy and the requirement to protect those with diminished autonomy" (prisoners, for example).

25. Beauchamp and Childress, *Principles*, 59. Italics added.

26. Tom L. Beauchamp and James F. Childress, *Principles of Biomedical Ethics* (New York: Oxford University Press, 2nd ed., 1983), 64.

27. Beauchamp and Childress, *Principles of Biomedical Ethics*, 62.

28. Lysaught, *Respect*, 675–676. Quotation marks added around "respect."

29. Karen Lebacqz, interview with Karen Lebacqz, by Leroy B. Walters, *Oral History of the Belmont Report and the National Commission*, 2004, 2–4. See also K. Lebacqz, *We Sure Are Older but Are We Wiser*, in James F. Childress, Eric M. Meslin, and Harold T. Shapiro, eds., *Belmont Revisited: Ethical Principles for Research with Human Subjects* (Washington, DC: Georgetown University Press, 2005), 99–110.

30. One reason for this was that the implementing bodies envisioned by the *Report*'s authors were either short-lived or never founded. See James F. Childress, Eric M. Meslin, and Harold T. Shapiro, eds., *Belmont Revisited: Ethical Principles for Research with Human Subjects* (Washington, DC: Georgetown University Press, 2005), vi, ix.

31. Thomas Beauchamp, *Belmont Oral History Project Interview*, September 22, 2004. Interviewer: Dr. Bernard A. Schwetz.

32. Beauchamp, *Belmont Oral History Project Interview*.

33. Albert Jonsen, *Oral History of the Belmont Report and the National Commission for the Protection of Human Subjects of Biomedical and Behavioral Research*.

Interview with Albert R. Jonsen, PhD, Professor of Medical Ethics, University of California at San Francisco Medical School, San Francisco, CA. Belmont Oral History Project, interview with Dr. Albert Jonsen, May 14, 2004. Interviewer: Dr. Bernard A. Schwetz, DVM, PhD, Director, Office for Human Research Protections.

34. Jonathan Moreno, back cover of Beauchamp and Childress's eighth edition of *Principles of Biomedical Ethics*, 2019.

35. Cedric M. Smith, "Origin and Uses of *Primum Non Nocere*—Above All, Do No Harm!" *Journal of Clinical Pharmacology* 45, no. 4 (April 2005): 371–377.

36. Albert B. Jonsen, Mark Siegler, and William J. Winslade, *Clinical Ethics: A Practical Approach to Ethical Decisions in Clinical Medicine* (New York: Macmillan Publishing Co., 1983). The book's ninth edition was published in 2022.

37. Appiah, *Honor Code*, xii.

38. Kuhn, *Structure of Scientific Revolutions*, 156–158.

39. Google Scholar, "Ethics and Clinical Research," accessed October 8, 2022, https://scholar.google.com/scholar?q=beecher+ethics+and+clinical+research&hl=en&as_sdt=0&as_vis=1&oi=scholart.

40. "Tom L. Beauchamp," WorldCat Identities, accessed October 8, 2022, https://www.worldcat.org/identities/lccn-n78091263/.

41. "Principles of Biomedical Ethics," Google Scholar, accessed March 13, 2022, https://scholar.google.com/scholar?hl=en&as_sdt=0%2C33&as_vis=1&q=Principles+of+Biomedical+Ethics&btnG=.

42. Pope Pius XII, "Address to an International Congress of Anesthesiologist, Official Documents," *L'Osservatore Romano*, November 25–26, 1957, accessed March 13, 2022, http://lifeissues.net/writers/doc/doc_31resuscitation.html.

43. Pope Pius XII, "Address."

44. Pope Pius XII, "Address."

45. *In re Quinlan*, 70 N.J. 10 (1976). 355 A.2d 647. In the matter of Karen Quinlan, an alleged incompetent, Supreme Court of New Jersey.

46. Stevens, *Bioethics in America*, 142.

47. Mid-twentieth-century American hospitals had a variety of committees diffusing moral responsibility by offering ethical advice or warranting potentially controversial decisions. However, these committees dealt with a wide range of official policies (e.g., a dean's committee) or with specific issues (e.g., abortion, sterilization, quarantine). Teel recommended a separate standing hospital committee focused on ethics, instead of using ad hoc committee or committees that dealt with other issues (e.g., referring a non-resuscitation issue to an abortion committee). For an example of how a hospital's specialized abortion committee responded to a 1960s request for an abortion by Sherri Finkbine, a pregnant thalidomide mother, see Baker, *Structure of Moral Revolutions*, 133–135.

48. Karen Teel, "The Physician's Dilemma: A Doctor's View—What the Law Should Be." *Baylor Law Review* 27, no. 1 (1975): 6–9.

49. Clinical Care Committee of the Massachusetts General Hospital, "Optimum Care for Hopelessly Ill Patients. A Report of the Clinical Care Committee of the

Massachusetts General Hospital," *New England Journal of Medicine* 295, no. 7 (August 12, 1976): 362–364.

50. Stuart Younger et al., "A National Survey of Hospital Ethics Committees," in President's Commission for the Study of Ethical Problems in Medicine and Biomedical and Behavioral Research, *Deciding to Forego Life-Sustaining Treatment: A Report on the Ethical, Medical and Legal Issues in Treatment Decisions* (Washington, DC: US Government Printing Office, 1983), 443–457.

51. Jonsen, *Birth of Bioethics*, 363–364.

52. Kuhn, *Structure of Scientific Revolutions*, 92–93.

53. A second factor was that specialist societies had peeled off from the AMA in large measure because the AMA had abused its ethics code to create conditions inimical to specialized medical practice. Consequently, specialists' societies were formed outside the AMA and initially abjured ethics codes. See Baker, *Before Bioethics*, 183–185.

54. AMA, *Principles of Medical Ethics*, 1957. In Baker et al., *American Medical Ethics Revolution*, 335–336. Italics added.

55. AMA Code of Ethics 1903, chap. 1, sec. 1, Baker et al., *American Medical Ethics Revolution*, 335–336. Italics added.

56. AMA Code of Ethics 1912, chap. 1, "The Physician's Responsibility," Baker et al., *American Medical Ethics Revolution*, 346. Italics added.

57. Jonsen, *Birth of Bioethics*, 338–339.

58. American Association of Medical Colleges, *Report I: Learning Objectives for Medical Student Education: Guidelines for Medical Schools* (Washington, DC: American Association of Medical Colleges, 1998), 4–5.

59. Govind Presad et al., "The Current State of Medical School Education in Bioethics, Health Law, and Health Economics," *Journal of Law and Medical Ethics*, 36, no. 1 (Spring 2008): 89–94 at 94.

60. "Principles of Biomedical Ethics," Google Scholar, accessed March 13, 2022, https://scholar.google.com/scholar?hl=en&as_sdt=0%2C33&as_vis=1&q=Princ iples+of+Biomedical+Ethics&btnG=(this link is no longer active).

61. The innovations in CPR that became available in the 1960s and 1970s were developed by several innovators: the team of James Elam (1918–1995), Archer Gordon (1921–1973), and Peter Safar (1924–2003) and, independently, by William B. Kouwenhoven (1886–1975).

62. Sherwin Nuland, *How We Die: Reflections on Life's Final Chapter* (New York: Alfred A. Knopf, 1994), 224. The statement in internal quotation marks was attributed to William Bean, MD (1909–1969), of the University of Iowa.

63. National Conference Steering Committee, "Standards for Cardiopulmonary Resuscitation (CPR) and Emergency Cardiac Care," *JAMA: The Journal of the American Medical Association* 227 (Suppl., 1974): 837, 864.

64. Mitchell T. Rabkin, Gerald Gillerman, and Nancy R. Rice, "Orders Not to Resuscitate," *New England Journal of Medicine* 295, no. 7 (August 12, 1976): 363–366.

65. Clinical Care Committee of the Massachusetts General Hospital, "Optimum Care for Hopelessly Ill Patients. A Report of the Clinical Care Committee of the Massachusetts General Hospital," *New England Journal of Medicine* 295, no. 7 (August 12, 1976): 362–364.

66. Robert Baker, "The Legitimation and Regulation of DNR Orders," in *Legislating Medical Ethics*, ed. Robert Baker and Martin Strosberg (Dordrecht: Kluwer Academic Publishers, 1995), 33–101 at 37.

67. Neither the notations nor the meaning of orders not to resuscitate were standard in the 1970s or early 1980s. For example, one New York hospital used purple dots as a proxy for "DNR," and another used the odd notation "OBP"—which, I surmise, meant "on the banana peel" (i.e., slipping away and hence not a candidate for CPR). Furthermore, there was no consensus on what a DNR order involved: Did it just mean no cardiopulmonary resuscitation? Or did the order to decease interventions extend to additional therapeutic interventions (e.g., no antibiotics)? For a case study of confusion over DNR orders in the 1980s written during that period, see Robert Baker, "The Patient Who Wants to Fight," in *The Machine at the Bedside: Strategies for Using Technology in Patient Care*, ed. Stanley Reiser and Michael Anbar (Cambridge: Cambridge University Press, 1984), 213–221. For an excellent overview of the state of DNR policies in the late 1970s and 1980s, see President's Commission for the Study of Ethical Problems in Medicine, *Deciding to Forego Life-Sustaining Treatment* (Washington, DC: US Government Printing Office, 1983), 248–252.

68. Writing DNR orders in medical records made them subject to the same standards of review and accountability as other medical orders. An incident illustrating why this is important involved an elderly homeless woman suffering from pneumonia who brought head lice into an ICU, infecting some staff members. An intern (a licensed physician receiving advanced training) had entered a DNR order into the woman's record. It was retracted by an attending physician, who then delivered a withering lecture on the irrelevance of age, social class—or head lice—to DNR status. Since the DNR order was recorded, the attending could discover and override it, and make it the subject of an impromptu ethics lecture.

69. Kuhn, *Structure of Scientific Revolutions*, 85. Bracketed terms added.

70. President's Commission, *Deciding to Forego*, 64, 65.

71. President's Commission, *Deciding to Forego*, 44.

72. President's Commission, *Deciding to Forego*, 44.

73. President's Commission, *Deciding to Forego*, 89–90.

74. President's Commission, *Deciding to Forego*, 89–90.

75. Bradford H. Gray, "Bioethics Commissions: What Can We Learn from Past Successes and Failures?" in *Society's Choices: Social and Ethical Decision Making in Biomedicine*, ed. Ruth Ellen Bulger, Elizabeth Meyer Bobby, and Harvey V. Fineberg (Washington, DC: National Academy Press, 1995), 261–306 at 286.

76. American Medical Association, House of Delegates, *139th Annual Meeting Proceedings*, June 24–28, 1990 (Chicago: American Medical Association, 1990), 237.

77. American Medical Association, *139th Annual Meeting Proceedings*, 237–238.

78. American Medical Association, House of Delegates, December 3–6, 1989, Honolulu, Hawaii, 43rd Interim Meeting (Chicago: American Medical Association, 1989), 159.

79. American Medical Association, *Code of Medical Ethics: Current Opinions with Annotations, 1998–1999* (Chicago: American Medical Association 1999), 45–46.

Chapter 11

1. Michel Foucault, *The Birth of the Clinic: An Archeology of Medical Perception*, trans. A. M. Sheridan Smith (New York: Vintage Books, 1975), xi.

2. Benjamin Freedman, "Where Are the Heroes of Bioethics?" *Journal of Clinical Ethics* 7, no. 4 (Winter 1996): 297–299, accessed October 8, 2022, https://philpapers.org/rec/FREWAT.

3. Rothman, *Strangers at the Bedside*, 145, 146.

4. Rothman, *Strangers at the Bedside*, 5–7, 71–73, 216.

5. Rothman, *Strangers at the Bedside*, 31.

6. Rothman, *Strangers at the Bedside*, 62–63.

7. As Rothman properly points out, the Nazi scientists "were university-trained and university-appointed researchers [who] possessed first rate credentials and had pursued notable careers." Rothman, *Strangers at the Bedside*, 63.

8. Jonsen, *Birth of Bioethics*, one paragraph divided between, 368–369.

9. Jonsen, *Birth of Bioethics*, 368.

10. *In the Matter of William A. Hyman, Appellant, v. Jewish Chronic Disease Hospital, Respondent*, 15 N.Y.2d 317 (1965), accessed October 8, 2022, https://law.justia.com/cases/new-york/court-of-appeals/1965/15-n-y-2d-317-0.html. This case may be the earliest US legal proceeding that refers to the ten principles expounded in *US v. Brandt* as "The Nuremberg Code."

11. Federal Food and Drug and Cosmetics Act of 1962, Federal Food and Cosmetic Act of 1962, PL 87–781, 76 Stat., 780–788, accessed October 8, 2022, https://www.govinfo.gov/content/pkg/STATUTE-76/pdf/STATUTE-76-Pg780.pdf, cited in Jonsen, *Birth of Bioethics*, 141, note 64.

12. Federal Food and Drug and Cosmetics Act of 1962, cited in Jonsen, *Birth of Bioethics*, 141, note 64. Italics added.

13. Rothman, *Strangers at the Bedside*, 92, 93.

14. Carpenter, *Reputation and Power*, 481.

15. Jonsen, *Birth of Bioethics*, 142.

16. Jonsen, *Birth of Bioethics*, 65.

17. Rothman, *Strangers at the Bedside*, 74.

18. Jonsen, *Birth of Bioethics*, 136.

19. World Medical Association, Declaration of Helsinki 1964, accessed October 8, 2022, https://www.wma.net/what-we-do/medical-ethics/declaration-of -helsinki/doh-jun1964/.

20. Beecher, *Research*, 231.

21. WMA, Declaration of Helsinki 1964. Italics added.

22. Jonsen, *Birth of Bioethics*, 368, 369; Rothman, *Strangers at the Bedside*, 145.

23. Jonsen, *Birth of Bioethics*, 368.

24. Alden Whitman, "Dr. George Wiley Feared Drowned," *New York Times*, August 10, 1973, accessed October 8, 2022, https://www.nytimes.com/1973/08/10 /archives/dr-george-wiley-feared-drowned-civil-rights-leader-42-who-headed.html.

25. William J. Curran, "The Patient's Bill of Rights Becomes Law," *New England Journal of Medicine* 290 (January 3, 1974): 32–33.

26. Jonsen, *Birth of Bioethics*, 369.

27. Pollner, *Off Our Backs*, 3, cited in Ruzek, *The Women's Health Movement*, 191–192.

28. George A. Wiley, "Health Care in the Inner City: Like It Is: Point of View of a Consumer," 1969, cited from d'Oronzio, "A Human Right to Healthcare," note 23, 294.

29. Paul Starr, 1982. *The Social Transformation of American Medicine* (New York: Basic Books), 390.

30. Jonsen, *Birth of Bioethics*, 390.

31. Jonsen, *Birth of Bioethics*, 390.

32. Jonsen, *Birth of Bioethics*, 392.

33. Jonsen, *Birth of Bioethics*, 392.

34. Jonsen, *Birth of Bioethics*, 397.

35. Jonsen, *Birth of Bioethics*, 397.

36. National Advisory Commission on Civil Disorders, *Report of the National Advisory Commission on Civil Disorders Summary of Report* (Washington, DC: US Government Printing Office, 1968), accessed October 8, 2022, https://www .hsdl.org/?abstract&did=35837.

37. CDC (Centers for Disease Control and Prevention), "The Tuskegee Timeline," April 22, 2021, accessed October 8, 2022, https://www.cdc.gov/tuskegee /timeline.htm.

Epilogue

1. Michigan Governor's Commission on Victimless Crime (i.e., opiate addiction). Paul Lowinger's papers are at the library of the University of Pennsylvania, accessed October 8, 2022, https://findingaids.library.upenn.edu/records/UPENN_RBML _PUSP.MS.COLL.635.

2. *Kaimowitz v. Department of Mental Health for the State of Michigan*, No. 73·19434·AW (Mich. Cir. Ct., Wayne County, July 10, 1973), accessed October 8,

2022, https://psychrights.org/States/Michigan/Kaimowitz(WayneCtMichCirCt1973).pdf%20rgetown.edu/handle/10822/765840.

3. Baker, "Metaethicist to Bioethicist," 369–379.

4. The numbers in parentheses correspond to numbered stages described in *Structure of Moral Revolutions*.

5. Frankel, *Public Health Service Guidelines*, cited in Rothman, *Strangers at the Bedside*, 86.

6. One reviewer's rejection note remarked that they disliked the word "paradigm" and simply listed "feminists?" "Welfare recipients?" A notable exception was the *American Journal of Bioethics*, which published "Erasing Blackness from Bioethics," in 2022.

7. See https://www.worldcat.org/identities/lccn-n84214331/ (accessed September 21, 2022). My academic appointment was in philosophy, but I also belong to the ASBH and the American Association of Historians of Medicine (AAHM). Before my retirement, I was a lifelong member of the American Philosophical Association (APA).

8. George Santayana, *Reason in Common Sense*, 1995, vol. 1, 284, Project Gutenberg ebook, accessed October 8, 2022, https://www.gutenberg.org/files/15000/15000-h/15000-h.htm.

9. Rachel Bluth, "'My Body, My Choice': How Vaccine Foes Co-Opted the Abortion Rallying Cry," *KHN*, July 6, 2022, accessed October 8, 2022, https://khn.org/news/article/my-body-my-choice-slogan-abortion-rights-anti-vaccine/; Jacob Jarvis, "GOP Lawmaker Appropriates 'My Body, My Choice' Slogan to Reject Mask Use," *Newsweek*, June 24, 2020, accessed October 8, 2022, https://www.newsweek.com/gop-lawmaker-my-body-my-choice-facemasks-1513121.

10. However, the Association of Bioethics Program Directors forthrightly came out against antivax movement. https://www.bioethicsdirectors.net/wp-content/uploads/2021/09/ABPD-Statement-in-Support-of-COVID-19-Vaccine-Mandates_FINAL9.22.2021.pdf (accessed August 11, 2022).

11. Lord Byron, *Personal Lyric, and Elegaic: Ode to Napoleon Buonaparte* (1881), accessed October 8, 2022, https://www.bartleby.com/205/31.html.

Appendix

1. Ivy to Pappworth, letter, April 6, 1966, London: PP/MHP/C5 Wellcome Library, London (UK).

2. Grodin papers, Appendix C, 4, cite 151, cited in Schmidt, *Justice at Nuremberg*, 136; see also Weindling, *Nuremberg Trials*, 263.

3. Schmidt, *Justice at Nuremberg*. See also Weindling, *Nuremberg Trials*, 263.

4. Schmidt, *Justice at Nuremberg*, 2004, 136–137.

5. Christian Bonah and Florian Schmaltz, "From Nuremberg to Helsinki: The Preparation of then Declaration of Helsinki in the Light of the Prosecution of

Medical War Crimes at the Struthof Medical Trials, France, 1952–1955," in *Ethical Research: The Declaration of Helsinkim and the Past, Present, and Future of Human Experimentation*, ed. Ulf Schmidt, Andreas Frewer, and Dominique Sprumont (New York: Oxford University Press, 2002), 72.

6. See also Schmidt, *Justice at Nuremberg*, 137, which draws on Grodin papers, Appendix B.

7. Andrew C. Ivy, "Report on War Crimes of a Medical Nature Committed in Germany and Elsewhere on German Nationals and the Nationals of Occupied Countries by the Nazi Regime During World War II," Document JC 9218, AMA Archives, 1946, 9–11. A shortened version of these guidelines was adopted by the AMA on the opening day of the Nuremberg Doctors Trial. Schmidt, *Justice at Nuremberg*, 13.

8. Ivy, "Report on War Crimes," 13.

9. "Permissible Medical Experiments," *Trials of War Criminals before the Nuremberg Military Tribunals under Control Council Law No. 10*. Nuremberg, October 1946–April 1949. Washington, DC: U.S. Government Printing Office, 1949–1953. This document was retrospectively designated "The Nuremberg Code" in the 1960s, when critics began to challenge the ethics of accepted practice in research on human subject. See Paul S. Weindling, "From the Nuremberg 'Doctors Trial' to the 'Nuremberg Code,'" in "Medical Ethics in the 70 Years after the Nuremberg Code, 1947 to the Present," *Wien Klin Wochenschr: The Central European Journal of Medicine* 130 (June 2018): 159–253.

10. Weindling, *Nuremberg Trials*, Table 12, 357–358.

11. Sue Pickard and I provided materials that Paul Lowinger and Gabe Kaimowitz used in fashioning arguments challenging whether an incarcerated criminal-sexual psychopath had the capacity to give informed voluntary consent to behavior-modifying psychosurgery.

12. This summary is based on D'Oronzio, "A Human Right to Healthcare," 296–298.

13. American Hospital Association, 1972, "Statement on A Patient's Bill of Rights," in Beauchamp and Childress, *Principles*, 285–287.

Bibliography

Ad Hoc Committee of the Harvard Medical School to Examine the Definition of Brain Death. "A Definition of Irreversible Coma." *JAMA* 205, no. 6 (August 5, 1968): 337–340.

Advisory Committee on Human Radiation Experiments. *Final Report of the Advisory Committee on Human Radiation Experiments*. New York: Oxford University Press, 1996.

Altman, Lawrence. *Who Goes First? The Story of Self-Experimentation*. Berkeley: University of California Press, 1986.

Alving, Alf S., Branch Craige Jr., Theodore N. Pullman, C. Merrill Whorton, Ralph Jones, Jr., and Lillian Eichelberger. "Procedures Used at Stateville Penitentiary for the Testing of Potential Antimalarial Agents." *Journal of Clinical Investigation* 27, no. 3 (1948): 2–5. Accessed October 8, 2022. https://doi.org/10.1172/JCI101956.

American Association of Medical Colleges. *Report I: Learning Objectives for Medical Student Education: Guidelines for Medical Schools*. Washington, DC: American Association of Medical Colleges, 1998.

American Hospital Association. "About the AHA (website)." Accessed October 8, 2022. https://www.aha.org/about.

American Medical Association. "Code of Ethics (1847)." In *The American Medical Ethics Revolution: How the AMA's Code of Ethics Has Transformed Physicians' Relationships to Patients, Professionals and Society*, edited by Robert Baker, Arthur Caplan, Linda Emanuel, and Stephen Latham, 324–334. Baltimore: The Johns Hopkins University Press, 1999.

American Medical Association. *Code of Medical Ethics: Current Opinions with Annotations, 1998–1999*. Chicago: American Medical Association, 1999.

American Medical Association. House of Delegates (Proceedings). *139th Annual Meeting*, June 24–28, 1990, Chicago, Illinois. Accessed October 8, 2022. https://ama.nmtvault.com/jsp/PsImageViewer.jsp?doc_id=1ee24daa-2768-4bff-b792-e4859988fe94%2Fama_arch%2FHOD00001%2F00000130.

American Medical Association. "Principles of Medical Ethics (1903)." In *The American Medical Ethics Revolution: How the AMA's Code of Ethics Has Transformed Physicians' Relationships to Patients, Professionals and Society*, edited by

Robert Baker, Arthur Caplan, Linda Emanuel, and Stephen Latham, 346–355. Baltimore: The Johns Hopkins University Press, 1999.

American Medical Association. "Principles of Medical Ethics (1912)." In *The American Medical Ethics Revolution: How the AMA's Code of Ethics Has Transformed Physicians' Relationships to Patients, Professionals and Society*, edited by Robert Baker, Arthur Caplan, Linda Emanuel, and Stephen Latham, 346–355. Baltimore: The Johns Hopkins University Press, 1999.

American Nurses Association. "A Suggested Code." *American Journal of Nursing* 26, no. 8 (1926): 599–601.

Andrews, Robert. *Famous Lines: A Columbia Dictionary of Familiar Quotations*. New York: Columbia University Press, 1996. Accessed October 8, 2022. https://books.google.com/books?id=MtciwlIG3sMC&pg=PA53#v=onepage&q&f=false.

Annas, George J., and Michael A. Grodin, eds. *The Nazi Doctors and the Nuremberg Code: Human Rights in Human Experimentation*. New York: Oxford University Press, 1992.

Annas, George J., and Joseph M. Healey Jr. "The Patient Rights Advocate: Redefining the Doctor-Patient Relationship in the Hospital Context." *Vanderbilt Law Review* 27, no. 2 (March 1974): 243–269.

Appelbaum, Paul S., and Charles W. Lidz, "Twenty-Five Years of Therapeutic Misconception." *The Hastings Center Report* 38, no. 2 (March–April 2008): 5–6; reply, 6–7.

Appiah, Kwame Anthony. *Honor Code: How Moral Revolutions Happen*. New York: W. W. Norton & Company, 2010.

Aristotle. *Politics 1.1252b*. Gregory R. Crane, ed. *Perseus Digital Library*. Accessed October 8, 2022. http://www.perseus.tufts.edu/hopper/text?doc=Perseus:abo:tlg,0086,035:1:1252b.

Avdeev, Alexandre, Alain Blum, and Irina Troitskaya. "The History of Abortion Statistics in Russia and the USSR from 1900 to 1991." *Population: An English Selection* 7 (1995): 39–66. Accessed April 12, 2023. https://www.academia.edu/1758149/The_History_of_Abortion_Statistics_in_Russia_and_the_USSR_from_1900_to_1991.

Bacon, Francis. 1605. "On the Dignity and Advancement of Learning." Project Gutenberg Section VII, no. 7. Accessed October 8, 2022. https://www.gutenberg.org/files/5500/5500-h/5500-h.htm.

Baicus, Anda. "History of Polio Vaccination." *World Journal of Virology* 1, no. 4 (2012): 108–114.

Baker, Robert. *Before Bioethics: A History of American Medical Ethics from the Colonial Period to the Bioethics Revolution*. New York: Oxford University Press, 2013.

Baker, Robert. "Bioethics and Human Rights: A Historical Perspective." *Cambridge Quarterly of Healthcare Ethics* 10, no. 3 (Summer 2001): 241–252.

Baker, Robert. "Erasing Blackness from Bioethics." *The American Journal of Bioethics* 22, no. 3 (2022): 33–35.

Baker, Robert. "From Metaethicist to Bioethicist." *Cambridge Quarterly of Healthcare Ethics* 11, no. 4 (Fall 2002): 369–379.

Baker, Robert. "Kübler-Ross and Bioethics: A Cautionary Tale." *The American Journal of Bioethics* 19, no. 12 (2019): 48–49.

Baker, Robert. "The Declaration of Helsinki and the Foundations of Global Bioethics." In *Ethical Research: The Declaration of Helsinki and the Past, Present, and Future of Human Experimentation*, edited by Ulf Schmidt, Andreas Frewer, and Dominique Sprumont, 47–69. New York: Oxford University Press, 2020.

Baker, Robert. "The Legitimation and Regulation of DNR Orders." In *Legislating Medical Ethics*, edited by Robert Baker and Martin Strosberg, 33–101. Dordrecht: Kluwer Academic Publishers, 1995.

Baker, Robert. "The Patient Who Wants to Fight." In *The Machine at the Bedside: Strategies for Using Technology in Patient Care*, edited by Stanley Reiser and Michael Anbar, 213–221. Cambridge, UK: Cambridge University Press, 1984.

Baker, Robert. *The Structure of Moral Revolutions: Studies of Changes in the Morality of Abortion, Death, and the Bioethics Revolution*. Cambridge, MA: The MIT Press, 2019.

Baker, Robert, Arthur Caplan, Linda Emanuel, and Stephen Latham, eds. *The American Medical Ethics Revolution: How the AMA's Code of Ethics Has Transformed Physicians' Relationships to Patients, Professionals and Society*. Baltimore: The Johns Hopkins University Press, 1999.

Baker, Robert, and Laurence McCullough, eds. *Cambridge World History of Medical Ethics*. New York: Cambridge University Press, 2009.

Baker, Robert, and Laurence McCullough. "Medical Ethics through the Life Cycle in Europe and the Americas." In *The Cambridge World History of Medical Ethics*, edited by Robert Baker and Laurence McCullough, 137–162. New York: Cambridge University Press, 2009.

Baron-Cohen Simon, Ami Klin, Steve Silberman, and Joseph D. Buxbaum. "Did Hans Asperger Actively Assist the Nazi Euthanasia Program?" *Molecular Autism* 9, no. 28 (April 19, 2018). Accessed October 8, 2022. https://molecularautism.biomedcentral.com/articles/10.1186/s13229-018-0209-5.

Baumgartner, Fritz. "Hippocrates and the Dignity of Human Life." *American Journal of Obstetrics and Gynecology* 186, no. 6 (June 2002): 1378–1379.

Baxter, Richard, "A Christian Directory: Or a Sum of Practical Theology and Cases of Conscience." In *The Practical Works of the Rev. Richard Baxter: With a Life of the Author and a Critical Examination of His Writings*, edited by William Orme, vol. V. London: Mills, Jowett, and Mills, [1672–1673] 1830.

Beauchamp, Tom. *Oral History of the Belmont Report and the National Commission for the Protection of Human Subjects of Biomedical and Behavioral Research*. Interview with Tom Lamar Beauchamp, Washington, DC, September 22, 2004.

Beauchamp, Tom. "The Origins, Goals, and Core Commitments of the *Belmont Report* and *Principles of Biomedical Ethics*." In *The Story of Bioethics: From*

Seminal Works to Contemporary Explorations, edited by Jennifer K. Walter and Eran P. Klein, 17–46. Washington, DC: Georgetown University Press.

Beauchamp, Tom, and James Childress. *Principles of Biomedical Ethics.* New York: Oxford University Press, 1979.

Beauchamp, Tom, and James Childress. *Principles of Biomedical Ethics.* 2nd ed. New York: Oxford University Press, 1983.

Beauchamp, Tom, and James Childress. *Principles of Biomedical Ethics.* 8th ed. New York: Oxford University Press, 2019.

Beecher, Henry K. "Consent in Clinical Experimentation: Myth and Reality." *JAMA* 195, no. 1 (January 1966): 34–35.

Beecher, Henry K. "Ethics and Clinical Research." *New England Journal of Medicine* 274, no. 24 (June 16, 1966): 1354–1360.

Beecher, Henry K. "Letter to Dean George P. Berry," February 1958. Cited in Vincent J. Kopp, "Henry Knowles Beecher and the Development of Informed Consent in Anesthesia Research." *Anesthesiology* 90 (June 1999): 1756–1765.

Beecher, Henry K. "Letters to John Talbott," August 20 and August 30, 1965. Countway Library (Boston, MA). Cited in David J. Rothman, *Strangers at the Bedside: A History of How Law and Bioethics Transformed Medical Decision Making,* 72. New York: Basic Books, 1991.

Beecher, Henry K. "Letter to Maurice Pappworth," January 7, 1965. The Wellcome Library, London. Accessed October 8, 2022. https://wellcomecollection.org/works/sn8mqmuy.

Beecher, Henry K. "Letter to George Burch," June 27, 1966. Cited in Vincent J. Kopp, "Henry Knowles and the Development of Informed Consent in Anesthesia Research." *Anesthesiology* 90 (Special Article, June 1999): 1756–1765.

Beecher, Henry K. "Letter to MH Pappworth," July 11, 1966. PP/MHP/C5. Wellcome Library, London. Cited in Allan Gaw, "Exposing Unethical Human Research: The Transatlantic Correspondence of Beecher and Pappworth." *Annals of Internal Medicine* 156, no. 2 (January 17, 2012): 150–155.

Beecher, Henry K. "Letter to MH Pappworth," July 20, 1966. PP/MHP/C5. Wellcome Library, London. Cited in Allan Gaw, "Exposing Unethical Human Research: The Transatlantic Correspondence of Beecher and Pappworth." *Annals of Internal Medicine* 156, no. 2 (January 17, 2012): 152–153.

Beecher, Henry K. *Research and the Individual: Human Studies.* Boston: Little Brown and Company, 1970.

Beecher, Henry K. "Scarce Resources and Medical Advancement." *Daedalus: Journal of the American Academy of Arts and Science* 98 (Spring 1969): 248–274.

Beecher, Jonathan. "Personal Communication," March 10, 2006. Cited in Michael Gionfriddo, "Enduring Contributions of Henry K. Beecher to Medicine, Science, and Society: Part I." *International Anesthesiology Clinics* 45, no. 4 (Fall 2007): 18–19.

Bentham, Jeremy. *Benthamiana: Select Abstracts from the Works of Jeremy Bentham: With an Outline of His Opinions on the Principal Subjects Discussed in*

His Works, edited by John Hill Burton. Philadelphia: Lea & Blanchard, 1844; reprinted, London: Forgotten Books, 2018.

Berger, Melvin. "A History of Immune Globulin Therapy, from the Harvard Crash Program to Monoclonal Antibodies." *Current Allergy and Asthma Reports* 2, no. 5 (September 2002): 368–378.

Bernard, Claude. *An Introduction to the Study of Experimental Medicine.* Translated by Henry Copley Green. Reprint, New York: Dover Publications, 1957.

Binding, Karl, and Alfred Hoche. *Die Freigabe der Vernichtung lebensunwerten Lebens.* Leipzig: Felix Meiner, 1920.

Blaha, Franz. "Testimony of Dr. Franz Blaha, January 11, 1946." Document Number 3249-PS, Exhibit USA-66, 166–195. Accessed October 8, 2022. http://avalon.law.yale.edu/imt/01-11-46.asp.

Bluth, R. "'My Body, My Choice': How Vaccine Foes Co-opted the Abortion Rallying Cry." *KHN.* July 6, 2022. Accessed October 8, 2022. https://khn.org/news/article/my-body-my-choice-slogan-abortion-rights-anti-vaccine/.

Bonah, Christian, and Florian Schmaltz. "From Nuremberg to Helsinki: The Preparation of Then Declaration of Helsinki in the Light of the Prosecution of Medical War Crimes at the Struthof Medical Trials, France, 1952–5." In *Ethical Research: The Declaration of Helsinki and the Past, Present, and Future of Human Experimentation*, edited by Ulf Schmidt, Andreas Frewer, and Dominique Sprumont, 69–100. New York: Oxford University Press, 2020.

Booth, Christopher. "Obituary: M. H. Pappworth." *British Medical Journal* 309, no. 6968 (1994): 1577–1578.

Boralevi, Lea Campos. *Bentham and the Oppressed.* Berlin: de Gruyter. 1984.

Boston Women's Health Book Collective. "Women, Medicine and Capitalism." In *Women and Their Bodies: A Course*, 6–7. Boston: Self-published, 1970. Accessed February 13, 2022. https://www.ourbodiesourselves.org/wp-content/uploads/2020/04/Women-and-Their-Bodies-Free-Press.pdf.

Brandt, Allan M. "Racism and Research: The Case of the Tuskegee Syphilis Study." *The Hastings Center Report* 22, no. 6 (November–December 1978): 21–29.

Bren, Linda. "Frances Oldham Kelsey: FDA Medical Reviewer Leaves Her Mark on History." US Food and Drug Administration, *FDA Consumer Magazine*, March–April 2001. Accessed October 9, 2022. https://web.archive.org/web/20061020043712/https://www.fda.gov/fdac/features/2001/201_kelsey.html.

Brent, R. L. "The Contributions of Widukind Lenz to Teratology and Science: Comments on 'Thalidomide Retrospective: What Did the Clinical Teratologist Learn?'" *Teratology* 46, no. 5 (November 1992): 45–46.

Bruenig, Elizabeth. "The Undeserving Poor: A Very Tiny History," June 6, 2019. Accessed April 12, 2023. https://medium.com/@ebruenig/the-undeserving-poor-a-very-tiny-history-96c3b9141e13.

Büchner, Franz. November 18, 1941. "*Der Eid des Hippokrates: Wortlaut des am 18. November 1941 in der Aula der Universität Freiburg gehalten öffentlichen*

Vortages," November 18, 1941. In *Der Mensch in der Sicht moderner Medizin*, edited by Franz Büchner, 131–151. Freiburg: Herder, 1985.

Buxtun, Peter. "Letter to William Brown," November 24, 1968. Cited in *Tuskegee's Truths: Rethinking the Tuskegee Syphilis Study*, edited by Susan Reverby, 79. Chapel Hill: University of North Carolina Press, 2009.

Buxton, Peter. 1973. "Testimony by Peter Buxton from the United States Senate Hearings on Human Experimentation." Cited in *Tuskegee's Truths: Rethinking the Tuskegee Syphilis Study*, edited by Susan Reverby. Chapel Hill: University of North Carolina Press, 2009.

Callahan, Daniel. "Bioethics as a Discipline." *The Hastings Center Studies* 1, no. 1 (1973): 66–73.

Callahan, Daniel. "Institute of Society, Ethics, and the Life Sciences: A Survey of Goals, Plans, and Budgetary Needs." From a folder marked ISELS, p. 1. Rockefeller Archives, Rockefeller Brothers Fund, Sleepy Hollow, New York. Cited in Tina M. L. Stevens, *Bioethics in America: Origins and Cultural Contexts*. Baltimore: Johns Hopkins University Press, 2000.

Callahan, Daniel. Interview conducted by Allan Brandt in March 17, 1977. Cited in David K. Rothman, *Strangers at the Bedside: A History of How Law and Bioethics Transformed Medical Decision Making*, 209. New York: Basic Books, 1991.

Callahan, Daniel. "Religion and the Secularization of Bioethics." *Hastings Center Report* 20, no. 4 (Suppl., July–August 1990): 2–4.

Callahan, Daniel, Bernard Haring, and Charles E. Curran. *The Catholic Case for Contraception: Leading Catholic Authorities Oppose Pope Paul's Position on Birth Control*. London: Macmillan, 1969.

Carpenter, Daniel. *Reputation and Power: Organizational Image and Pharmaceutical Regulation at the FDA*. Princeton, NJ: Princeton University Press, 2010.

Carroll, James. *Report to Surgeon General Sternberg*, August 18, 1906. Cited in Baker, Robert. 2013. *Before Bioethics: A History of American Medical Ethics from the Colonial Period to the Bioethics Revolution*, 258. New York: Oxford University Press.

Catania, Joseph A., Bernard Lo, Leslie E. Wolf, Margaret Dolcini, Lance M. Pollack, Judith C. Barker, Stacy Wortlieb, and Jeffe Henne. "Survey of U.S. Human Research Protection Organizations: Workload and Membership." *Journal of Empirical Research on Human Research Ethics* 3, no. 4 (December 2008): 57–69.

Cavanaugh, Thomas A. *Hippocrates' Oath and Asclepius Snake: The Birth of the Medical Profession*. New York: Oxford University Press, 2018.

Childress, James F., Eric M. Meslin, and Harold T. Shapiro, eds. *Belmont Revisited: Ethical Principles for Research with Human Subjects*. Washington, DC: Georgetown University Press, 2005.

Cicero. *De Legibus*. (Circa 52–44 BCE). Translated by C. D. Yonge. "On the Laws" in *Treatises of M. T. Cicero*. London: H. G. Bohn, 1853. Accessed October 9, 2022. https://books.google.com/books?id=joyeLp3Ok-oC&pg=PA398&dq=cicero&as_brr=1&hl=de#v=onepage&q=cicero&f=false.

Clegg, Hugh. 1962. [Unsigned with prefatory editorial comment] "Draft Code of Ethics on Human Experimentation." *British Medical Journal* 2 (October 27, 1962): 119.

Clinical Care Committee of the Massachusetts General Hospital. "Optimum Care for Hopelessly Ill Patients. A Report of the Clinical Care Committee of the Massachusetts General Hospital." *New England Journal of Medicine* 295, no. 7 (August 12, 1976): 362–364.

Cohen, I. Bernard. *Revolution in Science*. Cambridge, MA: The Belknap Press of Harvard University Press, 1985.

Cumming, H. S. "Letter to R. R. Moton, NAWNRC," September 20, 1932. Cited in Allan M. Brandt, "Racism and Research: The Case of the Tuskegee Syphilis Study." *The Hastings Center Report* 8, no. 6 (1978): 21–29.

Curran, William J. "Governmental Regulation of the Use of Human Subjects in Medical Research: The Approach of Two Federal Agencies." *DÆDALUS* 98, no. 2 (Spring 1969): 542–594.

Czech, H. 2018. "Hans Asperger, National Socialism, and 'Race Hygiene' in Nazi-era Vienna." *Molecular Autism* 9, no. 29 (April 19, 2018). Accessed October 9, 2022. https://molecularautism.biomedcentral.com/articles/10.1186/s13229-018-0208-6.

Dadrian, Vahakn, N. "The Role of Turkish Physicians on the World War I Genocide of Ottoman Armenians." *Holocaust and Genocide Studies* 1, no. 2 (1986): 169–192.

Dale, James B., and Alan L. Bisno. "In Memoriam: Gene H. Stollerman." *Clinical Infectious Diseases* 59, no. 12 (September 2014): 1805–1806.

Davenport, Mary L., Jennifer Lahl, and Evan C. Rosa. "Right of Conscience for Health-Care Providers." *Linacre Quarterly* 79, no. 2 (May 2012): 169–191.

Dickens, Charles. 1859. *A Tale of Two Cities*. London: Chapman and Hall. Accessed October 9, 2022. https://etc.usf.edu/lit2go/22/a-tale-of-two-cities/108/book-the-first-recalled-to-lifechapter-1-the-period/.

Dimbleby, Richard. BBC News, April 15, 1945. Cited from transcript of a radio report by Charlie Beckett, "75 Years On: Richard Dimbleby's BBC Report on the Liberation of Belsen Concentration Camp." Accessed April 8, 2023. https://blogs.lse.ac.uk/polis/2020/04/15/75-years-on-richard-dimblebys-bbc-report-on-the-liberation-of-belsen-concentration-camp/.

Donnini Macciò, Daniela. "On Economics and Philosophy; G.E. Moore's Imprint on the Cambridge Apostle's: From *Principia Ethica* to *My Early Beliefs*." PhD Thesis, Università Degli Studi Di Macerata, Italy, 2013.

D'Oronzio, Joseph C. "A Human Right to Healthcare Access: Returning to the Origins of the Patients' Rights Movement." *Cambridge Quarterly of Healthcare Ethics* 10, no. 3 (Summer 2001): 285–298.

Dossabhoy, Sheenaz S., Jessica Feng, and Manisha S. Desai. "The Use and Relevance of the Hippocratic Oath in 2015—A Survey of US Medical Schools." *Journal of Anesthesia History* 4, no. 2 (April 2018): 139–146.

Downer, Carol. "Covert Discrimination against Women as Medical Patients." Address to the American Psychological Association, Honolulu (mimeographed). Cited in Sheryl Burt Ruzek, *The Women's Health Movement, Feminist Alternative to Medical Control*, 1–2. New York: Prager Publishers, 1972.

Downie, Robert S., and Elizabeth Telfer. *Respect for Persons*. New York: Schocken Books, 1966.

Edelson, Paul J. "Henry Beecher and Maurice Pappworth: Honor in the Development of the Ethics of Human Experimentation." In *Twentieth Century Ethics of Human Subjects Research*, edited by Volker Roelcke and Giovanni Maio, 219–233. Stuttgart: Franz Steiner Verlag, 2004.

Edelstein, Ludwig. *The Hippocratic Oath: Text, Translation, and Interpretation*. Baltimore: The Johns Hopkins University Press, 1943.

Edwards, Marc A., Carol Yang, and Siddhartha Roy, "Who Dares to Speak Up? A Federal Agency Allowed Unethical Experimentation on Black Men for Four Decades before Someone Finally Decided to Blow the Whistle." *American Scientist* 109, no. 4 (July–August 2021): 238–242.

Eisenhower, General Dwight D. "The Horrifying Discovery of Dachau Concentration Camp—And Its Liberation by US Troops," April 12, 1945. History.com. Accessed October 9, 2022. https://www.history.com/news/dachau-concentration-camp-liberation.

Elkeles, Barbara. "The German Debate on Human Experimentation between 1880 and 1914." In *Twentieth Century Ethics of Human Subjects Research: Historical Perspectives on Values, Practices, and Regulations*, ed. Volker Roelcke and Giovanni Maio, 19–33. Stuttgart: Franz Steiner Verlag, 2004.

Elliot, Carl. "Tuskegee Truth Teller." *The American Scholar* 1 (January 2018): 44–56.

Emanuel, Eziekiel. "The History of Euthanasia Debates in the United States and Britain." *Annals of Internal Medicine* 121, no. 10 (November 15, 1994): 793–802.

Emerson, Ralph W. "Self-Reliance." In *Essays: First Series*, 1847. Accessed October 9, 2022. https://en.wikisource.org/wiki/Essays:_First_Series/Self-Reliance.

Engelhardt, H. Tristram, Jr. "Basic Ethical Principles in the Conduct of Biomedical and Behavioral Research Involving Human Subjects." In The National Commission for the Protection of Human Subjects of Biomedical and Behavioral Research, *Appendix: The Belmont Report: Ethical Principles and Guidelines for the Protection of Human Subjects of Research*, vol. I, section 8, 2–4. U.S. Department of Health, Education, and Welfare, DHEW Publication No. (OS) 78–0013. Washington, DC: US Government Printing Office, 1978.

Engelhardt Jr., H. Tristram. *The Foundations of Bioethics*. New York: Oxford University Press, 1986 (2nd edition, 1996).

Epstein, Beth, and Martha Turner. "The Nursing Code of Ethics: Its Value, Its History." *Online Journal of Issues in Nursing* 20, no. 2 (May 2015). Accessed April 12, 2023. https://ojin.nursingworld.org/MainMenuCategories/ANAMarketplace/ANAPeriodicals/OJIN/TableofContents/Vol-20-2015/No2-May-2015/The-Nursing-Code-of-Ethics-Its-Value-Its-History.html.

Eriksen, Cecilie. "The Dynamics of Moral Revolutions—Prelude to Future Investigations and Interventions." *Ethical Theory and Moral Practice* 22, no. 3 (2019): 779–792.

Evans, John. *The History and Future of Bioethics: A Sociological View.* New York: Oxford University Press, 2012.

Farias, Victor. *Heidegger and Nazism.* Philadelphia: Temple University Press, 1987.

Foucault, Michel. *The Birth of the Clinic: An Archeology of Medical Perception.* Translated by A. M. Sheridan Smith. New York: Vintage Books, 1963.

Fox, Daniel M. "Who Are We: The Political Origins of the Medical Humanities." *Theoretical Medicine* 6 (1985): 327–341.

Frankel, Mark S. *The Public Health Service Guidelines Governing Research Involving Human Subjects.* Washington, DC: George Washington University Program of Policy Studies in Science and Technology, Monograph no. 10, 1972.

Frewer, Andreas. "Debates on Human Experimentation in Weimar and Early Nazi Germany as Reflected in the Journal 'Ethik.'" In *Twentieth Century Ethics of Human Subjects Research: Historical Perspectives on Values, Practices, and Regulations,* edited by Volker Roelcke and Giovanni Maio, 137–150. Stuttgart: Franz Steiner Verlag, 2004.

Friedlander E. "Rudolf Virchow on Pathology Education." *Journal of Epidemiological Community Health* 60, no. 8 (2006): 671.

Freedman, Benjamin. 1996. "Where Are the Heroes of Bioethics?" *Journal of Clinical Ethics* 7, no. 4 (Winter): 297–299.

Gamble, Vanessa Northington. "A Legacy of Distrust: African Americans and Medical Research." *American Journal of Preventive Medicine* 9, no. 6 (Suppl., November 1, 1993): 35–38.

Gamble, Vanessa Northington. "Under the Shadow of Tuskegee: African Americans and Health Care." *American Journal of Public Health* 87, no. 11 (November 1997): 1773–1778.

Garland, Joseph. "Letter to Henry K. Beecher," March 3, 1955. Beecher Papers, Countway Library. Cited in Vincent J. Kopp, "Henry Knowles Beecher and the Development of Informed Consent in Anesthesia Research." *Anesthesiology* 90, no. 6 (June 1999): 1756–1765.

Gaw, Allan. "Exposing Unethical Human Research: The Transatlantic Correspondence of Beecher and Pappworth." *Annals of Internal Medicine* 156, no. 2 (January 17, 2012): 150–155.

Gaylin, William. "The Patient's Bill of Rights Becomes Law." *New England Journal of Medicine* 290 (January 3, 1974): 32–33.

Geraghty, Karen. "Protecting the Public: Profile of Dr. Frances Oldham Kelsey." *Virtual Mentor: American Medical Association Journal of Ethics* 3, no. 7 (July 2001): 252–254.

Gibson, Count, Jr. "Letter to Dr. Sidney Olansky," May 28, 1955. In Susan M. Reverby, *Examining Tuskegee: The Infamous Syphilis Study and Its Legacy.* Chapel Hill: University of North Carolina Press, 2009, 75.

Gibson, Count, Jr. "Obituary." *Hartford Courant*, July 25, 2002. Accessed October 9, 2022. https://www.courant.com/news/connecticut/hc-xpm-2002-07-25-0207242365-story.html.

Gionfriddo, Michael. "Enduring Contributions of Henry K. Beecher to Medicine, Science, and Society: Part I." *International Anesthesiology Clinics*, 45, no. 4 (Fall 2007): 18–19.

Goffman, Erving. "The Characteristics of Total Institutions." In *Symposium on Preventive and Social Psychiatry*, 43–84. Washington, DC: Walter Reed Army Institute of Research.

Gray, Bradford H. "Bioethics Commissions: What Can We Learn from Past Successes and Failures?" In *Society's Choices: Social and Ethical Decision Making in Biomedicine*, edited by Ruth Ellen Bulger, Elizabeth Meyer Bobby, and Harvey V. Fineberg, 261–306. Washington, DC: National Academy Press, 1995.

Green, Dwight H. "Ethics Governing the Service of Prisoners as Subjects in Medical Experiments." *JAMA* 136, no. 7 (February 14, 1948): 457–458.

Grodin, Michael. "Historical Origins of the Nuremberg Code." In *The Nazi Doctors and the Nuremberg Code: Human Rights in Human Experimentation*, edited by George J. Annas and Michael A. Grodin, 121–144. New York: Oxford University Press.

Gustafson, James M. "Theology Confronts Technology and the Life Sciences." *Commonweal* 105, no. 12 (June 16, 1978): 386–392.

Halberstrom, Michael. July 14, 1967. *The New York Times*. Cited in Joanna Seldon, *The Whistle-Blower*, 147. Buckingham: University of Buckingham Press, 2017.

Halpern, Sydney A. *Dangerous Medicine: The Story behind Human Experiments with Hepatitis*. New Haven, CT: Yale University Press, 2021.

Harcourt, Bernard E. "Making Willing Bodies: The University of Chicago Human Experiments at Stateville Penitentiary." *Journal of Social Research* 78, no. 2 (June 2011): 443.

Hargrove, Jenny. "British Research Ethics after the Second World War." In *Twentieth Century Ethics of Human Subjects Research*, edited by Volker Roelcke and Giovanni Maio, 181–197. Stuttgart: Franz Steiner Verlag, 2004.

Harris-Wehling, Jo, Jill C. Feasley, and Carroll L. Estes. *An Evaluation of the Long-Term Care Ombudsman Programs of the Older Americans Act*. Washington, DC: Division of Healthcare Services, Institute of Medicine, 1995.

Hastings Center. "Daniel Callahan, 1930–2019," July 18, 2019. Accessed October 9, 2022. https://www.thehastingscenter.org/news/daniel-callahan-1930-2019/.

Hellegers, Andre E. "A Scientists Analysis." In *Contraception, Authority, and Dissent*, edited by Charles E. Curran, 216–217. New York: Herder and Herder, 1960.

Heller, Jean. "Syphilis Victims in U.S. Study Went Untreated for 40 Years." *The New York Times*, July 26, 1972, 1. Accessed October 9, 2022. http://www.nytimes.com/1972/07/26/archives/syphilis-victims-in-us-study-went-untreated-for-40-years-syphilis.html.

Hillel, R. (Circa 110 BCE–10 CE). *Pirkei Avot 1*. Translated by Dr. Joshua Kulp. https://www.sefaria.org/Pirkei_Avot.1.14?ven=Mishnah_Yomit_by_Dr._Joshua _Kulp&vhe=Torat_Emet_357&lang=bi.

Hitler, Adolf. *Mein Kampf* [My Battle]. Translated by Ralph Manheim. Boston: Houghton Mifflin Company, 1943.

Hobbes, Thomas. 1651. *Leviathan, The Second Part: Of Commonwealth*. Chapter 17 *Of the Causes, Generation, and Definition of a Commonwealth*. Excerpts from *Sections 15–20*, Eris Project at Virginia Tech. Accessed October 9, 2022. https://history.hanover.edu/courses/excerpts/161hob.html.

Hufeland, Christoph Wilhelm. "The Results of Fifty Years' Experience." *Manual of the Practice of Medicine*. Translated by C. Bruchhausen and R. Nelson. London: Hippolyte Bailliere, [1836] 1844.

Ingelfinger, Franz J. *Yearbook of Medicine*. Chicago: Yearbook Medical Publishers, 1967–1968.

In re Quinlan, 70 N.J. 10, 355 A.2d 647 (1976).

International Conference on Harmonisation of Technical Requirements of Pharmaceuticals for Human Use (ICH). Glossary. (1995) 2016. 1.12. Accessed October 9, 2022. https://ichgcp.net/.

Ivy, Andrew C. "Letter to Pappworth," April 6, 1966. PP/MHP/C5 Wellcome Library, London.

Ivy, Andrew C. *Report on War Crimes of a Medical Nature Committed in Germany and Elsewhere on German Nationals and the Nationals of Occupied Countries by the Nazi Regime During World War II*. Document JC 9218. Chicago: AMA Archives, 1946, 9–11.

Jahr, Fritz. "Bio-Ethik," *Kosmos* 24, no. 4 (1927): 2–4.

Jarvis, Jacob. "GOP Lawmaker Appropriates 'My Body, My Choice' Slogan to Reject Mask Use." *Newsweek*, June 24, 2020. Accessed October 9, 2022. https://www.newsweek.com/gop-lawmaker-my-body-my-choice-facemasks-1513121.

Jones, David S., Christine Grady, and Susan E. Lederer. "Ethics and Clinical Research—The 50th Anniversary of Beecher's Bombshell." *New England Journal of Medicine* 374, no. 24 (June 16, 2016): 2393–2398.

Jones, James H. *Bad Blood: The Tuskegee Syphilis Experiment*. New York: The Free Press, Macmillan Publishing Co. 1981.

Jost, Adolf. 1895. *Das Recht auf den Tod*. Göttingen: Dietrich, 1895.

Jouanna, Jacques. *Hippocrates*. Translated by M. B. DeBevoise. Baltimore: The Johns Hopkins University Press, 1999.

Jonsen, Albert B. *The Birth of Bioethics*. New York: Oxford University Press, 1998.

Jonsen, Albert B. "Interview with Albert B. Jonsen," June 19 and 22, 1998. Department of Medical History and Ethics, School of Medicine, University of Washington, Seattle Interviewed by Judith Swazey at the Professor Jonsen's office at the University of Washington, June 19 and 22, 1998.

Jonsen, Albert B., Mark Siegler, and William J. Winslade. *Clinical Ethics: A Practical Approach to Ethical Decisions in Clinical Medicine.* New York: Macmillan Publishing Co., 1983.

Kaimowitz v. Department of Mental Health for The State of Michigan, no. 73 19434·AW (Michigan Circuit Court, Wayne County, July 10, 1973).

Kao, Audey C., and Kayh P. Parsi. "Content Analyses of Oaths Administered at U.S. Medical Schools in 2000." *Academic Medicine* 79, no. 9 (September 2004): 882–887.

Kassels, Austin Connor, and Jon F. Merz. "The History and Policy Evolution of Waivers of Informed Consent in Research." *Journal of Legal Medicine* 41, nos. 1–2 (January–June 2021): 1–28.

Katz, Jay, with Alexander M. Capron and Eleanor S. Glass. *Experimentation with Human Beings: The Authority of the Investigator, Subject, Professions, and State in the Human Experimentation Process.* New York: Russell Sage Foundation, 1972.

Kean, Sam. "Leo Alexander's Unflinching Pursuit: In the Waning Days of World War II, a Psychiatrist Raced across Germany to Uncover the Harrowing Abuses of Nazi Doctors," November 17, 2020. *Distillations,* Science History Institute. Accessed October 9, 2022. https://www.sciencehistory.org/distillations /leo-alexanders-unflinching-pursuit#:~:text=Leo%20Alexander's%20Unflinching%20Pursuit%20In%20the%20waning%20days,doctors.%20By%20 Sam%20Kean%20%7C%20November%2017%2C%202020.

Kelsey, Frances Oldham. *Autobiographical Reflections.* U.S. Food and Drug Administration. Accessed October 9, 2022. https://www.fda.gov/media/89162 /download.

Kelsey, Frances Oldham. "Patient Consent Provisions of the Federal Food, Drug, and Cosmetic Act." In *Clinical Investigation in Medicine: Legal, Ethical, and Moral Aspects: An Anthology and Bibliography,* edited by Irving Ladimer and Roger W. Newman, 336–338. Boston: Law-Medicine Research Institute of Boston University, 1963.

Klugman, Craig. "Bioethics: The Revolution Is Over," November 10, 2015. Accessed October 9, 2022. http://www.bioethics.net/2015/11/bioethics-the-revolution-is-over/.

Kopp, Vincent J. 1999. "Henry Knowles Beecher and the Development of Informed Consent in Anesthesia Research." *Anesthesiology* 90, no. 6 (June 1999): 1756–1765.

Kotz, Nick, and Mary Lynn Kotz. *A Passion for Equality: George Wiley and the Movement.* New York: W. W. Norton & Company, 1977.

Kübler-Ross, Elisabeth. *On Death and Dying: What the Dying Have to Teach Nurses, Clergy, and Their Own Families.* New York: Macmillan, 1969.

Kuhn, Thomas S. *The Structure of Scientific Revolutions.* 4th ed. Chicago: University of Chicago Press, 2012.

Lancaster, John, and James P. Rathwell. "The Moralist." *The Delacorte Review,* July 14, 2016. Accessed October 9, 2022. https://medium.com/thebigroundtable /the-moralist-ad8159ebe6be.

Langer, Elinor. "Human Experimentation: Cancer Studies at Sloan-Kettering Stir Public Debate on Medical Ethics." *Science*, 143, no. 3606 (February 7, 1964): 551–553.

Lauritzen, Paul. "Daniel Callahan & Bioethics: Where the Best Arguments Take Him." *Commonweal*, June 1, 2007. Accessed October 9, 2022. https://www .commonwealmagazine.org/daniel-callahan-bioethics.

Lebacqz, Karen. Interview with Karen Lebacqz, Robert Gordon Sproul Professor of Theological Ethics Emeritus, Pacific School of Religion, Berkeley, California, by Leroy B. Walters, October 26, 2004. *Oral History of the Belmont Report and the National Commission for the Protection of Human Subjects of Biomedical and Behavioral Research*. Accessed October 9, 2022. https://www.hhs.gov/ohrp /education-and-outreach/luminaries-lecture-series/belmont-report-25th-anniversary -interview-klebacqz/index.html.

Lebacqz, Karen. 2005. "We Sure Are Older but Are We Wiser." In *Belmont Revisited: Ethical Principles for Research with Human Subjects*, edited by James F. Childress, Eric M. Meslin, and Harold T. Shapiro, 99–110. Washington, DC: Georgetown University Press, 2005.

Lederer, Susan E. "'Ethics and Clinical Research' in Biographical Perspective." *Perspectives in Biology and Medicine 59*, no. 1 (Winter, 2019): 18–36.

Lederer, Susan E. "Research without Borders: The Origins of the Declaration of Helsinki." In *Twentieth Century Ethics of Human Subjects Research*, edited by Volker Roelcke and Giovanni Maio, 199–2018. Stuttgart: Franz Steiner Verlag, 2004.

Lederer, Susan E. 2000. "The Tuskegee Syphilis Study in the Context of American Medical Research." In *Tuskegee's Truths: Rethinking the Tuskegee Syphilis Study*, edited by Susan Reverby, 266–275. Chapel Hill: University of North Carolina Press, 2000.

Leopold, Nathan. *Life Plus 99 Years*. Garden City, NY: Doubleday, 1958.

Leven, Karl-Heinz. "The Invention of Hippocrates: Oath, Letters and Hippocratic Corpus." In *Ethics Codes in Medicine: Foundations and Achievements of Codification since 1947*, edited by Ulrich Tröhler, Stella Reiter-Theil, and Eckhard Herych, 111–129. Ashgate: Aldershot, 1998.

Ley, Herbert L., Jr. "Federal Law and Patient Consent." *Food, Drug, Cosmetic Law Journal 24*, no. 11 (November 1969): 520–526.

Lifton, Robert Jay. *The Nazi Doctors: Medical Killing and the Psychology of Genocide*. New York: Basic Books, 1986.

Linnaeus, Carl. Systema naturæ per regna tria naturæ, secundum classes, ordines, genera, species, cum characteribus & differentiis. 1766–1768. Accessed October 9, 2022. https://archive.org/stream/systemanaturaepe01linn/systemanaturaepe01linn _djvu.txt.

Lo Presti, Robert. 2014. "History of Science: The First Scientist." *Nature* 512: 250–251. Accessed October 9, 2022. https://www.nature.com/articles/512250a.

Luther, Martin. *Commentary on the Sermon on the Mount.* Translated by Charles A. Hay. Berks County, PA: Lutheran Publication Society, January 1, 1892. Accessed October 9, 2022. http://www.godrules.net/library/luther/37luther0.htm.

Lysaught, M. Therese. "Respect: Or, How Respect for Persons Became Respect for Autonomy," *Journal of Medicine and Philosophy* 29, no. 6 (December 2004): 665–680.

Marcuse, Herbert. *One-Dimensional Man: Studies in the Ideology of Advanced Industrial Society.* New York: Beacon Press, 1964.

Macklin, Ruth. Interview conducted by Drs. Renee C. Fox and Carla Messikomer at Dr. Macklin's apartment May 18, 1999. Acadia Institute Project on Bioethics in American Society. Accessed October 18, 2022. https://repository.library .georgetown.edu/bitstream/handle/10822/557039/MacklinR.pdf?sequence=6.

Maehle, Andreas-Holger. *Doctors, Honour, and the Law: Medical Ethics in Imperial Germany.* New York: Palgrave MacMillan, 2009.

Maehle, Andreas-Holger. "'God's Ethicist': Albert Moll and His Medical Ethics in Theory and Practice." *Medical History* 56, no. 2 (April 2012): 217–236.

Mashour, George A. 2009. "Altered States: LSD and the Anesthesia Laboratory of Henry Knowles Beecher." *Bulletin of Anesthesia History* 23, no. 3 (July 2005): 11–14.

McBride, William G. "Thalidomide and Congenital Abnormalities." Letter to the Editor. *The Lancet* 2 (December 16, 1961): 1358.

McClory, Robert. 1995. *Turning Point: The Inside Story of the Papal Birth Control Commission, and How Humanae Vitae Changed the Life of Patty Crowley and the Future of the Church.* New York: Crossroad, 1995; revised March 1, 1997.

McCoy, Alfred E. "Science in Dachau's Shadow: Hebb, Beecher, and The Development of CIA Psychological Torture and Modern Medical Ethics." *Journal of the History of the Behavioral Sciences* 43, no. 4 (Fall 2007): 401–417.

Mellanby, Kenneth. "A Moral Problem." Letter to the Editor. *The Lancet* 248, no. ii (December 7, 1946): 850.

Mellanby, Kenneth. *Human Guinea Pigs, by Kenneth Mellanby: A Reprint with Commentaries.* 3rd ed. Edited by Lisa M. Rasmussen. Cham: Springer Nature, 2020.

"Membership in the World Medical Association for the Japan and Western German Medical Associations." *World Medical Association Bulletin* 3 (1951): 204, 210.

Miles, Steven. *The Hippocratic Oath and the Ethics of Medicine.* New York: Oxford University Press, 2004.

Mill, John Stuart. "A Few Observations on the French Revolution: Review of Alison's *History of Europe.*" *Monthly Repository*, August 1833. Accessed October 10, 2022. http://www.laits.utexas.edu/poltheory/jsmill/diss-disc/french-rev.html.

Mill, John Stuart. "Bentham." *London and Westminster Review*, August 1838. In *Utilitarianism and On Liberty: Including 'Essay on Bentham' and Selections from the Writings of Jeremy Bentham and John Austin*, edited by Mary Warnock, 52–87. Malden, MA: Blackwell Publishing, 2003.

Miller, Franklin G. "Clinical Research before Informed Consent." *Kennedy Institute of Ethics Journal* 24, no. 2 (June 2014): 141–157.

Miller, Franklin G. "Henry Beecher and Consent to Research." *Perspectives in Biology and Medicine* 59, no. 1 (Winter 2016): 78–94.

Miller, Franklin G. "The Stateville Penitentiary Malaria Experiments: A Case Study in Retrospective Ethical Assessment." *Perspectives in Biology and Medicine* 56, no. 4 (Autumn 2013): 548–567.

Mintz, Morton. "'Heroine' of FDA Keeps Bad Drug Off Market." *Washington Post*, July 15, 1962. Accessed October 10, 2022. https://www.washingtonpost.com/wp-srv/washtech/longterm/thalidomide/keystories/071598drug.htm.

Moll, Albert. 1902. *Ärztliche Ethik: Die Pflichten des Arztes in allen Beziehungen seiner Thätigkeit*. Stuttgart: Ferdinand Enke.

Moore, J. E. "Letter to Taliaferro Clark of the United States Public Health Service," September 28, 1932, Clark, NA-WNRC. Cited in Allan M. Brandt, "Racism and Research: The Case of the Tuskegee Syphilis Study," *The Hastings Center Report* 8, no. 6 (December 1978): 21–29.

Moore, J. E., H. N. Cole, P. A. O'Leary, J. H. Stokes, U. J. Wile, Clark, T. Parran, and J. H. Usilton. "Cooperative Clinical Studies in the Treatment of Syphilis: Latent Syphilis." *Venereal Disease Information* 13 (September 20, 1932): 371–379, 389–401.

Moreno, Jonathan. "Acid Brothers Henry Beecher, Timothy Leary, and the Psychedelic of the Century." *Perspectives in Biology and Medicine* 59, no. 1 (Winter 2016): 107–112.

Moreno, Jonathan. "The Declaration of Helsinki and the 'American Stamp.'" In *Ethical Research: The Declaration of Helsinki and the Past, Present, and Future of Human Experimentation*, edited by Ulf Schmidt, Ulf, Andreas Frewer, and Dominique Sprumon, 351–365. New York: Oxford University Press, 2020.

Mount Sinai Hospital. "About the Hospital." Accessed October 10, 2022. https://www.mountsinai.org/locations/mount-sinai/about/history.

Myerson, Abraham, James B. Ayer, Tracy J. Putnam, Clyde E. Keeler, and Leo Alexander. *Eugenic Sterilization: A Reorientation of the Problem*. New York: The Macmillan Company, 1936.

National Advisory Commission on Civil Disorders. 1968. *Report of the National Advisory Commission on Civil Disorders Summary of Report*. Washington, DC: U.S. Government Printing Office. Accessed October 10, 2022. https://www.hsdl.org/?abstract&did=35837.

National Commission for the Protection of Human Subjects of Biomedical and Behavioral Research. *The Belmont Report: Ethical Principles and Guidelines for the Protection of Human Subjects of Research*. U.S. Department of Health, Education, and Welfare, DHEW Publication No. (OS) 78–0012. Washington, DC: U.S. Government Printing Office, 1978.

National Commission for the Protection of Human Subjects of Biomedical and Behavioral Research. 1978. *Ethical Guidelines for the Delivery of Health Services by DHEW*. U.S. Department of Health, Education, and Welfare, DHEW Publication No. (OS) 1978–0010. Washington, DC: U.S. Government Printing Office. Accessed October 10, 2022. https://repository.library.georgetown.edu/bitstream /handle/10822/559347/ethical_guidelines_health_services_min.pdf?sequence =1&isAllowed=y.

National Commission for the Protection of Human Subjects of Biomedical and Behavioral Research. 1978. *Report and Recommendations Institutional Review Boards*. Accessed October 10, 2022. https://videocast.nih.gov/pdf/ohrp _institutional_review_boards.pdf.

National Conference Steering Committee. "Standards for Cardiopulmonary Resuscitation (CPR) and Emergency Cardiac Care." *JAMA: The Journal of the American Medical Association* 227, no. 7 (February 18, 1974): 837, 864.

National Constituent Assembly (France). *Déclaration des droits de l'homme et du citoyen*. July 9, 1789. Accessed October 10, 2022. https://avalon.law.yale.edu /18th_century/rightsof.asp.

Neisser, Albert. 1898. "Was wissen wir von einer Serumtherapie bei Syphilis und was haben wir von ihr zu erhoffen?" *Archiv für Dermatologie und Syphilis* 44 (December 1898): 431–539.

Nietzsche, Friedrich. "Twilight of the Idols." In *The Complete Works of Friedrich Nietzsche*, edited by Oscar Levy and Robert Guppy, vol. 16. New York: Russell and Russell, [1889] 1964.

Nowell-Smith, Patrick H. 1954. *Ethics*. Harmondsworth: Penguin Books.

Nuland, Sherwin. 1994. *How We Die: Reflections on Life's Final Chapter*. New York: Alfred A. Knopf.

Nuremberg Tribunal. 1946. "Permissible Medical Experiments." *Trials of War Criminals before the Nuremberg Military Tribunals under Control Council Law No. 10*. Nuremberg, October 1946–April 1949. Washington, D.C.: U.S. Government Printing Office, 1949–1953.

Offit, Paul A. *Vaccinated: One Man's Quest to Defeat the World's Deadliest Diseases*. Washington, DC: Smithsonian Books, 2007.

Oxford English Dictionary. "New Words Notes: June 2017." Accessed October 10, 2022. https://public.oed.com/blog/june-2017-update-new-words-notes/.

Packianathan, Satyaseelan, Srinivasan Vijayakumar, and Paul Russell Roberts III. "Reflections on the Hippocratic Oath and Declaration of Geneva in Light of the COVID-19 Pandemic." *Southern Medical Journal* 113, no. 7 (July 2020): 326–329.

Palazzo, Guido, Franciska Krings, and Ulrich Hoffrage. "Ethical Blindness." *Journal of Business Ethics* 109, no. 3 (September 2012): 323–338.

Pappworth, Maurice. "'Human Guinea Pigs'—A History." *British Medical Journal* 301, no. 6766 (December 22–29, 1990), 1456–1460.

Pappworth, Maurice. "Human Guinea Pigs: A Warning." *Twentieth Century Magazine* 50, no. 4 (1962): 66–75.

Pappworth, Maurice. "Letter to HK Beecher, January 19, 1965." Henry K. Beecher Papers (H MS c64), Harvard Medical Library in the Francis A. Countway Library of Medicine. Cited in Allan Gaw, "Exposing Unethical Human Research: The Transatlantic Correspondence of Beecher and Pappworth." *Annals of Internal Medicine* 156, no. 2 (January 17, 2012): 150–155.

Pappworth, Maurice. "Letter to HK Beecher, February 20, 1965." Henry K. Beecher Papers (H MS c64), Harvard Medical Library in the Francis A. Countway Library of Medicine. Cited in Allan Gaw, "Exposing Unethical Human Research: The Transatlantic Correspondence of Beecher and Pappworth." *Annals of Internal Medicine* 156, no. 2 (January 17, 2012): 152.

Pappworth, Maurice. *Human Guinea Pigs: Experimentation on Man.* Boston: Beacon Press, 1968.

Parsons (Addelson), Kathryn Pyne. "Nietzsche and Moral Change." *Feminist Studies* 2, no. 1 (1974): 57–76.

Parsons, Talcott. *The Social System.* London: Collier-Macmillan Company, The Free Press of Glencoe, 1951.

Paul, Diane B., and James Moore. "The Darwinian Context: Evolution and Inheritance." In *The Oxford Handbook of the History of Eugenics*, edited by Alison Bashford and Philippa Levine, 27–42. Oxford: Oxford University Press, 2010.

Percival, Thomas. *Medical Ethics; or, a Code of Institutes and Precepts, Adapted to the Professional Conduct of Physicians and Surgeons.* London, UK: J. Johnson, 1803.

Percival, Thomas. *Medical Jurisprudence or a Code of Ethics and Institutes, Adapted to the Professions of Physic and Surgery.* Manchester, UK, 1794.

Phillips, Stephen. "How a Courageous Physician-Scientist Saved the U.S. from a Birth-Defects Catastrophe." 2011, reissued March 9, 2020. Accessed October 10, 2022. https://www.uchicagomedicine.org/forefront/biological-sciences-articles /courageous-physician-scientist-saved-the-us-from-a-birth-defects-catastrophe.

Paul VI, Pope. "Encyclical Letter *Humanae Vitae* of the Supreme Pontiff Paul VI to His Venerable Brothers the Patriarchs, Archbishops, Bishops and Other Local Ordinaries in Peace and Communion with the Apostolic See, to the Clergy and Faithful of the Whole Catholic World, and to All Men of Good Will, on the Regulation of Birth," July 25, 1968. Accessed October 10, 2022. https://www.vatican.va/content /paul-vi/en/encyclicals/documents/hf_p-vi_enc_25071968_humanae-vitae.html.

Pius XII, Pope. "The Moral Limits of Medical Research and Treatment, Proposition 12." In *Papal Encyclicals Online.* September 14, 1952. Accessed October 10, 2022. https://www.papalencyclicals.net/pius12/p12psych.htm.

Pius XII, Pope. "Address to an International Congress of Anesthesiologist. Official Documents." *L'Osservatore Romano.* November 25–26, 1957. Accessed October 12, 2022. http://lifeissues.net/writers/doc/doc_31resuscitation.html.

Plato. *The Republic of Plato.* 3rd ed. Translated by Benjamin Jowett. Oxford: Clarendon Press, 2017. Stephanus numbering from 1908, Project Gutenberg. Accessed October 12, 2022. https://www.gutenberg.org/files/55201/55201-h/55201-h.htm.

Plotz, David. "The Greatest Magazine Ever Published: What I Learned Reading All of *Life* Magazine from the Summer of 1945." *Slate*, December 27, 2013. Accessed

October 12, 2022. https://slate.com/human-interest/2013/12/life-magazine-1945 -why-it-was-the-greatest-magazine-ever-published.html.

Pollner, Fran. "NWRO Convention: Health Care." *Off Our Backs* 3 (July/ August 1973): 8.

Poore, Carol. *Disability in Twentieth-Century German Culture.* Ann Arbor: University of Michigan Press, 2009.

Pow, Stephan, and Frank W. Stahnisch. "Ludwig Edelstein (1902–1965): A German Historian of Medicine in North American Exile and the Emergence of the Modern Hippocratic Oath." *Journal of Medical Biography* 24, no. 4 (November 2016): 527–537.

President's Commission for the Study of Ethical Problems in Medicine and Biomedical and Behavioral Research. *Deciding to Forego Life-Sustaining Treatment.* Washington, DC: U.S. Government Printing Office, 1983.

Pridham, J. A. "Founding of the World Medical Association." *World Medical Association Bulletin* 3 (1951): 207.

"Prison Malaria: Convicts Expose Themselves to Disease so Doctors Can Study It." *Life Magazine* (June 1945): 43–45. Accessed October 12, 2022. https://books .google.com/books?id=h0gEAAAAMBAJ&pg=PA43&lpg=PA43&dq=Prison +malaria;+convicts+expose+themselves+to+disease+so+doctors+can+study+it .&source=bl&ots=RcmbOYMzd1&sig=2WrKyi-xWkS1Astv3IdXHCRoc8&hl =en&sa=X&ved=2ahUKEwj5-eTm5YvdAhUkiOAKHVBjBKUQ6AEwAXoECAk QAQ#v=onepage&q=Prison%20malaria%3B%20convicts%20expose%20them selves%20to%20disease%20so%20doctors%20can%20study%20it.&f=false.

Potter, Van Rensselaer. *Bioethics: Bridge to the Future.* Englewood Cliffs, NJ: Prentice-Hall, 1971.

Proctor, Robert N. *Racial Hygiene: Medicine under the Nazis.* Cambridge, MA: Harvard University Press, 1988.

Proctor, Robert N. "Nazi Doctors, Racial Medicine, and Human Experimentation." In *The Nazi Doctors and the Nuremberg Code: Human Rights in Human Experimentation,* edited by George J. Annas and Michael A. Grodin, 17–31. New York: Oxford University Press, 1992.

Proctor, Robert N. "Nazi Science and Nazi Medical Ethics: Some Myths and Misconceptions." *Perspectives in Biology and Medicine* 43, no. 3 (Spring 2000): 335–346.

Rabkin, Mitchell T., Gerald Gillerman, and Nancy R. Rice. "Orders Not to Resuscitate." *New England Journal of Medicine* 295, no. 7 (August 12, 1976): 363–366.

Ramm, Rudolf. *Ärztlizche Rechts-und Standeskunde Der Artz als Gesundheitserzieher.* Berlin: De Gruyter, 1943. English edition: Rudolf Ramm, *Medical Jurisprudence and Rules of the Medical Profession.* Translated by Melvin Wayne Cooper. New York: Springer International Publishing, 2019.

Ramsey, Paul. *The Patient as Person: The Lyman Beecher Lectures at Yale University.* New Haven, CT: Yale University Press, 1970.

Ravich, Ruth, and Lucy Schmolka. "Patient Representation as a Quality Improvement Tool." *The Mount Sinai Journal of Medicine* 60, no. 5 (October 1993): 374–378.

Reich, Warren T.S.T.D. [Doctor of Sacred Theology], Professor Emeritus and Senior Research Scholar, Kennedy Institute of Ethics, Washington, DC. Interview conducted by Dr. Judith P. Swazey at the Kennedy Institute of Ethics, Georgetown University. Acadia Institute Project on Bioethics in American Society, Archives of the Bioethics Library, Kennedy Institute of Ethics, March 29, 2000.

Reverby, Susan M. *Examining Tuskegee: The Infamous Syphilis Study and Its Legacy.* Chapel Hill: University of North Carolina Press, 2009.

Reverby, Susan M. *Tuskegee's Truths: Rethinking the Tuskegee Syphilis Study.* Chapel Hill: University of North Carolina Press, 2000.

Rockwell, Donald H., Anne R. Yobs, and M. Brittain Moore Jr. "The Tuskegee Study of Untreated Syphilis: The 30th Year of Observation." *Archives of Internal Medicine* 114, no. 6 (December 1964): 792–798.

Rodman, Marc A. *Conflicts of Interest and the Future of Medicine.* New York: Oxford University Press, 2012.

Roelcke, Volker, and Giovanni Maio. *Twentieth Century Ethics of Human Subjects Research: Historical Perspectives on Values, Practices, and Regulations.* Stuttgart: Franz Steiner Verlag, 2004.

Rothman, David J. *Strangers at the Bedside: A History of How Law and Bioethics Transformed Medical Decision Making.* New York: Basic Books, 1991.

Rutten, Thomas. *Hippokrates im Gespräch. Katalog der Ausstellung des Institutes für Theorie und Geschicte der Medizin und der Universitäts- und Landesbibliothek Münster,* 10 December 1993–8 January 1994, Münster. (Catalogue).

Rütten, Thomas. "Ludwig Edelstein at the Crossroads of 1933. On the Inseparability of Life, Work, and Their Reverberations." *Early Science and Medicine* 11, no. 1 (January 2006): 50–99.

Ruzek, Sheryl Burt. *The Women's Health Movement: Feminist Alternatives to Medical Control.* New York: Praeger Publishers, 1978.

Santayana, George. *The Life of Reason: The Phases of Human Progress.* Vol. 1. 1905. Project Gutenberg ebook. Accessed October 16, 2022. https://www.gutenberg.org/files/15000/15000-h/15000-h.htm.

Sass, Hans-Martin. "Fritz Jahr's 1927 Concept of Bioethics." *Kennedy Institute of Ethics Journal* 17, no. 4 (December 2008): 279–295.

Schilling, Klaus Karl. 1945. "Dr. Klaus Karl Schilling Testifies at the Trial of Former Camp Personnel and Prisoners from Dachau, United States Holocaust Memorial Museum." Accessed October 16, 2022. https://collections.ushmm.org/search/catalog/pa1069345.

Schleiner, Winfried. *Medical Ethics in the Renaissance.* Washington, DC: Georgetown University Press, 2007.

Schmidt, Ulf. *Justice at Nuremberg: Leo Alexander and the Nazi Doctors' Trial.* Houndmills: Palgrave Macmillan, 2004.

296   Bibliography

Schmidt, Ulf. *Karl Brandt: The Nazi Doctor, Medicine and Power in the Third Reich*. London: Continuum, 2007.

Schmidt, Ulf, Andreas Frewer, and Dominique Sprumont, eds. 2020. *Ethical Research: The Declaration of Helsinki and the Past, Present, and Future of Human Experimentation*. New York: Oxford University Press.

Schoenfield, Philip, Catherine Pease-Watkin, and Michael Quinn, eds. 2014. *Of Sexual Irregularities and Other Writings on Sexual Morality: The Collected Works of Jeremy Bentham*. Oxford: The Clarendon Press, 2014.

Seldon, Joanna. *The Whistle-Blower*. Buckingham: The University of Buckingham Press, 2017.

Shuster, Evelyn. "Fifty Years Later: The Significance of the Nuremberg Code." *New England Journal of Medicine* 337, no. 20 (November 13, 1997): 1436–1440.

Smith, Cedric M. "Origin and Uses of *Primum Non Nocere*—Above All, Do No Harm!" *Journal of Clinical Pharmacology* 45, no. 4 (April 2005): 371–377.

Smith, Wesley J. *Culture of Death: The Age of "Do Harm" Medicine*. New York: Encounter Books, 2002.

Smith, Wesley J. "The War on the Hippocratic Oath." *NRL News Today*, February 2, 2018. Accessed October 18, 2022. https://www.nationalrighttolifenews.org/2018/02/war-hippocratic-oath/.

Spinelli, A. *Report of the Committee on Medical Ethics*. 44th Council Session, Chicago, Illinois. New York: World Medical Association, May 6–12, 1962.

Stanger, Allison. *Whistleblowers: Honesty in America from Washington to Trump*. New Haven, CT: Yale University Press, 2019.

Stanley, Jenn. "CHOICE/LESS: The Backstory, Episode 4: Tuskegee Was the 'Tip of the Iceberg,'" June 21, 2017. Accessed October 16, 2022. https://rewirenewsgroup.com/multimedia/podcast/choiceless-backstory-episode-4-tuskegee-tip-iceberg/.

Starr, Paul. *The Social Transformation of American Medicine*. New York: Basic Books, 1982.

Stevens, M. L. Tina. *Bioethics in America: Origins and Cultural Contexts*. Baltimore: The Johns Hopkins University Press, 2000.

Swift, Johnathan. *Polite Conversation in Three Dialogues by Jonathan Swift with Introduction and Notes by George Saintsbury*. 1738. Project Gutenberg edition 2019. Accessed October 16, 2022. https://www.gutenberg.org/files/60186/60186-h/60186-h.htm.

Teel, Karen. "The Physician's Dilemma: A Doctor's View—What the Law Should Be: *Baylor Law Review* 27, no. 1 (1975): 6–9.

Tröhler, Ulrich, Stella Reiter-Theil, and Eckhard Herych, eds. *Ethics Codes in Medicine: Foundations and Achievements of Codification since 1947*. Aldershot: Ashgate, 1998.

Tsuchiya, Kazuyo. National Welfare Rights Organization (1966–1975). Posted January 23, 2007. Accessed October 16, 2022. https://www.Blackpast.Org/African-American-History/National-Welfare-Rights-Organization-1966-1975/.

United Nations. "Universal Declaration of Human Rights." 183rd Meeting of United Nations General Assembly (Paris, December 10, 1948). Accessed October 16, 2022. https://www.un.org/en/about-us/universal-declaration-of-human-rights.

United States Department of Health, Education, and Welfare: Public Health Service. *Final Report of the Tuskegee Syphilis Study Ad Hoc Advisory Panel.* Washington, DC: Government Printing Office, April 28, 1973.

United States Department of Health, Education, and Welfare. "Summary: The Belmont Report of The National Commission for the Protection of Human Subjects of Biomedical and Behavioral Research." April 18, 1979. Accessed October 16, 2022. https://www.hhs.gov/ohrp/regulations-and-policy/belmont-report/read-the-belmont-report/index.html.

United States Department of the Interior, National Park Service. "Letter to Tuskegee Syphilis Study Subjects," National Register of Historic Places: U.S. Public Health Service Syphilis Study, Macon County AL. 1932–1973, Section E, p. 4.

United States Holocaust Memorial Museum. Hadamar Trial. In *Holocaust Encyclopedia.* Accessed October 16, 2022. https://encyclopedia.ushmm.org/content/en/article/the-hadamar-trial.

United States Holocaust Memorial Museum. "Recollections of a Member of a British Army Film and Photographic Unit. United States Holocaust Museum, Bergen-Belsen," T. J. Stretch, British Army Chaplin. *Liberation of Bergen Belsn* (April 15, 1954). Accessed October 16, 2022. https://www.ushmm.org/learn/timeline-of-events/1942-1945/liberation-of-bergen-belsen.

United States Public Health Service. "Minutes, April 5, 1965," unpublished typescript, Tuskegee Syphilis Study-National Library of Medicine. Cited in Alan Brandt, "Racism and Research: The Case of the Tuskegee Syphilis Study," *The Hastings Center Report* 8, no. 6 (December 1978): 21–29.

Urban Dictionary. "Unwoke" (January 21, 2021 characterization). Accessed February 11, 2022. https://www.urbandictionary.com/define.php?term=Unwoke. (Note this link has changed because the Urban dictionary is crowd-defined.)

Veatch, Robert M. *A Theory of Medical Ethics.* New York: Basic Books, 1981.

Veatch, Robert M. *Disrupted Dialogue: Medical Ethics and the Collapse of Physician-Humanist Communication (1770–1980).* New York: Oxford University Press, 2005.

Veatch, Robert M. "Henry Beecher's Contributions to the Ethics of Clinical Research." *Perspectives in Biology and Medicine* 95, no. 1 (Winter 2016): 3–17.

Veatch, Robert M. Interview by Dr. Renee C. Fox and Dr. Judith P. Swazey at Professor Veatch's office. March 26, 1999. Acadia Institute Project of Bioethics in American Society.

Vollmann, Jochen, and Rolf Winau. "The Prussian Regulation of 1900: Early Ethical Standards for Human Regulation in Germany," *IRB: Ethics & Human Research* 18, no. 4 (July–August 1996): 9–11.

Von Staden, Heinrich. "'In a Pure and Holy Way': Personal and Professional Conduct in the Hippocratic Oath." *Journal of the History of Medicine and Allied Sciences* 51, no. 4 (October 1996): 406–408.

Walter, Jennifer K., and Eran P. Klein, eds. *The Story of Bioethics: From Seminal Works to Contemporary Explorations*. Washington, DC: Georgetown University Press, 2003.

Walters, Leroy. Interview with LeRoy Walters, PhD, Director, Kennedy Institute of Ethics, and Joseph P. Kennedy, Sr. Professor of Christian Ethics, Kennedy Institute of Ethics, Washington, DC, March 13, 2000. Conducted by Dr. Judith P. Swazey at the Kennedy Institute of Ethics, Georgetown University. Acadia Institute Project on Bioethics in American Society, Archives of the Bioethics Library, Kennedy Institute of Ethics.

Wenger, O. C. "Letter to R. A. Vonderlehr," July 21, 1933. Center for Disease Control Papers, Tuskegee Syphilis Study Administrative Records, 1930–1980, Box 5, Folder Correspondence, National Archives–Southeast Region. Cited in *Tuskegee's Truths: Rethinking the Tuskegee Syphilis Study*, edited by Susan Reverby, 85. Chapel Hill: University of North Carolina Press, 2000.

Weindling, Paul S. "From the Nuremberg 'Doctors Trial' to the 'Nuremberg Code.'" In "Medical Ethics in the 70 Years after the Nuremberg Code, 1947 to the Present." *Wien Klin Wochenschr: The Central European Journal of Medicine* 130 (Suppl. 3, June 2018): 159–253.

Weindling, Paul S. "Human Guinea Pigs and the Ethics of Experimentation: The *BMJ*'s Correspondent at the Nuremberg Medical Trial." *British Medical Journal* 313 (December 7, 1996): 1467–1470.

Weindling, Paul S. *Nazi Medicine and the Nuremberg Trials: From Medical War Crimes to Informed Consent*. Basingstoke, UK: Palgrave Macmillan, 2008.

Whewell, William. *History of the Inductive Sciences from the Earliest to the Present Day*. London: John W. Parker, 1837.

Whitman, Alden. "Dr. George Wiley Feared Drowned." *New York Times*, August 10, 1973. Accessed October 16, 2022. https://www.nytimes.com/1973/08/10/archives/dr-george-wiley-feared-drowned-civil-rights-leader-42-who-headed.html.

Wiley, George A. "Health Care in the Inner City: Like It Is: Point of View of a Consumer." In *Health Care Problems of the Inner City: Report of the 1969 National Health Forum*. New York: National Health Forum, edited by H. Milt, 1969, 12. Cited in Joseph C. D'Oronzio, "A Human Right to Healthcare Access: Returning to the Origins of the Patients' Rights Movement." *Cambridge Quarterly of Healthcare Ethics* 10, no. 3 (Summer 2001): 285–298.

Williford, Miriam. "Bentham on the Rights of Women." *Journal of the History of Ideas* 36, no. 1 (January–March 1975): 167–176.

Wilson, Duncan. 2014. *The Making of British Bioethics*. Manchester: Manchester University Press.

Wittgenstein, Ludwig. *Tractatus Logico-Philosophicus*. Translated by C. K. Ogden. Project Guttenberg edition (1922) 2021. Accessed October 16, 2022. https://www.gutenberg.org/files/5740/5740-pdf.pdf.

Wittgenstein, Ludwig. *Philosophical Investigations*. Translated by G. E. M. Anscombe, P. M. S. Hacker, and Joachim Schulte. Hoboken, NJ: Wiley-Blackwell Publishing Co., [1953] 2001.

Wittgenstein, Ludwig. *Culture and Value*. Translated by Peter Winch. *Vermischte Bemerkungen*, e.d., G. H. Von Wright, and Hekki Nyman. Chicago: University of Chicago Press, 1984.

World Medical Association. Declaration of Geneva. World Medical Association, 2017. Declaration of Geneva as amended by the 68th WMA General Assembly, Chicago, United States, October 2017. Accessed February 9, 2022. https://www.wma.net/policies-post/wma-declaration-of-geneva/.

World Medical Association. "*Serment de Geneve*, Declaration of Geneva, *Declaracion en Gemebra*." *World Medical Association Bulletin* 1, no. 2 (July 1949): 35–37.

World Medical Association. Declaration of Helsinki, 1964. Accessed October 16, 2022. https://www.wma.net/wp-content/uploads/2018/07/DoH-Jun1964.pdf.

World Medical Association. "Principles for Those in Research and Experimentation (approved by The General Assembly of the World Medical Association in 1954)." *World Medical Journal* 2, no. 1 (1955): 14–15.

World Medical Association. "Proceedings," *World Medical Association Bulletin* 1, no. 1 (1949): 6–12.

Yoshioka, Alan. "Use of Randomisation in the Medical Research Council's Clinical Trial of Streptomycin in Pulmonary Tuberculosis in the 1940s." *British Medical Journal* 317, (October 31, 1998): 1220–1223.

Younger, Stuart, David Jackson, Claudia Coulton, B. W. Juknialls, and Erin Murphy Smith. "A National Survey of Hospital Ethics Committees, in President's Commission for the Study of Ethical Problems." In Medicine and Biomedical and Behavioral Research, *Deciding to Forego Life-Sustaining Treatment: A Report on the Ethical, Medical and Legal Issues in Treatment Decisions*, 443–457. Washington DC: US Government Printing Office, 1983.

Index

Ethics, defined as justifications or critiques of morality, 13

Ethik (1922–1939), German medical journal, first journal dedicated to medical ethics, 14

Eugenics, 2, 20, 29–30, 46, 49, 243n57
Galton's characterization of, 50

Euthanasia (English/German, *see also,* "*euthanasie*") 157, 173, 178, 242n41
Blanchard's 1708 English physicians' dictionary on, 15
Christoph Hufeland's German palliative care characterization of, 15
Francis Bacon's view of euthanasia as palliative care for the dying, 15
Hippocratic Oath's prohibition on using deadly medicines, 50
Samuel Williams' English redefinition of euthanasia as "mercy killing," 15–16
William Munk on palliative euthanasia, 50

Euthanasie (German, pre-1945), 15–17, 20, 23, 25, 40, 41
Adolf Jost, *euthanasie* as the right to death, 16

Evans, John H. (1965–) US sociologist, studied bioethics lexicon, 154–155

Experimentation with Human Beings (1972), Jay Katz and Alex Capron's pioneering anthology on the ethics of research on humans, 221

FD&C, *see* United States Food, Drug, and Cosmetic Act

FDA, *see* United States Food and Drug Administration

Fletcher, Joseph F. (1905–1991), US moral theologian of situational ethics, 77

Fox, Renée (1928–2020), US sociologist of medicine and bioethics,156, 163, 165, 184

Führer (charismatic leader *see also* Adolf Hitler), 12, 39, 104, 231n4

Garland, Joseph (1893–1973), transformational editor of *New England Journal of Medicine* (1947–1967), 93

Gaylin, Willard (1925–1922), cofounder of The Hastings Center, 148, 152–153, 156, 189, 192, 202–203

German Health Council, (*Reichgesundheitsrat*)
1931 regulations on using experimental treatments on patients, 32

Genocide, 1, 21, 22, 235n14 (*see also,* Holocaust)

Georgetown University, 149, 203

Gibson, Count (1921–2002), US Army veteran, first Tuskegee Syphilis Study whistleblower, civil rights activist, 103, 105, 108–110, 119–121, 192, 194,195, 209, 210

Gorovitz, Samuel (1938–), US philosopher, organized 1974 Institute on Morals and Medicine, 152, 208

Green, Governor Dwight (1897–1958), commissioned study of experiments conducted on prisoners, 27, 28, 236n22,

Gustafson, James (1925–2021), influential Yale moral theologian, 156

Hastings Center, The, founding bioethics think tank, 78, 100, 106, 148–151, 163, 184, 186, 192, 203, 214, 254n143
Hastings Center Studies, 151

HEC, *see,* hospital ethics committee

Heidegger, Martin (1889–1976), German philosopher, supported Nazis, 12, 156, 230n4, 243n56

Hellegers, André (1926–1979), Dutch American obstetrician, founder of Kennedy Institute of Ethics, 147–153, 156, 189, 192
Pope's Biologist, 147

Heller, Jean (1943–), AP reporter who reported out the Tuskegee Syphilis Study scandal, 90, 114, 118, 121, 122, 159

Basic Bioethics

Arthur Caplan, editor

Books Acquired under the Editorship of Glenn McGee and
Arthur Caplan

Peter A. Ubel, *Pricing Life: Why It's Time for Health Care Rationing*

Mark G. Kuczewski and Ronald Polansky, eds., *Bioethics: Ancient Themes in Contemporary Issues*

Suzanne Holland, Karen Lebacqz, and Laurie Zoloth, eds., *The Human Embryonic Stem Cell Debate: Science, Ethics, and Public Policy*

Gita Sen, Asha George, and Piroska Östlin, eds., *Engendering International Health: The Challenge of Equity*

Carolyn McLeod, *Self-Trust and Reproductive Autonomy*

Lenny Moss, *What Genes Can't Do*

Jonathan D. Moreno, ed., *In the Wake of Terror: Medicine and Morality in a Time of Crisis*

Glenn McGee, ed., *Pragmatic Bioethics, 2d edition*

Timothy F. Murphy, *Case Studies in Biomedical Research Ethics*

Mark A. Rothstein, ed., *Genetics and Life Insurance: Medical Underwriting and Social Policy*

Kenneth A. Richman, *Ethics and the Metaphysics of Medicine: Reflections on Health and Beneficence*

David Lazer, ed., *DNA and the Criminal Justice System: The Technology of Justice*

Harold W. Baillie and Timothy K. Casey, eds., *Is Human Nature Obsolete? Genetics, Bioengineering, and the Future of the Human Condition*

Robert H. Blank and Janna C. Merrick, eds., *End-of-Life Decision Making: A Cross-National Study*

Norman L. Cantor, *Making Medical Decisions for the Profoundly Mentally Disabled*

Margrit Shildrick and Roxanne Mykitiuk, eds., *Ethics of the Body: Post-Conventional Challenges*

Alfred I. Tauber, *Patient Autonomy and the Ethics of Responsibility*

David H. Brendel, *Healing Psychiatry: Bridging the Science/Humanism Divide*

Jonathan Baron, *Against Bioethics*

Michael L. Gross, *Bioethics and Armed Conflict: Moral Dilemmas of Medicine and War*

Karen F. Greif and Jon F. Merz, *Current Controversies in the Biological Sciences: Case Studies of Policy Challenges from New Technologies*

Deborah Blizzard, *Looking Within: A Sociocultural Examination of Fetoscopy*

Ronald Cole-Turner, ed., *Design and Destiny: Jewish and Christian Perspectives on Human Germline Modification*

Holly Fernandez Lynch, *Conflicts of Conscience in Health Care: An Institutional Compromise*

Mark A. Bedau and Emily C. Parke, eds., *The Ethics of Protocells: Moral and Social Implications of Creating Life in the Laboratory*

Jonathan D. Moreno and Sam Berger, eds., *Progress in Bioethics: Science, Policy, and Politics*

Eric Racine, *Pragmatic Neuroethics: Improving Understanding and Treatment of the Mind-Brain*

Martha J. Farah, ed., *Neuroethics: An Introduction with Readings*

Jeremy R. Garrett, ed., *The Ethics of Animal Research: Exploring the Controversy*

Books Acquired under the Editorship of Arthur Caplan

Sheila Jasanoff, ed., *Reframing Rights: Bioconstitutionalism in the Genetic Age*

Christine Overall, *Why Have Children? The Ethical Debate*

Yechiel Michael Barilan, *Human Dignity, Human Rights, and Responsibility: The New Language of Global Bioethics and Bio-Law*

Tom Koch, *Thieves of Virtue: When Bioethics Stole Medicine*

Timothy F. Murphy, *Ethics, Sexual Orientation, and Choices about Children*

Daniel Callahan, *In Search of the Good: A Life in Bioethics*

Robert Blank, *Intervention in the Brain: Politics, Policy, and Ethics*

Gregory E. Kaebnick and Thomas H. Murray, eds., *Synthetic Biology and Morality: Artificial Life and the Bounds of Nature*